Transport Theory

PROCEEDINGS OF A SYMPOSIUM
IN APPLIED MATHEMATICS
OF THE AMERICAN MATHEMATICAL SOCIETY
AND THE SOCIETY FOR INDUSTRIAL AND
APPLIED MATHEMATICS

HELD IN NEW YORK CITY
APRIL 5-8, 1967

Richard Bellman
Garrett Birkhoff
Ibrahim Abu-Shumays

EDITORS

VOLUME I
SIAM-AMS PROCEEDINGS

Transport Theory

AMERICAN MATHEMATICAL SOCIETY
PROVIDENCE, RHODE ISLAND
1969

Prepared by the American Mathematical Society under Air Force Office of Scientific Research Contract No. F44620-67-C-0010, and U.S. Army Research Office (Durham) Contract No. DAHC04-67-C-0074.

Standard Book Number 821-81320-X
Library of Congress Catalog No. 68-23112
AMS 1968 Subject Classifications 8260, 8220, 8265

Contents

IV. Kinetic Theory and Plasma Transport

Introduction

The industrial and military applications of atomic energy have stimulated much mathematical research in neutron transport theory. The possibility of controlled thermonuclear processes has similarly focussed attention upon plasmas, sometimes called the "fourth state of matter." Independently, many classical aspects of kinetic theory and radiative transfer theory have been studied both because of their basic mathematical interest and because of their physical applications to areas such as upper-atmosphere meteorology.

The mathematical difficulties in all of the areas cited are notoriously formidable. As a consequence, ingenious techniques have been developed to handle particular questions. Since the equations of mathematical physics are fortunately limited in number, there is a strong possibility that methods developed in one field on the basis of physical reasoning can be applied to other fields as well. An example of this is provided by the use of methods of radiative transfer in neutron transport theory.

There is thus strong motivation to collect together a number of experts in the field of transport theory and to give them an opportunity to compare methods and results. This volume is dedicated to that purpose. There are four main parts, (I) Analytical Neutron Transport, (II) Numerical Neutron Transport, (III) Stochastic Aspects, (IV) Kinetic Theory and Plasma Transport, each containing four expository accounts of recent developments. We believe that these accounts provide an excellent and comprehensive survey of the current state of transport theory and that it should serve as a stimulus for research in a fascinating subject where very much more remains to be done.

Richard Bellman
Garrett Birkhoff
Ibrahim Abu-Shumays

I. Analytical Neutron Transport

Multiple scattering of partially polarized light

T. W. Mullikin

1. **Introduction.** In his paper presented at the Symposium on Nuclear Reactor Theory in 1961, Wigner [**33**] briefly discusses transport theory which takes into account the quantum mechanical phenomena of polarization of neutrons as particles with spin. Such a transport equation is formulated and studied by Bell and Goad [**3**].

A class of problems in which the mathematical formulation is very similar to that for polarization of neutrons is in the study of multiple scattering of light when account is taken of the polarization as well as intensity. Consideration of polarization of light in scattering problems goes back to Rayleigh [**24**], Stokes [**32**] and others. A proper formulation of a transfer equation to account for multiple scattering seems to have been given first by Chandrasekhar [**5**], [**6**].

From the mathematical viewpoint the different nature of problems in neutron transport and in radiative transfer can dictate different objectives in the numerical or analytic study of the equations. Bell and Goad seem to find that the influence of polarization on criticality of fast reactors can be satisfactorily accounted for by P_1-approximations.

In the study of radiative transfer in planetary atmospheres certain detailed features of the polarization are of interest. For example, one well-known observed phenomenon in the earth atmosphere is the appearance of neutral points, of zero polarization, relative to the sun's position [**5**], [**25**]. A prediction of these from a mathematical model requires a rather accurate computation of a vector function in its dependence on position and direction [**5**, Chapter X]. Hopefully, such detailed structure can be of use in studying the inverse problems where one wants to specify a class of atmospheres compatible with observations [**1**], [**5**], [**17**], [**27**].

Chandrasekhar was the first to solve successfully a complete multiple scattering problem accounting for polarization. This work, leading to accurate numerical results, was possible before the advent of large computers by the clever analytic reduction to simpler equations.

One approach to any problem which has been formulated mathematically is to put the equations on a computer in some approximate form. Of course, this is not always the most efficient or most reliable way to proceed. We use below a slight generalization of the problem studied by Chandrasekhar to discuss various alternative formulations of the transport equation accounting for polarization of light.

2. **The transport equation.** We refer to [**5**], [**6**] and other papers in this volume for a discussion of the physical model and derivation of the following equation. We have a time-independent vector Boltzmann equation

$$\boldsymbol{\Omega} \cdot \nabla \boldsymbol{\Psi} + \boldsymbol{\sigma} \boldsymbol{\Psi} = \int \boldsymbol{P}(x, \boldsymbol{\Omega}, \boldsymbol{\Omega}')\boldsymbol{\Psi}(x, \boldsymbol{\Omega}')\, d\boldsymbol{\Omega}' + \boldsymbol{S} \tag{2.1}$$

where $\boldsymbol{\Psi}$ is a 4-vector, $\boldsymbol{\sigma}$ is a physical scalar function, $\boldsymbol{S}$ is a source, and $\boldsymbol{P}$ is a 4×4 phase matrix. This is a system of linear integro-differential equations which, of course, must be supplemented by a specification of the geometric boundary of the scattering medium together with boundary conditions.

The vector $\boldsymbol{\Psi}$ can be represented in several ways. The Stokes representation used by Chandrasekhar [**5**, Chapter I] gives $\boldsymbol{\Psi}$ relative to a basis so that each component is real valued. The components all have the dimension of intensity and serve to determine the total intensity as well as the orientation and major and minor axes of any elliptically polarized component of the field.

The phase matrix $\boldsymbol{P}$ is determined by physical properties of the scattering medium. For a molecular atmosphere the Rayleigh law is used [**5**, Chapter I]. If aerosol particles are accounted for, the phase matrix is obtained by adding to the Rayleigh phase matrix a phase matrix determined from solutions for scattering of electromagnetic radiation by homogeneous spheres with an appropriate average taken with respect to a distribution of sphere radii [**11**], [**17**, Part 2]. The elements of the phase matrices for aerosols vary by orders of magnitude over the range of scattering angle. Phase matrices for scattering by anisotropic particles and for resonance line scattering are discussed by Chandrasekhar [**5**] (see also [**14**], [**17**]).

For the case of conservative Rayleigh scattering in a homogeneous plane-parallel atmosphere, Chandrasekhar was able to reduce certain functionals of the solution to equation (2.1) to scalar equations for the boundary value problem of a uniform unidirectional radiation field on one face of the atmosphere. These functionals will be given later and this reduction explained.

With the wide variety of phase matrices which are of interest there does not seem to be any analytic method for simplifying equation (2.1) which will be applicable in general. In fact, we will show that the mere introduction of imperfect scattering into the Rayleigh law (cf. [**30**]) seems to deny the nice reduction to scalar equations. We will see this as the mathematical impossibility of diagonalizing a system of singular integral equations.[1]

For plane-parallel atmospheres with inhomogeneities only in the direction normal to the plane of stratification, numerical studies have been made of equation (2.1) for quite general phase matrices. One approach [**13**] is to replace integrals by quadratures and derivatives by differences and solve by a relaxation procedure the boundary value problem of a uniform unidirectional field of radiation on one boundary. This gives a rather general, but slow, numerical procedure.

There are several motivations for attempting to reduce equation (2.1) to other equations. Aside from the mathematical challenge, there are problems of asymptotics for thick atmospheres. Also in many situations it is only the reflected or transmitted radiation which is observed. It is possible that these quantities can be computed more easily than by determining the radiation field throughout the atmosphere. Also, in comparison of computations with such experiments a range of incident and emergent angles is needed. Since incident angles appear parametrically in the boundary conditions for equation (2.1), this means many repetitions of a direct computational program for solving this equation. We will discuss in §4 various alternative formulations of equation (2.1) in a special context.

3. **Rayleigh scattering.** We consider now a homogeneous plane-parallel atmosphere of finite thickness τ in which local scattering is described by the Rayleigh law. If Stokes parameters are appropriately related to the plane of scattering, the phase matrix for Rayleigh scattering is given by

$$\boldsymbol{R} = \frac{3}{2}\begin{pmatrix} \cos^2\theta & 0 & 0 & 0 \\ 0 & 1 & 0 & 0 \\ 0 & 0 & \cos\theta & 0 \\ 0 & 0 & 0 & \cos\theta \end{pmatrix} \tag{3.1}$$

where θ is the scattering angle. The phase matrix which enters equation (2.1) is more complicated when directions are referred to a fixed Cartesian coordinate system. We do not display this but refer to Chandrasekhar [**5**, Chapter I].

We consider a steady state problem in which one face of the atmosphere is exposed to a uniform radiation field from one direction. With probability $\exp(-s)\,ds$ the field propagates a distance s without disturbance and then interacts with the medium in the distance ds. Upon interaction with the medium, radiation is absorbed with probability $(1 - c)$ and with probability c it is scattered according to the Rayleigh law. Reemission of absorbed radiation is not considered so we do not have to consider frequency changes due to absorption, scattering and reemission. The problem we consider is the same as that presented by Chandrasekhar in his book, Chapters I and X, except that we are not considering

[1] C.f. I. C. Gohberg and M. G. Krein, *Systems of integral equations on a half line with kernels depending on the difference of the arguments*, Uspehi Mat. Nauk **13** (1958), 3–72 = AMS Transl. **14** (1960), 217–287.

the scattering to be conservative. The case of $0 \leq c \leq 1$ has also been considered by Sobolev [**30**] in less detail than the case of $c = 1$ treated by Chandrasekhar.

We consider equation (2.1) for the diffuse radiation field which has suffered at least one collision. In this way the attenuated incident radiation on the bounding face enters the equation as the source term in equation (2.1). It is shown by Chandrasekhar [**5**, Chapter X] that the transport equation can be separated into three uncoupled equations by a harmonic expansion. Two of these equations can further be reduced immediately to scalar transport equations. We will not discuss these (cf. [**4**], [**18**], [**19**]).

The remaining equations lead to equations for the fundamental matrix of a two-vector system

$$\mu \frac{\partial}{\partial x} \boldsymbol{M} + \boldsymbol{M} = c \int_{-1}^{1} \boldsymbol{P}(\mu, \mu') \boldsymbol{M}(x, \mu', \nu)\, d\mu' + c\boldsymbol{P}(\mu, \nu) \exp\left(-\frac{x}{\nu}\right), \tag{3.2}$$

where

$$\boldsymbol{P}(\mu, \mu_0) = \frac{3}{8} \begin{pmatrix} 2(1-\mu^2)(1-\mu_0^2) + (\mu\mu_0)^2 & \mu^2 \\ \mu_0^2 & 1 \end{pmatrix}, \tag{3.3}$$

where $0 \leq x \leq \tau$, $\ -1 \leq \mu \leq 1$, $\ 0 \leq \nu \leq 1$. The parameters μ and ν are cosines of polar angles relative to the positive x-axis which is normal to the face of the atmosphere and measures optical depth into the atmosphere.

The boundary conditions for the diffuse field are that no radiation reenters the atmosphere, that is,

$$\boldsymbol{M}(0, \mu, \nu) = \boldsymbol{M}(\tau, -\mu, \nu) = 0 \quad (0 \leq \mu \leq 1). \tag{3.4}$$

Scattering and transmission matrices are defined by

$$\begin{aligned} \boldsymbol{S}(\mu, \nu) &= \mu \boldsymbol{M}(0, -\mu, \nu), \\ \boldsymbol{T}(\mu, \nu) &= \mu \boldsymbol{M}(\tau, \mu, \nu), \end{aligned} \qquad (0 \leq \mu \leq 1). \tag{3.5}$$

These determine the azimuth independent part of the Stokes vector for scattered and transmitted radiation by

$$\begin{pmatrix} \Psi_l \\ \Psi_n \end{pmatrix}(0, -\mu, \nu) = (1/\mu)\boldsymbol{S}(\mu, \nu)\boldsymbol{F}, \qquad \begin{pmatrix} \Psi_l \\ \Psi_n \end{pmatrix}(\tau, \mu, \nu) = (1/\mu)\boldsymbol{T}(u, \nu)\boldsymbol{F},$$

where $\boldsymbol{F}$ is a two vector determined by the incident field.

In the case $c = 1$, Chandrasekhar shows that the scattering and transmission matrices can be expressed in terms of various scalar X- and Y-functions which are functions of the angular variable μ and the parameter τ. This amounts to a separation of variables in the matrices $\boldsymbol{S}$ and $\boldsymbol{T}$, and an expression of these matrices in terms of simpler functions. This means that intensity vectors can be computed on the bounding faces of the atmosphere without having to compute the field throughout the atmosphere. This is in the spirit of the principles of invariance introduced into radiative transfer by Ambartsumian [**1**] and Chandrasekhar (see references in [**1**] and [**5**]).

We want now to discuss the cases $0 \leq c \leq 1$ from other viewpoints. We will see a reason for the possibility of separation of the variables μ and ν and the computation of the matrix $\boldsymbol{M}$ from equations in which the variable x enters parametrically. Two components of the Stokes vector for radiation interior to the atmosphere are, of course, given by

$$\begin{pmatrix}\Psi_l \\ \Psi_n\end{pmatrix}(x, \mu, \nu) = \boldsymbol{M}(x, \mu, \nu)\boldsymbol{F}.$$

This alternate computation of M is possible because the assumption of homogeneity of the atmosphere allows the fruitful use of Laplace transform methods (Wiener-Hopf methods). We will also see the reason that in the case $c = 1$ it is further possible to reduce to scalar equations modulo some 2×2 systems of linear algebraic equations.

We have previously discussed [**20**] the complete field throughout the atmosphere for the case $c = 1$. Recent numerical results for $\boldsymbol{S}$ and $\boldsymbol{T}$ matrices based on the representation in terms of scalar $\boldsymbol{X}$ and $\boldsymbol{Y}$ functions are given for the case of $c = 1$ in [**16**], [**29**].

4. **Methods of solution.** We wish now to discuss various methods for analyzing the matrix equation (3.2) with boundary conditions (3.4). Aside from numerical methods which replace (3.2) by a system of algebraic equations, there are three principal methods.

We first discuss separation of the variables μ and ν. It is observed by Sekera [**28**] that the matrix $\boldsymbol{P}$ of (3.3) can be factored as

$$\boldsymbol{P}(\mu, \mu_0) = \tfrac{3}{8}\boldsymbol{Q}(\mu)\boldsymbol{Q}'(\mu_0)$$

where $\boldsymbol{Q}'$ denotes the transpose of $\boldsymbol{Q}$ and

$$\boldsymbol{Q}(\mu) = \begin{pmatrix}\mu^2 & 2^{1/2}(1-\mu^2) \\ 1 & 0\end{pmatrix}. \tag{4.1}$$

The possibility of similar factorizations for more general phase matrices is discussed in [**26**].

With the matrix $\boldsymbol{N}$ defined by

$$\boldsymbol{M}(x, \mu, \nu) = \boldsymbol{Q}(\mu)\boldsymbol{N}(x, \nu), \tag{4.2}$$

we obtain from equation (3.2) the equation

$$\mu\, \partial \boldsymbol{N}/\partial x + \boldsymbol{N} = c\boldsymbol{J} \tag{4.3}$$

where

$$\boldsymbol{J} = \boldsymbol{I} \exp(-x/\nu) + \int_{-1}^{1} \psi(\mu')\boldsymbol{N}(x, \mu', \nu)\, d\mu' \tag{4.4}$$

and

$$\psi(\mu) = \frac{3}{8}\begin{pmatrix}1+\mu^4 & 2^{1/2}\mu^2(1-\mu^2) \\ 2^{1/2}\mu^2(1-\mu^2) & 2(1-\mu^2)^2\end{pmatrix}. \tag{4.5}$$

The boundary conditions are that

$$N(0, \mu) = N(\tau, -\mu) = 0, \quad (0 \leq \mu \leq 1).$$

The matrix $\boldsymbol{I}$ is the 2×2 identity matrix. The 2×2 matrix $\boldsymbol{\psi}$ should not be confused in the remainder of this paper with the Stokes vector $\boldsymbol{\Psi}$ in equation (2.1). The other notation and equations follow closely those of our paper [**20**]; however, a slight change has been made in the definitions of $\boldsymbol{Q}$, $\boldsymbol{\psi}$, $\boldsymbol{N}$ and $\boldsymbol{J}$ to preserve certain matrix symmetries sacrificed in [**20**] to avoid the appearance throughout of a factor of $2^{1/2}$.

(a) *Integral equations.* By use of the boundary conditions the matrix $\boldsymbol{N}$ is expressed in terms of the matrix $\boldsymbol{J}$ by

$$\text{(4.6)} \qquad \begin{aligned} N(x, \mu, \nu) &= \frac{c}{\mu}\int_0^x J(y, \nu)\exp\left(\frac{y - x}{\mu}\right) dy, \\ N(x, -\mu, \nu) &= \frac{c}{\mu}\int_x^\tau J(y, \nu)\exp\left(\frac{x - y}{\mu}\right) dy. \end{aligned} \qquad 0 \leq \mu \leq 1,$$

This leads to a matrix integral equation

$$\text{(4.7)} \qquad J(x, \nu) = I\exp(-x/\nu) + c\Lambda(J)(x, \nu)$$

where ν appears as a parameter and the operator Λ is defined by

$$\text{(4.8)} \qquad \Lambda(J)(x, \nu) = \int_0^1 \psi(\mu)\int_0^\tau J(y, \nu)\exp\left(-\frac{|x - y|}{\mu}\right) dy\,\frac{d\mu}{\mu}.$$

Existence and uniqueness of solutions to equation (3.2) and boundary conditions (3.4) follow from a similar result for equation (4.7). This is, of course, a familiar approach to the study of scalar transport equations [**12**]. The usual proof is to show that the operator $(I - c\Lambda)^{-1}$ exists and is representable by a Neumann series

$$(I - c\Lambda)^{-1} = \sum_{n=0}^{\infty}(c\Lambda)^n,$$

when some function space has been selected for the discussion of these operators.

If we consider Λ as an operator in the Hilbert space of vector functions whose two components are square integrable functions of x $(0 \leq x \leq \tau)$, it is easy to see that Λ is positive-definite, selfadjoint, and compact (if $\tau < \infty$). We obtain the upper bounds [**20**]

$$\|\Lambda\| \leq \tfrac{1}{2}\{\rho_1 + \rho_3 + ((\rho_1 - \rho_3)^2 + 8\rho_2^2)^{1/2}\} < 1 \quad (0 \leq \tau < \infty),$$

where

$$\rho_i = 2\int_0^1 \frac{\chi_i(\mu)}{1 + [\mu\beta(\mu)]^2}\, d\mu,$$

$\beta(\mu)$ is the root of the equation

$$\mu\beta = \cot(\tau\beta/2) \quad (0 \le \tau\beta \le \pi),$$

and

$$\chi_1(\mu) = \tfrac{3}{8}(1 + \mu^4), \qquad \chi_2(\mu) = \tfrac{3}{8}\mu^2(1 - \mu^2), \qquad \chi_3(\mu) = \tfrac{3}{4}(1 - \mu^2)^2.$$

These estimates are obtained by considering Λ as an integral of a one-parameter family of simpler selfadjoint operators and interchanging the usual Rayleigh-Ritz maximization and the integration [**21**].

If c is a nonnegative function of x, $0 \le c(x) \le 1$, it appears under the integral sign in equation (4.7). The resulting operator can be symmetrized, and the above estimate can be used to prove existence and uniqueness of a solution and to estimate the rate of convergence of iterations of equation (4.7) for inhomogeneous atmospheres as well.

Iteration in equation (4.7) defines one computational method which also applies to inhomogeneous atmospheres where c is a function of optical depth. Numerical results for inhomogeneous atmospheres have been obtained by these methods [**10**]. It should be noticed that x is the active variable in these integral equations and that ν is a parameter. For each μ and ν, by this iterative procedure $\boldsymbol{J}(x, \nu)$ must be computed for all $0 \le x \le \tau$ to determine the matrix $\boldsymbol{S}(\mu, \nu)$ of equation (3.5) even though, for constant c, we obtain in equation (4.22) the following representation

$$\boldsymbol{S}(\mu, \nu) = (\mu\nu/(\mu + \nu))\boldsymbol{Q}(\mu)[\boldsymbol{J}'(0, \mu)\boldsymbol{J}(0, \nu) - \boldsymbol{J}'(\tau, \mu)\boldsymbol{J}(\tau, \nu)]. \tag{4.9}$$

(b) *Linear singular integral equations.* Even though the above results establish existence and uniqueness of the solution to equation (4.7), the computation by iteration is not always the most efficient method. This is especially true if the solution is desired for a range of values of the parameters μ and ν at particular values of x, for example if the matrix S is to be computed.

We obtain from equation (4.7) an equation in which the variable x enters parametrically as follows. We apply the operator Λ to both sides of equation (4.7) and determine an equation for the matrix function $\Lambda(J)$ in which the $\boldsymbol{J}$-independent terms are integrals with respect to the parameter ν of combinations of tne exponential terms in equation (4.7).

Omitting the details [**20**], we give the equations which result for complex z ($z \notin [-1, 1]$):

$$\begin{aligned} \boldsymbol{J}(x, z)\boldsymbol{\lambda}(z) &= \boldsymbol{I}\exp(-x/z) + cz\int_0^1 \boldsymbol{J}(x, \mu')\boldsymbol{\psi}(\mu')\frac{d\mu'}{\mu' - z} \\ &\quad - cz\exp(-\tau/z)\int_0^1 \boldsymbol{J}(\tau - x, \mu')\boldsymbol{\psi}(\mu')\frac{d\mu'}{\mu' + z}, \end{aligned} \tag{4.10}$$

and a similar equation with x replaced by $(\tau - x)$. The matrix function $\boldsymbol{\lambda}$ is given by

$$\boldsymbol{\lambda}(z) = \boldsymbol{I} - 2cz^2\int_0^1 \boldsymbol{\psi}(\mu')\frac{d\mu'}{z^2 - (\mu')^2}. \tag{4.11}$$

The variable x enters parametrically in equation (4.10).

The connection with Wiener-Hopf methods is clear from the fact that

$$\boldsymbol{\lambda}(z) = \boldsymbol{I} - c\hat{\boldsymbol{K}}(1/z)$$

where $\hat{K}$ denotes the Laplace transform of the matrix kernel of the operator Λ.

From the form of the equation (4.7) and its unique solution, it follows that $\boldsymbol{J}(x, z)$ is analytic in the complex variable z, for $z \neq 0$. Suppose the matrix function $\boldsymbol{\lambda}(z)$ is singular for some z_0 ($z_0 \notin [-1, 1]$). Then $\boldsymbol{\lambda}(z_0)$ annihilates at least one vector $\boldsymbol{v}_0$. This gives linear constraints on the matrix $\boldsymbol{J}$

(4.12)

$$\boldsymbol{v}_0 \exp(-x/z_0) = cz \int_0^1 \left[\frac{\exp(-\tau/z_0)}{\mu' + z_0} \boldsymbol{J}(\tau - x, \mu') - \frac{1}{\mu' - z_0} \boldsymbol{J}(x, \mu') \right] \psi(\mu') \boldsymbol{v}_0 \, d\mu'.$$

For example, for $c = 1$, we can compute [20] that

$$\det \boldsymbol{\lambda}(z) \neq 0, \quad |z| < \infty \quad \text{and} \quad z \notin [-1, 1],$$

but that for $|z| \to \infty$

$$\boldsymbol{\lambda}(z)\begin{pmatrix} 2 \\ 1 \end{pmatrix} = O(1/z^2), \qquad \boldsymbol{\lambda}(z)\begin{pmatrix} 1 \\ -1 \end{pmatrix} = O(1).$$

So $\det \boldsymbol{\lambda}(z)$ vanishes at $|z| = \infty$ to second order but the rank of $\boldsymbol{\lambda}(z)$ equals one at $|z| = \infty$. Combined with equation (4.10) for $c = 1$, we obtain two linear constraints by equating coefficients of powers of $1/z$, [20]. For $0 \leq c < 1$ the analysis is more complicated.

We also obtain inhomogeneous matrix Hilbert problems as follows. Define the matrix function by

$$\boldsymbol{\phi}(x, z) = \frac{1}{2\pi i} \int_0^1 [\boldsymbol{J}(x, \mu') + \boldsymbol{J}(\tau - x, \mu')] \boldsymbol{\psi}(\mu') \frac{d\mu'}{\mu' - z}. \tag{4.13}$$

Then equation (4.10) requires the determination of $\boldsymbol{\phi}$ which vanishes at $|z| = \infty$, is holomorphic for $z \notin [0, 1]$, and satisfies on $[0, 1]$ the condition

$$\begin{aligned} \boldsymbol{\phi}^+ \boldsymbol{\psi}^{-1} \boldsymbol{\lambda}^- = \boldsymbol{\phi}^- \boldsymbol{\psi}^{-1} \boldsymbol{\lambda}^+ &+ 2\pi i c \mu \boldsymbol{\phi}(-\mu) \exp(-\tau/\mu) \\ &+ \boldsymbol{I}[\exp(-x/\mu) + \exp(x - \tau)/\mu)], \end{aligned} \tag{4.14}$$

where

$$\boldsymbol{\lambda}^{\pm} = \boldsymbol{\lambda}_0 \pm c\pi i \mu \boldsymbol{\psi} \tag{4.15}$$

and $\pm$ denotes nontangential limits from above and below the cut $[0, 1]$ in the z-plane. The constraints must also be considered with this equation. A corresponding problem is defined for the difference $[\boldsymbol{J}(x, \mu') - \boldsymbol{J}(\tau - x, \mu')]$.

In order to study the equation (4.14) we must first study the homogeneous problem of finding a matrix $\boldsymbol{E}$, holomorphic for $z \notin [-1, 1]$, and for which

$$\boldsymbol{E}^+(\mu)\boldsymbol{\lambda}^-(\mu) = \boldsymbol{E}^-(\mu)\boldsymbol{\lambda}^+(\mu), \quad (0 \leq \mu \leq 1).$$

This matrix Hilbert problem we attack in the standard manner (say in Muskhelishvili) by converting it to the form

$$E^+ = E^- G \tag{4.16}$$

where $G = \lambda^+(\lambda^-)^{-1}$.

We obtain for G the expression

$$G(\mu) = \frac{1}{\Delta(\mu)}[\det \lambda_0(\mu) + \tfrac{3}{2}(\pi c\mu(1-\mu^2))^2]I + \frac{3}{8}\frac{\pi i c\mu}{\Delta(\mu)}\begin{pmatrix} a(\mu) & b(\mu) \\ (1-c)\, d(\mu) & -a(\mu)\end{pmatrix} \tag{4.17}$$

where

$$a(\mu) = -1 + (4-3c)\mu^2 - (1-c)\mu^4, \qquad b(\mu) = 2^{1/2}[(2-c)\mu^2 - 2(1-c)\mu^4],$$
$$d(\mu) = 2\cdot 2^{1/2}\mu^2(1-\mu^2), \text{ and } \Delta = \det(\lambda^-).$$

We see that G is a full matrix if $c < 1$, and that the determination of a fundamental solution matrix E is not simple.

For $c = 1$ we see, however, that G is triangular

$$G(\mu) = \text{diagonal} + \frac{3\pi i\mu}{8\Delta(\mu)}\begin{pmatrix} \mu^2 - 1 & 2^{1/2}\mu^2 \\ 0 & 1-\mu^2\end{pmatrix}.$$

It then follows that the change of basis

$$E = FA \tag{4.18}$$

with

$$A(z) = \frac{1}{1-z^2}\begin{pmatrix} 1-z^2 & -z^2/2^{1/2} \\ 0 & 1\end{pmatrix}$$

gives the diagonalization of the system as $F^+ = F^- H$, where

$$H(\mu) = \text{diagonal} + \frac{3\pi i}{8\Delta(\mu)}\mu(1-\mu^2)\begin{pmatrix} -1 & 0 \\ 0 & 1\end{pmatrix}.$$

In fact, it has been shown [20] that

$$H = \begin{pmatrix} \lambda_1^+/\lambda_1^- & 0 \\ 0 & \lambda_2^+/\lambda_2^-\end{pmatrix}$$

with

$$\lambda_i(z) = 1 - 2z^2\int_0^1 \frac{\psi_i(\mu)}{z^2-\mu^2}\,d\mu$$

and

$$\psi_1(\mu) = \tfrac{3}{8}(1-\mu^2), \qquad \psi_2(\mu) = \tfrac{3}{4}(1-\mu^2).$$

A transformation similar to that in equation (4.18) was used [**20**] to express the matrix function $\boldsymbol{J}$ in terms of solutions to scalar singular equations

$$\lambda_i(\mu)J_i(x,\mu) = \exp(-x/\mu) + \mu\int_0^1 \psi_i(\mu')\left[\frac{1}{\mu'-\mu}J_i(x,\mu') - \frac{\exp(-\tau/\mu)}{\mu'+\mu}J_i(\tau-x,\mu')\right]d\mu'. \tag{4.19}$$

The various characteristic functions ψ_i are those introduced by Chandrasekhar [**5**, Chapter X]. These scalar singular integral equations can be transformed to rapidly convergent scalar Fredholm equations for which a computer program exists and has been used for computations [**4**], [**29**]. From these equations we have obtained asymptotics for large τ [**4**], [**20**]. The results given in [**20**] reduce the computation of the matrix function $\boldsymbol{J}$ of equation (4.3) to the solving of scalar Fredholm equations and some systems of algebraic equations. It is also shown that nonlinear combinations of these results give solutions to the original matrix equation (3.2) and boundary conditions (3.4). In all of these computations the variable x enters parametrically.

The above diagonalization, and the representation in equation (4.22) below, give an explanation of Chandrasekhar's success in reducing scattering and reflection matrix computations to computations of scalar functions X_i and Y_i, where one choice of these functions is

$$X_i(\mu) = J_i(0,\mu), \qquad Y_i(\mu) = J_i(\tau,\mu)$$

for the scalar functions J_i in equation (4.19).

When the constant c satisfies $0 < c < 1$, it is, of course, possible to diagonalize the matrix $\boldsymbol{G}$ of equation (4.16). The transformation needed, however, is not a meromorphic matrix function of the complex variable z. This method does not work and one seems forced to resort to the theory of systems of singular integral equations [**23**, Chapter 18] to reduce equations (4.12) and (4.14) to more desirable Fredholm equations. We know of no complete analysis in this particular context [**30**]. Similar problems arise in the study of multi-group neutron transport [**17′**].

(c) *Nonlinear equations—principles of invariance.* As an illustration of the nonlinear combinations of results of subsection (b) used to give solutions $\boldsymbol{M}(x,\mu,\nu)$ to equation (3.2) we discuss only the special matrix $\mu\boldsymbol{M}(0,\mu,\nu)$, the scattering matrix $\boldsymbol{S}$ of equation (3.5). For the case $c = 1$, we refer to [**20**, Theorem 10] for results for $0 \le x \le \tau$.

From equation (4.7) it is easy to show that the x-derivative of the matrix $\boldsymbol{J}$ has matrix elements which are square integrable functions of x $(0 \le x \le \tau)$ and that

$$\left(\frac{\partial}{\partial x}+\frac{1}{\nu}\right)\boldsymbol{J}(x,\nu) = c\int_0^1 \boldsymbol{J}(x,\mu)\boldsymbol{\psi}(\mu)\frac{d\mu}{\mu}\boldsymbol{J}(0,\nu) - c\int_0^1 \boldsymbol{J}(\tau-x,\mu)\boldsymbol{\psi}(\mu)\frac{d\mu}{\mu}\boldsymbol{J}(\tau,\nu). \tag{4.20}$$

We have from equations (3.5), (4.2) and (4.6) that

$$\text{(4.21)} \qquad \boldsymbol{S}(\mu, \nu) = \boldsymbol{Q}(\mu)\int_0^\tau \boldsymbol{J}(x, \nu)\exp(-x/\mu)\,dx.$$

It is easy to show that a computation of $\boldsymbol{S}$ from (4.20), use of equation (4.7), and use of the selfadjointness of the operator Λ gives the following representation

$$\text{(4.22)} \qquad \boldsymbol{S}(\mu, \nu) = \boldsymbol{Q}(\mu)[\boldsymbol{J}'(0, \mu)\boldsymbol{J}(0, \nu) - \boldsymbol{J}'(\tau, \mu)\boldsymbol{J}(\tau, \nu)].$$

It is not obvious from this result, but it is shown in [**20**, Theorem 7] that, for $c = 1$ at least,

$$\text{(4.22)} \qquad \boldsymbol{S}(\mu, \nu) = \boldsymbol{S}'(\nu, \mu).$$

This is a statement of the "principle of reciprocity" discussed by Chandrasekhar [**5**, Chapter VII].

The above representation shows that variables separate in the matrix $\boldsymbol{S}$ and requires the computation of the matrix $\boldsymbol{J}(x, \nu)$ at the two values $x = 0$ and $x = \tau$ only. For the special case $c = 1$ it is possible, as mentioned above, to reduce the computation of these matrices to computations of solutions of scalar Fredholm equations.

For the case $0 < c < 1$, we have a separation of variables in $\boldsymbol{S}$. In this case, however, further reduction to scalar equations does not seem possible and an efficient method for solving the system of singular equations for the matrices $\boldsymbol{J}(0, \mu)$ and $\boldsymbol{J}(\tau, \mu)$ is not obvious.

There are other equations which can be used for the determination of matrices such as $\boldsymbol{S}$. In analogy with scalar equations we define matrices

$$\boldsymbol{X}(\mu) = \boldsymbol{J}(0, \mu), \qquad \boldsymbol{Y}(\mu) = \boldsymbol{J}(\tau, \mu).$$

Then from equations (4.21) and (4.22) and equation (4.7) for $\boldsymbol{J}$, evaluated at $x = 0$ and $x = \tau$, we obtain nonlinear matrix equations [**5**], [**17**], [**28**]

$$\text{(4.23)} \qquad \begin{aligned} \boldsymbol{X}(\nu) &= \boldsymbol{I} + c\nu\int_0^1 \boldsymbol{\psi}(\mu)[\boldsymbol{X}'(\mu)\boldsymbol{X}(\nu) - \boldsymbol{Y}'(\mu)\boldsymbol{Y}(\nu)]\frac{d\mu}{\mu + \nu}, \\ \boldsymbol{Y}(\nu) &= \boldsymbol{I}\exp(-\tau/\nu) - c\nu\int_0^1 \boldsymbol{\psi}(\mu)[\boldsymbol{X}'(\mu)\boldsymbol{Y}(\nu) - \boldsymbol{Y}'(\mu)\boldsymbol{X}(\nu)]\frac{d\mu}{\mu - \nu}. \end{aligned}$$

One motivation for using these nonlinear equations as opposed to the linear singular equations (4.10) would be the hope that they have a unique solution which could be found by iteration. Such equations have been extensively used in the study of scalar X and Y functions [**1**], [**5**], [**7**], [**8**], [**9**], [**30**], [**31**]. In the scalar case we have given necessary and sufficient conditions for the unique solvability of similar equations [**18**], [**19**]. We have not carried out the analysis, but undoubtedly for the above matrix equations the condition for uniqueness is analogous to that for the scalar case, and for the linear singular matrix equations of subsection (b), namely

$$\det \boldsymbol{\lambda}(z) \neq 0 \quad \text{for } z \notin [-1, 1].$$

Any use of the above equations for computations must incorporate linear constraints (4.12).

There are other nonlinear equations which use the fact that the matrix function $\partial J/\partial\tau$ has entries which are square integrable functions of x $(0 \leq x \leq \tau)$. It follows from equation (4.7) that

$$\frac{\partial}{\partial\tau} J(x, \nu, \tau) = c\int_0^1 J(\tau - x, \mu, \tau)\psi(\mu)\frac{d\mu}{\mu} Y(\nu, \tau). \tag{4.24}$$

This equation can be used to obtain a nonlinear integro-differential equation for the associated scattering matrix defined by

$$s(\mu, \nu, \tau) = \int_0^\tau J(x, \nu) \exp(-x/\mu)\, dx. \tag{4.25}$$

Differentiating this with respect to τ and using (4.24), we find

(4.26)

$$\frac{\partial}{\partial\tau} s(\mu, \nu, \tau) = \left[\exp(-\tau/\mu)I + \int_0^1\int_0^\tau \exp(-x/\mu)J(\tau - x, \mu', \tau)\, dx\, \psi(\mu')\frac{d\mu'}{\mu'}\right] Y(\nu, \tau).$$

From equation (4.7) we see that the solution to the equation

$$(I - c\Lambda)K(x, \mu', \tau) = \exp(-(\tau - x)/\mu')I$$

is given by $K(x, \mu', \tau) = J(\tau - x, \mu', \tau)$. We can use this fact, together with the selfadjointness of the operator $(I - c\Lambda)^{-1}$ to rewrite the inner product

$$\int_0^\tau \exp(-x/\mu')J(\tau - x, \mu', \tau)\, dx = \int_0^x J(x, \mu', \tau) \exp\left(\frac{x - \tau}{\mu'}\right) dx$$

so that equation (4.26) becomes

$$\frac{\partial}{\partial\tau} s(\mu, \nu, \tau) = \left[\exp(-\tau/\mu)I + \int_0^1\int_0^\tau \exp\left(\frac{x - \tau}{\mu'}\right) J'(x, \mu, \tau)\, dx\, \psi(\mu')\frac{d\mu'}{\mu'}\right] Y(\nu, \tau).$$

By symmetry of the matrix ψ we recognize the bracket from equation (4.7) as the matrix $Y'(\mu, \tau)$ to obtain

$$\partial s(\mu, \nu, \tau)/\partial\tau = Y'(\mu, \tau)Y(\nu, \tau). \tag{4.27}$$

This can be combined with equations (4.22) and (4.23) to obtain the nonlinear integro-differential equation

(4.28)

$$\left(\frac{\partial}{\partial\tau} + \frac{\mu + \nu}{\mu\nu}\right) s(\mu, \nu, \tau) = \left[I + c\int_0^1 s(\mu, \mu')\psi(\mu')\frac{d\mu'}{\mu'}\right]\left[I + c\int_0^1 \psi(\mu'')s'(\mu'', \nu)\frac{d\mu''}{\mu''}\right].$$

The initial condition for $\tau = 0$, no atmosphere, is obviously

$$s(\mu, \nu, 0) = \mathbf{0}. \tag{4.29}$$

To take advantage of the separation of variables in the s matrix we can use the easily established relation

$$X(-\nu, \tau) = \exp(\tau/\nu) Y(\nu, \tau)$$

to obtain the system of nonlinear matrix integro-differential equations, for $0 \leq \nu \leq 1$,

$$\begin{aligned} \frac{\partial}{\partial \tau} X(\nu, \tau) &= c \int_0^1 Y(\mu, \tau) \psi(\mu) \frac{d\mu}{\mu} Y(\nu, \tau), \\ \left(\frac{\partial}{\partial \tau} + \frac{1}{\mu}\right) Y(\nu, \tau) &= c \int_0^1 Y(\mu, \tau) \psi(\mu) \frac{d\mu}{\mu} X(\nu, \tau). \end{aligned} \tag{4.30}$$

The initial conditions are

$$X(\nu, 0) = Y(\nu, 0) = I. \tag{4.31}$$

For scalar transport problems Chandrasekhar [**5**, Chapters VII and VIII] gives the above nonlinear equations. He mentions the extension to matrix equations but does not write them down (cf. [**26**], [**28**]).

For the scalar case corresponding to isotropic scattering these nonlinear equations have been used for numerical computations [**2**], [**15**]. We have discussed elsewhere some stability problems for these equations [**22**]. We have shown that for the conservative case of isotropic scattering the equation analogous to (4.28) is not uniformly stable. We show that for any $\epsilon > 0$ there are initial conditions for the s-equation which determine a solution which differs from the solution for the boundary condition $s = 0$ for $\tau = 0$ by less than ϵ for some τ_ϵ and is unbounded for some $T > \tau_\epsilon$. Similar studies have not been made of the matrix equations.

References

1. V. A. Ambartsumian, *Theoretical astrophysics*, Gosudstv. Isdat. Tehn-Teor. Lit., Moscow, 1952; English transl., Pergamon, London, 1958.

2. R. Bellman, R. Kalaba and M. Prestrud, *Invariant imbedding and radiative transfer in slabs of finite thickness*, American Elsevier, New York, 1963.

3. G. I. Bell and W. B. Goad, *Polarization effects on neutron transport*, Nucl. Sci. and Engrg. **23** (1965), 380–391.

4. J. L. Carlstedt and T. W. Mullikin, *Chandrasekhar's X- and Y-functions*, Astrophys. J. Suppl., XII, No. 113, 1966, pp. 449–586.

5. S. Chandrasekhar, *Radiative transfer*, Dover, New York, 1960 and Cambridge Univ. Press, Cambridge, 1950.

6. ———, *On the radiative equilibrium of a stellar atmosphere*. X and XI, Astrophys. J. **103** (1946), 351–370 and **104** (1946), 110–132.

7. S. Chandrasekhar and D. D. Elbert, *Illumination and polarization of the sunlit sky on Rayleigh scattering*, Trans. Amer. Philos. Soc. **44** (1954), 643–728.

8. ———, *The X- and Y-functions for isotropic scattering*. I, II, Astrophys. J. **115** (1952), 244–278.

9. K. L. Coulson, J. V. Dave and Z. Sekera, *Tables related to radiation emerging from a planetary atmosphere with Rayleigh scattering*, Univ. of California Press, Los Angeles, Calif., 1960.

10. J. V. Dave, *Multiple scattering in a non-homogeneous, Rayleigh atmosphere*, J. Atmospheric Sci. **22** (1965), 273–279.

11. D. Deirmendjian, *Scattering and polarization properties of polydispersed suspensions with partial absorption*, Proc. of Interdisciplinary Conf. on Electromagnetic Scattering, Pergamon Press, New York, 1963.

12. E. Hopf, *Mathematical problems of radiative equilibrium*, Cambridge Tracts, No. 31, 1934.

13. B. Herman and D. N. Yarger, *Multiple scattering from particles in the Mie region*, Proc. of the Interdisciplinary Conf. on Electromagnetic Scattering, Amherst, Mass., 1965, Gordon and Breach, New York, 1967.

14. H. C. van de Hulst, *Light scattering by small particles*, Wiley, New York, 1957.

15. H. Kagiwada and R. Kalaba, *Initial value methods for the basic boundary value problem and integral equation of radiative transfer*, The RAND Corporation, RM-4928-PR, 1966.

16. A. B. Kahle, *Global radiation emerging from a Rayleigh scattering atmosphere of large optical thickness*, The RAND Corporation, RM-5343-PR, 1967.

17. M. Kerker (Editor), *Electromagnetic scattering*, Proc. of First Interdisciplinary Conf. on Electromagnetic Scattering, Potsdam, N.Y., 1962, Pergamon Press, New York, 1963.

17′. A. Leonard and J. H. Ferziger, *Energy-dependent neutron transport in plane geometry.* II, III, Nucl. Sci. and Engrg. **26** (1966), 170–191.

18. T. W. Mullikin, *A complete solution of the X and Y equations of Chandrasekhar*, Astrophys. J. **136** (1962), 627–635.

19. ———, *Chandrasekhar's X and Y equations*, Trans. Amer. Math. Soc. **113** (1964), 316–332.

20. ———, *The complete Rayleigh-scattered field within a homogeneous plane-parallel atmosphere*, Astrophys. J. **145** (1966), 886–931.

21. ———, *Estimates of critical dimensions of spherical and slab reactors*, J. Math. Anal. Appl. **5** (1962), 184–199.

22. ———, *A nonlinear integrodifferential equation in radiative transfer*, J. Soc. Indust. Appl. Math. **13** (1965), 388–410.

23. N. I. Muskhelishvili, *Singular integral equations*, Noordhoff, Gröningen-Holland, 1953.

24. Lord Rayleigh, *Scientific papers of Lord Rayleigh.* I, Cambridge, 1899.

25. Z. Sekera, "Polarization of skylight," in *Handbuch der Physik*, vol. 48, Springer-Verlag, Berlin, 1957, pp. 288–328.

26. ———, *Reduction of the equations of radiative transfer for a plane-parallel planetary atmosphere*, Part I and II, The RAND Corporation, RM-4951-PR, RM-5056-PR, 1966.

27. ———, *Determination of atmospheric parameters from measurement of polarization of upward radiation by satellite or space probe*, The RAND Corporation, RM-5158-PR, 1966.

28. ———, *Radiative transfer in a planetary atmosphere with imperfect scattering*, The RAND Corporation, R-413-PR, 1963.

29. Z. Sekera and A. B. Kahle, *Scattering functions for Rayleigh atmospheres of arbitrary thickness*, The RAND Corporation, R-452-PR, 1966.

30. V. V. Sobolev, *A treatise on radiative transfer*, Moscow, 1956; English transl., Van Nostrand, Princeton, N.J., 1963.

31. Y. Sobouti, *Chandrasekhar's X- and Y- and related functions*, Astrophys. J. Suppl. (72) **7** (1963), 411.

32. G. Stokes, *Mathematical and physical papers of Sir George Stokes.* III, Cambridge, 1922, pp. 233–259.

33. E. P. Wigner, *Problems of nuclear reactor theory*, Proc. Sympos. Appl. Math. XI, Amer. Math. Soc., Providençe, R.I., 1961.

PURDUE UNIVERSITY
LAFAYETTE, INDIANA

On the boundary value problems of linear transport theory[1]

K. M. Case

I. **Introduction.** In many branches of physics one encounters linear transport equations which are required to be solved subject to certain prescribed conditions on various bounding surfaces. Here we will discuss some methods which appear to be rather generally applicable to such problems. Our essential limitation is to *linear* equations. This is not because nonlinear problems are less interesting—they are more so—but rather the nonlinear problems are much harder and do not seem amenable to as general a discussion.

The point of view to be followed arises so: one frequently says that the problems in transport theory are significantly different from the classical problems associated with partial differential equations. The question we want to raise is as to how closely one *can* parallel the methods used for the classical boundary value problems in transport theory. In particular we mean methods like those used in electromagnetic theory to reduce the boundary value problems to integral equations for the pertinent surface charges and currents. The advantages of that approach are, among others, that

(i) It permits us to gain some general insight into the nature of the solutions.

(ii) The method is particularly well adapted to approximation procedures.

We may expect similar advantages if transport problems are also so formulated.

Our main result is that the appropriate formulation is rather straightforward to obtain. Indeed, while the relevant kernels may be somewhat more complicated algebraically, the underlying mathematical structure of the resulting equations seem, if anything, to be simpler than for the classical problems. This, presumably, is because the transport equations are only first order differential equations in the position variables.

[1] The work reported here was supported in part by the National Science Foundation.

Two remarks may be made here:

(a) The relationship of the present approach to that of Invariant Imbedding is rather interesting. In both cases one obtains integral equations—for essentially the same quantities. Here, however, the equations to be solved are linear ones. While it is often claimed that nonlinear equations are easier to handle *numerically*, we think the present formulation is somewhat more desirable. It turns out to require much less ingenuity to pass from our linear equations to the nonlinear ones than vice versa.

(b) It will be seen that many earlier results—such as normal mode expansions—are readily incorporated in the present scheme. Indeed, generalizations of known completeness and orthogonality properties become trivially apparent.

In §II we develop the general formulation of the method and deduce some properties of the relevant kernels. As an illustration we treat (in §III) the simplest example—one velocity neutron transport in plane geometry. The modifications necessary for plasma problems are outlined in §IV.

II. **General formulation and properties.** The general form of equation we wish to consider is

$$\frac{\partial \Psi}{\partial t} + \boldsymbol{v} \cdot \nabla \Psi + \sigma' \Psi = \int f(\boldsymbol{v}, \boldsymbol{v}'; \boldsymbol{r}) \Psi(\boldsymbol{r}, \boldsymbol{v}', t)\, d^3v' + Q(\boldsymbol{r}, \boldsymbol{v}, t). \tag{1}$$

By taking the Fourier transform with respect to the time variable, we reduce this to

$$\boldsymbol{v} \cdot \nabla \Psi(\boldsymbol{r}, \boldsymbol{v}) + \sigma(\boldsymbol{r}, |\boldsymbol{v}|)\Psi = \int f(\boldsymbol{v}, \boldsymbol{v}'; \boldsymbol{r}) \Psi(\boldsymbol{r}, \boldsymbol{v}')\, d^3v' + Q(\boldsymbol{r}, \boldsymbol{v}). \tag{2}$$

The problem is to find the solution of (2) in a region V bounded by a surface S. Appropriate boundary conditions are presumed prescribed on S. Here $Q(\boldsymbol{r}, \boldsymbol{v})$ is some given function and we explicitly assume that σ depends only on $|\boldsymbol{v}|$. It may be noted that we have also assumed that the kernel f is *local* in $\boldsymbol{r}$. Temporarily, this will exclude plasma problems. However, it will be shown in §IV that only minor modifications suffice to adapt the procedure to that case.

It is convenient to introduce an *adjoint* equation. This we do by replacing the scattering kernel f by its time reversed form, i.e.

$$f(\boldsymbol{v}, \boldsymbol{v}'; \boldsymbol{r}) \to \tilde{f}(\boldsymbol{v}, \boldsymbol{v}'; \boldsymbol{r}) = f(-\boldsymbol{v}', -\boldsymbol{v}; \boldsymbol{r}). \tag{3}$$

Now let $\tilde{\Psi}(\boldsymbol{r}, \boldsymbol{v})$ be a solution of this adjoint equation with source $\tilde{Q}(\boldsymbol{r}, \boldsymbol{v})$. Then

$$-\boldsymbol{v} \cdot \nabla \tilde{\Psi}(\boldsymbol{r}, -\boldsymbol{v}) + \sigma \tilde{\Psi} = \int f(\boldsymbol{v}, \boldsymbol{v}'; \boldsymbol{r}) \tilde{\Psi}(\boldsymbol{r}, -\boldsymbol{v}')\, d^3v' + \tilde{Q}(\boldsymbol{r}, -\boldsymbol{v}). \tag{4}$$

Conventional manipulation with equations (2) and (4) yields the Green's type identity

$$\begin{aligned} &\iint_V \{\Psi(\boldsymbol{r}, \boldsymbol{v})\tilde{Q}(\boldsymbol{r}, -\boldsymbol{v}) - \tilde{\Psi}(\boldsymbol{r}, -\boldsymbol{v}) Q(\boldsymbol{r}, \boldsymbol{v})\}\, d^3r\, d^3v \\ &\qquad = \iint_S \boldsymbol{n}_i(\boldsymbol{r}_S) \cdot \boldsymbol{v} \tilde{\Psi}(\boldsymbol{r}_S, -\boldsymbol{v}) \Psi(\boldsymbol{r}_S, \boldsymbol{v})\, dS\, d^3v. \end{aligned} \tag{5}$$

(Here $\boldsymbol{n}_i(\boldsymbol{r}_S)$ denotes the inner normal at the point $\boldsymbol{r}_S$ on S.)

Specializing to various particular Q and $\tilde{Q}$, we obtain a number of useful results.

(a) Let $\tilde{\Psi}$ be the infinite medium Green's function for the adjoint equation, i.e.

$$\tilde{Q}(\boldsymbol{r}, \boldsymbol{v}) = \delta(\boldsymbol{r} - \boldsymbol{r}_0)\delta(\boldsymbol{v} - \boldsymbol{v}_0) \tag{6}$$

and the adjoint equation is to be satisfied everywhere.[2] Denote this solution by $\tilde{G}(\boldsymbol{r}, \boldsymbol{v}; \boldsymbol{r}_0, \boldsymbol{v}_0)$. Substituting in equation (5) we obtain

$$\begin{aligned}\int_V \Psi(\boldsymbol{r}, \boldsymbol{v}_0)\delta(\boldsymbol{r} - \boldsymbol{r}_0)\, d^3r &= \iint_V \tilde{G}(\boldsymbol{r}, -\boldsymbol{v}; \boldsymbol{r}_0, -\boldsymbol{v}_0)Q(\boldsymbol{r}, \boldsymbol{v})\, d^3r\, d^3v \\ &\quad + \iint_S \tilde{G}(\boldsymbol{r}_S, -\boldsymbol{v}; \boldsymbol{r}_0, -\boldsymbol{v}_0)\boldsymbol{n}_i(\boldsymbol{r}_S) \cdot \boldsymbol{v}\Psi(\boldsymbol{r}_S, \boldsymbol{v})\, dS\, d^3v.\end{aligned} \tag{7}$$

(b) In equation (7) we let V be all of space and Ψ be the infinite medium Green's function for the original transport equation, i.e.

$$Q(\boldsymbol{r}, \boldsymbol{v}) = \delta(\boldsymbol{r} - \boldsymbol{r}_1)\delta(\boldsymbol{v} - \boldsymbol{v}_1), \qquad \Psi = G(\boldsymbol{r}, \boldsymbol{v}; \boldsymbol{r}_1, \boldsymbol{v}_1).$$

This then becomes

$$\tilde{G}(\boldsymbol{r}_1, -\boldsymbol{v}_1; \boldsymbol{r}_0, -\boldsymbol{v}_0) = G(\boldsymbol{r}_0, \boldsymbol{v}_0; \boldsymbol{r}_1, \boldsymbol{v}_1). \tag{8}$$

We remark that this is the general *Reciprocity Principle*. The Green's function and its time reversed relative (the adjoint function) are related in a manner which is intuitively very clear. If in particular we have time reversal invariance, i.e.

$$f(\boldsymbol{v}, \boldsymbol{v}'; \boldsymbol{r}) = f(-\boldsymbol{v}', -\boldsymbol{v}; \boldsymbol{r}), \tag{9}$$

then equation (8) simplifies to

$$G(\boldsymbol{r}_0, \boldsymbol{v}_0; \boldsymbol{r}_1, \boldsymbol{v}_1) = G(\boldsymbol{r}_1, -\boldsymbol{v}_1; \boldsymbol{r}_0, -\boldsymbol{v}_0). \tag{10}$$

In any event, we can use the general principle (equation (8)) to eliminate adjoint quantities from equation (7). This then becomes

$$\begin{aligned}\int_V \Psi(\boldsymbol{r}', \boldsymbol{v})\delta(\boldsymbol{r} - \boldsymbol{r}')\, d^3r' &= \iint_V G(\boldsymbol{r}, \boldsymbol{v}; \boldsymbol{r}', \boldsymbol{v}')Q(\boldsymbol{r}', \boldsymbol{v}')\, d^3r'\, d^3v' \\ &\quad + \iint_S G(\boldsymbol{r}, \boldsymbol{v}; \boldsymbol{r}'_S, \boldsymbol{v}')\boldsymbol{n}_i(\boldsymbol{r}'_S) \cdot \boldsymbol{v}'\Psi(\boldsymbol{r}', \boldsymbol{v}')\, dS'\, d^3v'.\end{aligned} \tag{11}$$

(c) If $\boldsymbol{r} \in V$, we see that equation (11) becomes the integral representation

$$\begin{aligned}\Psi(\boldsymbol{r}, \boldsymbol{v}) &= \iint_V G(\boldsymbol{r}, \boldsymbol{v}; \boldsymbol{r}', \boldsymbol{v}')Q(\boldsymbol{r}', \boldsymbol{v}')\, d^3r'\, d^3v' \\ &\quad + \iint_S G(\boldsymbol{r}, \boldsymbol{v}; \boldsymbol{r}'_S, \boldsymbol{v}')\boldsymbol{n}_i(\boldsymbol{r}'_S) \cdot \boldsymbol{v}'\Psi(\boldsymbol{r}'_S, \boldsymbol{v}')\, dS'\, d^3v'.\end{aligned} \tag{12}$$

[2] In general some condition at infinity will be needed to make $\tilde{G}(G)$ unique. Here let us assume that vanishing at infinity is an appropriate requirement. (In other cases the modifications should be obvious.)

In particular, we see that the solution of the boundary value problem is known provided we can find the surface distribution $\Psi(\boldsymbol{r}_S, \boldsymbol{v})$.

(d) For $\boldsymbol{r} \notin V$, we find that equation (11) becomes the identity

$$(13)\quad \begin{aligned} 0 &= \iint_V G(\boldsymbol{r}, \boldsymbol{v}; \boldsymbol{r}', \boldsymbol{v}')Q(\boldsymbol{r}', \boldsymbol{v}')d^3r'\, d^3v' \\ &+ \iint_S G(\boldsymbol{r}, \boldsymbol{v}; \boldsymbol{r}', \boldsymbol{v}')\boldsymbol{n}_i(\boldsymbol{r}'_S) \cdot \boldsymbol{v}'\Psi(\boldsymbol{r}'_S, \boldsymbol{v}')\, dS'\, d^3v'. \end{aligned}$$

From equations (12) and (13) we readily deduce integral equations for the needed $\Psi(\boldsymbol{r}_S, \boldsymbol{v})$ by passing to the limit $\boldsymbol{r} \to \boldsymbol{r}_S$. These are

$$(14)\quad \begin{aligned} \begin{pmatrix} \Psi(\boldsymbol{r}_S, \boldsymbol{v}) \\ 0 \end{pmatrix} &= \iint_V G(\boldsymbol{r}_S, \boldsymbol{v}; \boldsymbol{r}', \boldsymbol{v}')Q(\boldsymbol{r}', \boldsymbol{v}')\, d^3r'\, d^3v' \\ &+ \iint_S G_\pm(\boldsymbol{r}_S, \boldsymbol{v}; \boldsymbol{r}'_S, \boldsymbol{v}')\boldsymbol{n}_i(\boldsymbol{r}'_S) \cdot \boldsymbol{v}'\Psi(\boldsymbol{r}'_S, \boldsymbol{v}')\, dS'\, d^3v', \end{aligned}$$

where

$$(15)\quad G_\pm(\boldsymbol{r}_S, \boldsymbol{v}; \boldsymbol{r}'_S, \boldsymbol{v}') = \lim_{r \to r_s \left\{\substack{\text{from within V} \\ \text{from without V}}\right\}} G(\boldsymbol{r}, \boldsymbol{v}; \boldsymbol{r}'_S, \boldsymbol{v}').$$

At first sight it may seem peculiar that we seem to have derived *two* integral equations for the surface distribution. Actually however, these two are identical. To see this, let us consider equation (2) with $Q = \delta(\boldsymbol{r} - \boldsymbol{r}'_S)\delta(\boldsymbol{v} - \boldsymbol{v}')$ in the vicinity of $\boldsymbol{r}'_S$. If we introduce an orthogonal coordinate system with one component in the direction of $\boldsymbol{n}_i(\boldsymbol{r}'_S)$, the other two components tangential to S at $\boldsymbol{r}'_S$, and integrate from a point slightly outside S to a point slightly inside S, we obtain

$$(16)\quad \{G_+(\boldsymbol{r}_S, \boldsymbol{v}; \boldsymbol{r}'_S, \boldsymbol{v}') - G_-(\boldsymbol{r}_S, \boldsymbol{v}; \boldsymbol{r}'_S, \boldsymbol{v}')\}\boldsymbol{n}_i(\boldsymbol{r}'_S) \cdot \boldsymbol{v}' = \delta^S(\boldsymbol{r}_S - \boldsymbol{r}'_S)\delta(\boldsymbol{v} - \boldsymbol{v}').$$

(Here $\delta^S(\boldsymbol{r}_S - \boldsymbol{r}'_S)$ denotes the two dimensional δ function on the surface S.) Using equation (16) we readily see that the lower of equation (14) implies the upper and vice versa.

Some other general properties of the kernels $G_\pm$ are readily found. Thus suppose we take for $\Psi(\boldsymbol{r}_S, \boldsymbol{v})$ in equation (14)

$$(17)\quad \Psi(\boldsymbol{r}_S, \boldsymbol{v}) = G_+(\boldsymbol{r}_S, \boldsymbol{v}; \boldsymbol{r}''_S, \boldsymbol{v}''),$$

then since the corresponding Q is zero within V we obtain the relations

$$(18)\quad G_+(\boldsymbol{r}_S, \boldsymbol{v}; \boldsymbol{r}''_S, \boldsymbol{v}'') = \iint_S G_+(\boldsymbol{r}_S, \boldsymbol{v}; \boldsymbol{r}'_S, \boldsymbol{v}')\boldsymbol{n}_i(\boldsymbol{r}'_S) \cdot \boldsymbol{v}'G_+(\boldsymbol{r}'_S, \boldsymbol{v}'; \boldsymbol{r}''_S, \boldsymbol{v}'')\, dS'\, d^3v'$$

and

$$(19)\quad 0 = \iint_S G_-(\boldsymbol{r}_S, \boldsymbol{v}; \boldsymbol{r}'_S, \boldsymbol{v}')\boldsymbol{n}_i(\boldsymbol{r}'_S) \cdot \boldsymbol{v}'G_+(\boldsymbol{r}'_S, \boldsymbol{v}'; \boldsymbol{r}''_S, \boldsymbol{v}'')\, dS'\, d^3v'.$$

Similarly by considering the *exterior* problem we get two more relations which differ from these in that G_+ and G_- are interchanged. These various properties are compactly described so. Let $P_\pm$ be operators defined on functions on the surface(s) S and all velocities by

$$(20)\qquad P_\pm \Psi(\boldsymbol{r}_S, \boldsymbol{v}) = \pm \iint_S G_\pm(\boldsymbol{r}_S, \boldsymbol{v}; \boldsymbol{r}'_S, \boldsymbol{v}')\boldsymbol{n}_i(\boldsymbol{r}'_S)\cdot \boldsymbol{v}'\Psi(\boldsymbol{r}'_S, \boldsymbol{v}')\, dS'\, d^3v',$$

then P_+ and P_- are mutually orthogonal projection operators that give a resolution of the identity, i.e.

$$(21)\qquad P_+ + P_- = 1,$$

and

$$(22)\qquad P_+^2 = P_+, \qquad P_-^2 = P_-, \qquad P_+P_- = P_-P_+ = 0.$$

For reasons that will be clear later, we call equation (21) the *generalized completeness* property while equations (22) are properly called the generalized orthogonality relations.

Our formulation of the boundary value problems can then be summarized in the following form:

(i) $\Psi(\boldsymbol{r}_S, \boldsymbol{v})$ is to be determined from equation (14). (The properties given by equations (21) and (22) may be expected to materially simplify the calculation.)

(ii) Given $\Psi(\boldsymbol{r}_S, \boldsymbol{v})$, we then compute Ψ everywhere in V using equation (12).

Some remarks are in order here:

(a) Clearly, to make any progress we must be able to construct G. This is hardly surprising—if we cannot handle the infinite medium problem we could not expect to be able to treat boundary value problems. It should be noted that in special cases of interest the construction of G is not difficult. Thus, suppose σ and f are independent of $\boldsymbol{r}$. Also suppose that f is a degenerate kernel. Then we can always write down immediately a Fourier integral representation for G.

(b) Our representation of the solution can be very convenient—for example, we readily read off asymptotic properties.

(c) In particular, if we can approximate $\Psi(\boldsymbol{r}_S, \boldsymbol{v})$, we will have a good representation everywhere. This is often easy to obtain in various limiting situations such as when V is very small or very large. Further it may be noted that in special cases, at least, our integral equation follows from a rather simple variational principle. Thus, in various neutron transport problems, we have in addition to the time reflection symmetry

$$(23)\qquad G_+(\boldsymbol{r}_S, \boldsymbol{v}; \boldsymbol{r}'_S, \boldsymbol{v}') = G_+(\boldsymbol{r}'_S, -\boldsymbol{v}'; \boldsymbol{r}_S, -\boldsymbol{v}),$$

also the symmetry

$$(24)\qquad G_+(\boldsymbol{r}_S, \boldsymbol{v}; \boldsymbol{r}'_S, \boldsymbol{v}') = G_+(\boldsymbol{r}'_S, \boldsymbol{v}'; \boldsymbol{r}_S, \boldsymbol{v}).$$

In this case, we readily see that

$$K = \left[2\iiint_V \int_S \Psi(\boldsymbol{r}_S, \boldsymbol{v})\boldsymbol{n}_i(\boldsymbol{r}_S)\cdot\boldsymbol{v}G(\boldsymbol{r}_S, \boldsymbol{v}'; \boldsymbol{r}', \boldsymbol{v}')Q(\boldsymbol{r}', \boldsymbol{v}')\, dS\, d^3r'\, d^3v\, d^3v' \right.$$

$$\left. + \iiint_S \int_S \Psi(\boldsymbol{r}_S, \boldsymbol{v})\boldsymbol{n}_i(\boldsymbol{r}_S)\cdot\boldsymbol{v}G_+(\boldsymbol{r}_S, \boldsymbol{v}; \boldsymbol{r}'_S, \boldsymbol{v}')\boldsymbol{n}_i(\boldsymbol{r}'_S)\cdot\boldsymbol{v}'\Psi(\boldsymbol{r}'_S, \boldsymbol{v}')\, dS\, dS'\, d^3v\, d^3v'\right] \tag{25}$$

$$\times \left(\iint_S \boldsymbol{n}_i(\boldsymbol{r}_S)\cdot\boldsymbol{v}\Psi^2(\boldsymbol{r}_S, \boldsymbol{v})\, dS\, d^3v\right)^{-1}$$

is stationary.

III. **Illustrations.**

A. *Simplifications.* To illustrate the approach and to see the general structure of the integral equations that are encountered let us consider some of the simplest problems of the type discussed in the previous section. These arise when one considers stationary one-velocity neutron transport with isotropic scattering [**1**].

The relevant Green's function satisfies the equation

$$\begin{aligned} &\boldsymbol{\Omega}\cdot\nabla G(\boldsymbol{r}, \boldsymbol{\Omega}; \boldsymbol{r}', \boldsymbol{\Omega}') + G(\boldsymbol{r}, \boldsymbol{\Omega}; \boldsymbol{r}', \boldsymbol{\Omega}') \\ &\qquad = \frac{c}{4\pi}\int G(\boldsymbol{r}, \boldsymbol{\Omega}''; \boldsymbol{r}', \boldsymbol{\Omega}')\, d\boldsymbol{\Omega}'' + \delta(\boldsymbol{r} - \boldsymbol{r}')\delta(\boldsymbol{\Omega}\cdot\boldsymbol{\Omega}'). \end{aligned} \tag{26}$$

(Here $\boldsymbol{\Omega}$'s are unit vectors specifying the *direction* of the neutron velocity and we have introduced units such as to remove extraneous constants. For simplicity, we restrict ourselves to the case $0 \leqq c < 1$.)

The integral representation for the distribution function then becomes

$$\begin{aligned} \Psi(\boldsymbol{r}, \boldsymbol{\Omega}) &= \iint_V G(\boldsymbol{r}, \boldsymbol{\Omega}; \boldsymbol{r}', \boldsymbol{\Omega}')Q(\boldsymbol{r}', \boldsymbol{\Omega}')\, d^3r'\, d\boldsymbol{\Omega}' \\ &\quad + \iint_S G(\boldsymbol{r}, \boldsymbol{\Omega}; \boldsymbol{r}'_S, \boldsymbol{\Omega}')\boldsymbol{n}_i(\boldsymbol{r}'_S)\cdot\boldsymbol{\Omega}'\Psi(\boldsymbol{r}'_S, \boldsymbol{\Omega}')\, dS'\, d\boldsymbol{\Omega}', \end{aligned} \tag{27}$$

while the integral equation for the needed surface distribution is

$$\begin{aligned} \Psi(\boldsymbol{r}_S, \boldsymbol{\Omega}) &= \iint_V G(\boldsymbol{r}_S, \boldsymbol{\Omega}; \boldsymbol{r}', \boldsymbol{\Omega}')Q(\boldsymbol{r}', \boldsymbol{\Omega}')\, d^3r'\, d\boldsymbol{\Omega}' \\ &\quad + \iint_S G_+(\boldsymbol{r}_S, \boldsymbol{\Omega}; \boldsymbol{r}'_S, \boldsymbol{\Omega}')\boldsymbol{n}_i(\boldsymbol{r}'_S)\cdot\boldsymbol{\Omega}'\Psi(\boldsymbol{r}'_S, \boldsymbol{\Omega}')\, dS'\, d\boldsymbol{\Omega}\,. \end{aligned} \tag{28}$$

Let us simplify even further to problems of plane symmetry, i.e. where

$$\Psi(x, y, z; \boldsymbol{\Omega}) = \Psi(x, 0, 0; \boldsymbol{\Omega}), \tag{29}$$

and the bounding surfaces are planes x = constant. The relevant Green's function then satisfies

(30)

$$\left(\mu \frac{\partial}{\partial x} + 1\right) G(x, \boldsymbol{\Omega}; x', \boldsymbol{\Omega}') = \frac{c}{4\pi} \int G(x, \boldsymbol{\Omega}''; x', \boldsymbol{\Omega}')\, d\boldsymbol{\Omega}'' + \delta(x - x')\delta(\boldsymbol{\Omega} \cdot \boldsymbol{\Omega}'),$$

where $\mu = \Omega_x$.

Further, equations (27) and (28) become

(31)

$$\Psi(x, \boldsymbol{\Omega}) = \iint_V G(x, \boldsymbol{\Omega}; x', \boldsymbol{\Omega}') Q(x', \boldsymbol{\Omega}')\, dx'\, d\boldsymbol{\Omega}' + \sum_{S'} \int G(x, \boldsymbol{\Omega}; x', \boldsymbol{\Omega}') \boldsymbol{n}_i(x'_S) \cdot \boldsymbol{\Omega}' \Psi(x'_S, \boldsymbol{\Omega}') d\boldsymbol{\Omega}'$$

and

(32)

$$\Psi(x_S, \boldsymbol{\Omega}) = \iint_V G(x_S, \boldsymbol{\Omega}; x', \boldsymbol{\Omega}') Q(x', \boldsymbol{\Omega}')\, dx'\, d\boldsymbol{\Omega}' + \sum_{S'} \int G_+(x_S, \boldsymbol{\Omega}; x'_S, \boldsymbol{\Omega}') \boldsymbol{n}_i(x'_S) \cdot \boldsymbol{\Omega}' \Psi(x'_S, \boldsymbol{\Omega}')\, d\boldsymbol{\Omega}'.$$

B. *Construction of the Green's function.* Clearly the first step is to construct G; while there are many simple ways to do this, we here choose a normal mode approach—it throws considerable light on the generalized orthogonality and completeness properties. Thus consider the homogeneous form of equation (30),

(33)
$$\left(\mu \frac{\partial}{\partial x} + 1\right) \Phi = \frac{c}{4\pi} \int \Phi(x, r)\, d\boldsymbol{\Omega}''.$$

The translational invariance and invariance with respect to rotations around the x-axis suggests we look for solutions of the form

(34)
$$\Phi = \Phi_\nu^{(m)}(\mu) e^{im\phi} e^{-x/\nu}.$$

Substituting in equation (33) yields

(35)
$$\left(1 - \frac{\mu}{\nu}\right) \Phi_\nu^{(m)}(\mu) = \delta_{m,0} \frac{c}{2} \int_{-1}^{1} \Phi_\nu^{(m)}(\mu'')\, d\mu''.$$

The solutions are well known. They are orthogonal in the sense that

(36)
$$\int_0^{2\pi} d\phi \int_{-1}^{1} d\mu \mu \Phi_\nu^{(m)}(\mu) e^{im\phi} \Phi_{\nu'}^{(m')}(\mu) e^{-im'\phi} = 0 \quad \text{unless } m = m' \text{ and } \nu = \nu'.$$

Further the $\Phi_\nu^{(m)} e^{im\phi}$ are complete in the sense that "any" $F(\mu, \phi)$ can be expanded in terms of these. (The details are to be found in [**1**].) For convenience we summarize the relevant formulae here:

(i) For $m \neq 0$ there is a continuum of functions with $-1 \leqq \nu \leqq 1$. Explicitly these are

(37)
$$\Phi_\nu^{(m)}(\mu) = \delta(\mu - \nu), \qquad -1 \leqq \nu \leqq 1.$$

(ii) For $m = 0$ there is firstly a continuum with ν in the same range with

$$\Phi_\nu^{(0)}(\mu) \equiv \phi_\nu(\mu) = \frac{c\nu}{2} P \frac{1}{\nu - \mu} + \lambda(\nu)\delta(\mu - \nu), \qquad -1 \leqq \nu \leqq 1, \tag{38}$$

where

$$\lambda(\nu) = \frac{\Lambda_+(\nu) + \Lambda_-(\nu)}{2}. \tag{39}$$

Here $\Lambda_\pm$ denote the boundary values of the analytic function

$$\Lambda(z) = 1 + \frac{cz}{2}\int_{-1}^{1} \frac{d\mu'}{\mu' - z}. \tag{40}$$

Secondly, there are two discrete modes corresponding to $\nu = \pm\nu_0$ where

$$\Lambda(\pm\nu_0) = 0 \tag{41}$$

and (with our present assumptions) such that ν_0 is real and greater than unity. The functions

$$\Phi_{\pm\nu_0}^{(0)}(\mu_0) \equiv \phi_{0\pm}(\mu) = \frac{c\nu_0}{2(\nu_0 \mp \mu)}. \tag{42}$$

Note: All our functions are normalized so that

$$\int_{-1}^{1} \Phi_\nu^{(m)}(\mu)\, d\mu = 1. \tag{43}$$

To determine G we now expand in this complete set of functions. In order that this vanish at infinity we write

$$G = \pm \sum_{m=-\infty}^{\infty} \int_{\nu \gtrless 0} A_\nu^{(m)} \Phi_\nu^{(m)}(\mu) e^{im\phi} e^{-(x-x')/\nu}\, d\nu, \qquad x \gtrless x'. \tag{44}$$

(Here $\int_\nu$ means integral over continuum plus a sum over the discrete modes.) From equation (30) we deduce that

$$G(x' + \epsilon, \mathbf{\Omega}; x', \mathbf{\Omega}') - G(x' - \epsilon, \mathbf{\Omega}; x', \mathbf{\Omega}') = \delta(\mu - \mu')\delta(\phi - \phi')/\mu'. \tag{45}$$

Inserting the form of G from equation (44) we get as an equation for the expansion coefficients $A_\nu^{(m)}$

$$\sum_{m=-\infty}^{\infty} \int d\nu' A_{\nu'}^{(m')} \Phi_{\nu'}^{(m')}(\mu) e^{im'\phi} = \frac{\delta(\mu - \mu')\delta(\phi - \phi')}{\mu'}. \tag{46}$$

Using the orthogonality relations we readily deduce that

$$A_\nu^{(m)} = e^{-im\phi'} \Phi_\nu^{m}(\mu')/2\pi N_\nu^{(m)}. \tag{47}$$

Here the $N_\nu^{(m)}$ for the continuum ν are defined by

$$\int_{-1}^{1} \mu \Phi_\nu^{(m)}(\mu)\Phi_{\nu}^{(m)}(\mu)\, d\mu = N_\nu^{(m)}\delta(\nu - \nu'), \tag{48}$$

while for the discrete modes

$$\int_{-1}^{1} \mu \Phi_{\pm\nu_0}^{(0)}(\mu)\Phi_{\pm\nu_0}^{(0)}(\mu)\, d\mu = N_0^{\pm}. \tag{49}$$

Explicitly we have

$$N_\nu^{(m)} = \nu, \qquad m \neq 0, \tag{50}$$

$$N_\nu^{(0)} = \nu\Lambda_+(\nu)\Lambda_-(\nu), \tag{51}$$

and

$$N_{\pm 0}^{(0)} = \frac{c}{2}\nu_0^2 \left.\frac{\partial \Lambda}{\partial z}\right|_{z=\pm\nu_0}. \tag{52}$$

Inserting these expressions into equation (44) we finally obtain

$$G(x, \mathbf{\Omega}; x', \mathbf{\Omega}') = \pm \sum_{m=-\infty}^{\infty} \int_{\nu \gtrless 0} \frac{\Phi_\nu^{(m)}(\mu)\Phi_\nu^{(m)}(\mu')}{2\pi N_\nu^{(m)}} e^{im(\phi-\phi')} e^{-(x-x')/\nu}\, d'\nu. \tag{53}$$

C. *Half-space problems.* The simplest situation arises when V is the region $x > 0$ and S is the surface $x = 0$. The pertinent kernels are then

$$\begin{aligned} G_\pm(\mathbf{\Omega}; \mathbf{\Omega}') &= \lim_{x \to 0^+, 0^-} G(x, \mathbf{\Omega}; 0, \mathbf{\Omega}') \\ &= \pm \sum_{m=-\infty}^{\infty} \int_{\nu>0;\, \nu<0} \frac{\Phi_\nu^{(m)}(\mu)\Phi_\nu^{(m)}(\mu')}{2\pi N_\nu^{(m)}} e^{im(\phi-\phi')}\, d\nu. \end{aligned} \tag{54}$$

We recognize here that the generalized completeness theorem which here takes the form

$$[G_+(\mathbf{\Omega}; \mathbf{\Omega}') - G_-(\mathbf{\Omega};\ \mathbf{\Omega}')]\mu' = \delta(\mathbf{\Omega} \cdot \mathbf{\Omega}') \tag{55}$$

is just the statement of the completeness of the functions $\Phi_\nu^{(m)}(\mu)e^{im\phi}$. Similarly, the generalized orthogonality relations for which the first two here take the form

$$G_+(\mathbf{\Omega}; \mathbf{\Omega}'') = \int G_+(\mathbf{\Omega}; \mathbf{\Omega}')\mu' G_+(\mathbf{\Omega}'; \mathbf{\Omega}'')\, d\mathbf{\Omega}' \tag{56a}$$

and

$$0 = \int G_-(\mathbf{\Omega}; \mathbf{\Omega}')\mu' G_+(\mathbf{\Omega}'; \mathbf{\Omega}'')\, d\mathbf{\Omega}' \tag{56b}$$

are simple consequences of the orthogonality relations for the normal modes. Thus, equation (56b) follows from the fact that all functions used in constructing G_+ are orthogonal to those used in constructing G_-. Similarly, the idempotent property of equation (56a) follows from the orthogonality of the different functions which are superposed to construct G_+.

As a simple application, let us see how the albedo problem for a half-space is formulated and solved with the present approach. We consider the situation with $Q(x, \mathbf{\Omega}) = 0$. Then the integral equation to be solved is

$$\Psi(0; \mathbf{\Omega}) = \int G_+(\mathbf{\Omega}; \mathbf{\Omega}')\mu'\Psi(0; \mathbf{\Omega}')\, d\mathbf{\Omega}'. \tag{57}$$

We remark that the properly phrased problem here is such that the incident distribution, i.e. $\Psi(0; \mathbf{\Omega})$ for $\mu > 0$ is given. Thus equation (57) is really two equations for $\Psi(0; \mathbf{\Omega})$ with $\mu < 0$. Let us see how these determine the unknown quantity. To treat what is essentially the most general case, consider

$$\Psi(0; \mathbf{\Omega}) = \delta(\mathbf{\Omega} \cdot \mathbf{\Omega}_0) \quad \text{for} \quad \mu, \mu_0 > 0. \tag{58}$$

Writing equation (57) for $\mu > 0$ we have

$$\delta(\mathbf{\Omega} \cdot \mathbf{\Omega}_0) = \mu_0 G_+(\mathbf{\Omega}; \mathbf{\Omega}_0) + \int_{\mu' < 0} G_+(\mathbf{\Omega}; \mathbf{\Omega}')\mu'\Psi(0; \mathbf{\Omega}')\, d\mathbf{\Omega}', \qquad \mu > 0. \tag{59a}$$

This can be materially simplified using our generalized completeness property which here states that

$$\delta(\mathbf{\Omega} \cdot \mathbf{\Omega}_0) = \mu_0\{G_+(\mathbf{\Omega}; \mathbf{\Omega}_0) - G_-(\mathbf{\Omega}; \mathbf{\Omega}_0)\}. \tag{60a}$$

With this result equation (59a) becomes

$$-\mu_0 G_-(\mathbf{\Omega}; \mathbf{\Omega}_0) = \int_{\mu' < 0} G_+(\mathbf{\Omega}; \mathbf{\Omega}')\mu'\Psi(0; \mathbf{\Omega}')\, d\mathbf{\Omega}', \qquad \mu > 0. \tag{59b}$$

For $\mu < 0$, equation (57) is

$$\Psi(0, \mathbf{\Omega}) = \mu_0 G_+(\mathbf{\Omega}; \mathbf{\Omega}_0) + \int_{\mu' < 0} G_+(\mathbf{\Omega}; \mathbf{\Omega}')\mu'\Psi(0; \mathbf{\Omega}')\, d\mathbf{\Omega}'. \tag{60b}$$

This clearly decomposes upon Fourier decomposition with respect to the variable ϕ. Thus suppose we multiply equation (59b) and (60b) by $e^{-im\phi}$ and integrate over all ϕ. We obtain

$$-\mu_0 G_-^{(m)}(\mu, \mu_0)e^{-im\phi_0} = \int_{\mu' < 0} \mu' G_+^{(m)}(\mu, \mu')e^{-im\phi'}\Psi(0; \mathbf{\Omega}')\, d\mathbf{\Omega}', \qquad \mu > 0, \tag{61}$$

and

$$\begin{aligned} \int_0^{2\pi} e^{-im\phi}\Psi(0; \mathbf{\Omega})\, d\phi &= \mu_0 G_+^{(m)}(\mu, \mu_0)e^{-im\phi_0} \\ &\quad + \int_{\mu' < 0} \mu' G_+^{(m)}(\mu, \mu_0)e^{-im\phi'}\Psi(0; \mathbf{\Omega}')\, d\mathbf{\Omega}', \qquad \mu < 0. \end{aligned} \tag{62}$$

Here

$$\begin{aligned} G_\pm^{(m)}(\mu, \mu_0) &= e^{im\phi_0}\int_0^{2\pi} e^{-im\phi}G_\pm(\mathbf{\Omega}; \mathbf{\Omega}_0)\, d\phi \\ &= \pm\int_{\nu \gtrless 0} \frac{\Phi_\nu^{(m)}(\mu)\Phi_\nu^{(m)}(\mu_0)}{N_\nu^{(m)}}\, d\nu. \end{aligned} \tag{63}$$

For $m \neq 0$, these are particularly simple. Thus

(64)
$$G_+^{(m)}(\mu, \mu_0) = \int_0^1 \frac{\delta(\mu - \nu)\delta(\mu_0 - \nu)}{\nu} \, d\nu = \frac{\delta(\mu - \mu_0)}{\mu} \Theta(\mu).$$

Here

(65)
$$\Theta(\mu) = 1, \quad \mu > 0$$
$$= 0, \quad \mu < 0.$$

Similarly

(66)
$$G_-^{(m)}(\mu, \mu_0) = \delta(\mu - \mu_0)\Theta(-\mu).$$

In particular then, we see that for $m \neq 0$, equation (61) is the trivial identity $0 = 0$, while equation (62) states that

(67)
$$\Psi_m(\mu) = \int_0^{2\pi} e^{-im\phi}\Psi(0; \mathbf{\Omega}) \, d\mathbf{\Omega} = 0, \qquad \mu < 0.$$

Thus only the equations with $m = 0$ remain. This is a rather pleasant feature of the approach. The ϕ asymmetry of the incident distribution is only mirrored in the *spatial dependence* of the solution. It plays no role in the integral equations we must solve. (This is, of course, due to our assumption of isotropic scattering.)

The only equations that remain are thus

(68a)
$$-\mu_0 G_-^{(0)}(\mu, \mu_0) = \int_{\mu' < 0} \mu' G_+^{(0)}(\mu, \mu')\Psi_0(\mu') \, d\mu', \qquad \mu > 0,$$

and

(69a)
$$\Psi_0(\mu) = \mu_0 G_+^{(0)}(\mu, \mu_0) + \int_{\mu' < 0} \mu' G_+(\mu, \mu')\Psi_0(\mu') \, d\mu',$$

where, to be more explicit,

(70)
$$\pm G_\pm^{(0)}(\mu, \mu') = \frac{\phi_{0\pm}(\mu)\phi_{0\pm}(\mu')}{N_{0\pm}} + \int_{\nu > 0, \nu < 0} \frac{\phi_\nu(\mu)\phi_\nu(\mu')}{N(\nu)} \, d\nu.$$

We note that equation (68a) is a singular integral equation of a rather standard form while equation (69a) is an equation of the second kind with a completely continuous kernel. The determination of Ψ_0 is as follows: Let

(71)
$$\Gamma_0 = \frac{1}{N_{0+}} \int_{-1}^0 \mu' \phi_{0+}(\mu')\Psi_0(\mu') \, d\mu',$$

and

(72)
$$\Gamma(\nu) = \frac{1}{N(\nu)} \int_{-1}^0 \mu' \phi_\nu(\mu')\Psi_0(\mu') \, d\mu'.$$

Then

(73)
$$\int_{-1}^0 \mu' G_+^{(0)}(\mu, \mu')\Psi_0(\mu') \, d\mu' = \Gamma_0\phi_{0+}(\mu) + \int_0^1 \Gamma(\nu)\phi_\nu(\mu) \, d\nu.$$

Hence equation (68a) is

$$-\mu_0 G_-^{(0)}(\mu, \mu_0) = \Gamma_0 \phi_{0+}(\mu) + \int_0^1 \Gamma(\nu)\phi_\nu(\mu)\, d\nu, \qquad \mu > 0. \tag{68b}$$

Inserting the explicit expressions for the $\phi_\nu(\mu)$ gives a familiar singular integral equation for the coefficients Γ_0 and $\Gamma(\nu)$. They are determined by a standard procedure (cf., for example, [**1**]). Having found these, we note that equation (69a) is

$$\Psi_0(\mu) = \mu_0 G_+^{(0)}(\mu, \mu_0) + \Gamma_0 \phi_{0+}(\mu) + \int_0^1 \Gamma(\nu)\phi_\nu(\mu)\, d\nu, \qquad \mu < 0. \tag{69b}$$

Hence, on inserting the known coefficients, we can use this to read off $\Psi_0(\mu)$. The result is

$$\Psi_0(\mu) = \frac{c}{2(1-c)} \frac{\mu_0}{(\nu_0 + \mu_0)(\nu_0 - \mu)(\mu_0 - \mu) X(\mu) X(-\mu_0)}, \qquad \mu < 0. \tag{74}$$

Here $X(z)$ is a function defined as follows. If $\Theta(\mu) = \arg \Lambda^+(\mu)$, then

$$X(z) = (1 - z)^{-1} \exp \frac{1}{\pi} \int_0^1 \frac{\Theta(\mu')\, d\mu'}{(\mu' - z)}. \tag{75}$$

Now we have $\Psi(0; \mathbf{\Omega})$. Indeed

$$\Psi(0; \mathbf{\Omega}) = \delta(\mathbf{\Omega} \cdot \mathbf{\Omega}_0), \quad \mu > 0, \qquad \text{and} \qquad \Psi(0; \Omega) = \frac{\Psi_0(\mu)}{2\pi}, \quad \mu < 0. \tag{76}$$

If we insert these into our integral representation, which now reads

$$\begin{aligned} \Psi(x, \mathbf{\Omega}) = &\int_{\mu' > 0} \mu' \Psi(0, \mathbf{\Omega}') G(x, \mathbf{\Omega}; 0, \mathbf{\Omega}')\, d\mathbf{\Omega}' \\ &+ \int_{\mu' < 0} \mu' \Psi(0, \mathbf{\Omega}') G(x, \mathbf{\Omega}; 0, \mathbf{\Omega}')\, d\mathbf{\Omega}', \end{aligned} \tag{77}$$

we can immediately find Ψ everywhere. First we note that in virtue of extreme simplicity of the eigenfunctions for $m \neq 0$, we can materially simplify the G of equation (53) by explicitly summing those terms. The result for $x > 0$ is that

$$G(x, \mathbf{\Omega}; 0, \mathbf{\Omega}') = \frac{G^{(0)}(x, \mu; 0, \mu')}{2\pi} + \left[\delta(\phi - \phi') - \frac{1}{2\pi}\right] \frac{\delta(\mu - \mu')}{\mu'} e^{-x/\mu'} \Theta(\mu'), \tag{78}$$

where

$$G^{(0)}(x, \mu; 0, \mu') = \int_{\nu > 0} \frac{\Phi_\nu^{(0)}(\mu) \Phi_\nu^{(0)}(\mu')}{N_\nu^{(0)}} e^{-x/\nu}\, d\nu. \tag{79}$$

Combining our results, we then have

$$\Psi(x, \mathbf{\Omega}) = \left[\delta(\phi - \phi_0) - \frac{1}{2\pi}\right]\delta(\mu - \mu_0)e^{-x/\mu_0}$$

(80)
$$+ \frac{\mu_0}{2\pi(\nu_0 + \mu_0)X(-\mu_0)}\left\{\frac{2\nu_0\phi_{0+}(\mu_0)\phi_{0+}(\mu)X(-\nu_0)e^{-x/\nu_0}}{N_{0+}}\right.$$
$$\left. + \int_0^1 (\nu_0 + \nu)\phi_\nu(\mu_0)\phi_\nu(\mu)\frac{X(-\nu)}{N(\nu)}e^{-x/\nu}\,d\nu\right\}.$$

Of interest below is the asymptotic form for large x. We readily see this is given by the term proportional to $\exp(-x/\nu_0)$. Thus

(81)
$$-\Psi(x, \mathbf{\Omega}) \sim \frac{\mu_0\phi_{0+}(\mu_0)\phi_{0+}(\mu)\exp(-x/\nu_0)}{\pi c(1 - c)(\nu_0 + \mu_0)\nu_0^2 X(-\mu_0)X(\nu_0)}.$$

(Here we have used some identities from [**1**] to simplify the coefficients.)

D. *Slab problems*. Let us take as V the slab between $x = 0$ and $x = l$. The general integral representation is then

(82)
$$\Psi(x, \mathbf{\Omega}) = \int \mu'\,d\mathbf{\Omega}'\Psi(0, \mathbf{\Omega}')G(x, \mathbf{\Omega}; 0, \mathbf{\Omega}') - \int \mu'\,d\mathbf{\Omega}'\Psi(l, \mathbf{\Omega}')G(x, \mathbf{\Omega}; l, \mathbf{\Omega}')$$

with G that of equation (53). (We omit sources Q for simplicity.)

Specializing to the boundaries, we have as integral equations for the surface values

(83)
$$\Psi(0, \mathbf{\Omega}) = \int \mu'\,d\mathbf{\Omega}'G(0 + \varepsilon, \mathbf{\Omega}; 0, \mathbf{\Omega}')\Psi(0, \mathbf{\Omega}')$$
$$- \int \mu'\,d\mathbf{\Omega}'G(0, \mathbf{\Omega}; l, \mathbf{\Omega}')\Psi(l, \mathbf{\Omega}')$$

and

(84)
$$\Psi(l, \mathbf{\Omega}) = \int \mu'\,d\mathbf{\Omega}'G(l, \mathbf{\Omega}; 0, \mathbf{\Omega}')\Psi(0, \mathbf{\Omega}')$$
$$- \int \mu'\,d\mathbf{\Omega}'G(l - \varepsilon, \mathbf{\Omega}; l, \mathbf{\Omega}')\Psi(l, \mathbf{\Omega}').$$

Note: $G(x, \mathbf{\Omega}; 0, \mathbf{\Omega}')$ for $x > 0$ is as given by equation (78), while for $x < l$ we have

(85)
$$G(x, \mathbf{\Omega}; l, \mathbf{\Omega}') = \frac{G^{(0)}(x, \mu; l, \mu')}{2\pi}$$
$$+ \left[\frac{1}{2\pi} - \delta(\phi - \phi')\right]\frac{\delta(\mu - \mu')\Theta(-\mu')e^{(l-x)/\mu'}}{\mu'}$$

with

$$G^{(0)}(x, \mu; l, \mu') = -\int_{\nu<0} \frac{\Phi_\nu^{(0)}(\mu)\Phi_\nu^{(0)}(\mu')}{N(\nu)} e^{(l-x)/\nu} d\nu. \tag{86}$$

It may be noted that the generalized orthogonality relations are essentially of the two types

$$\begin{aligned} G(0+\varepsilon, \mathbf{\Omega}; 0, \mathbf{\Omega}'') &= \int \mu' \, d\mathbf{\Omega}' G(0+\varepsilon, \mathbf{\Omega}; 0, \mathbf{\Omega}') G(0+\varepsilon, \mathbf{\Omega}'; 0, \mathbf{\Omega}'') \\ &\quad - \int \mu' \, d\mathbf{\Omega}' G(0, \mathbf{\Omega}; l, \mathbf{\Omega}') G(l, \mathbf{\Omega}'; 0, \mathbf{\Omega}'') \end{aligned} \tag{87}$$

and

$$\begin{aligned} G(0, \mathbf{\Omega}; l, \mathbf{\Omega}'') &= \int \mu' \, d\mathbf{\Omega}' G(0+\varepsilon, \mathbf{\Omega}; 0, \mathbf{\Omega}') G(0, \mathbf{\Omega}'; l, \mathbf{\Omega}'') \\ &\quad - \int \mu' \, d\mathbf{\Omega}' G(0, \mathbf{\Omega}; l, \mathbf{\Omega}') G(l-\varepsilon, \mathbf{\Omega}'; l, \mathbf{\Omega}''). \end{aligned} \tag{88}$$

Again, by direct substitution, one can verify that these are indeed satisfied as a result of the orthogonality of the $\Phi_\nu^{(m)}(\mu)e^{im\phi}$.

Of most interest here is the albedo-transmission problem, i.e. the problem where we have as prescribed

$$\Psi(0, \mathbf{\Omega}) = \delta(\mathbf{\Omega} \cdot \mathbf{\Omega}_0), \quad \mu, \mu_0 > 0 \quad \text{and} \quad \Psi(l, \mathbf{\Omega}) = 0, \quad \mu < 0. \tag{89}$$

Instead of using equations (83) and (84) directly, it is convenient to first perform the Fourier decomposition of equation (82). Thus multiplying equation (82) by $e^{-im\phi}$ and integrating yields:

a. For $m \neq 0$,

$$\Psi_m(x, \mu) = \Theta(\mu)e^{-x/\mu}\Psi_m(0, \mu) - \Theta(-\mu)e^{(l-x)/\mu}\Psi_m(l, \mu). \tag{90}$$

With the conditions of equation (89) we thus have

$$\Psi_m(x, \mu) = \delta(\mu - \mu_0)e^{-x/\mu_0}e^{-im\phi_0}. \tag{91}$$

b. For $m = 0$,

$$\begin{aligned} \Psi_0(x, \mu) &= \int_{-1}^{1} \mu' \, G^{(0)}(x, \mu; 0, \mu')\Psi_0(0, \mu') \, d\mu' \\ &\quad - \int_{-1}^{1} \mu' \, G^{(0)}(x, \mu; l, \mu')\Psi_0(l, \mu') \, d\mu'. \end{aligned} \tag{92}$$

For the albedo problem we thus need to determine $\Psi_0(0, \mu')$ for $\mu' < 0$ and $\Psi_0(l, \mu')$ for $\mu' > 0$. Inserting in equation (92) the boundary condition of equation (89) and letting x approach the boundaries thus yields as the explicit form of integral equations to be solved

(93)
$$\delta(\mu-\mu_0)=\mu_0G^{(0)}(0+\varepsilon,\mu;0,\mu_0)+\int_{-1}^{0}\mu'G^{(0)}(0+\varepsilon,\mu;0,\mu')\Psi_0(0,\mu')\,d\mu' -\int_0^1\mu'G^{(0)}(0,\mu;l,\mu')\Psi_0(l,\mu')\,d\mu',\qquad \mu>0,$$

(94)
$$\Psi_0(0,\mu)=\mu_0G^{(0)}(0+\varepsilon,\mu;0,\mu_0)+\int_{-1}^{0}\mu'G^{(0)}(0,\mu;0,\mu')\Psi_0(0,\mu')\,d\mu' -\int_0^1\mu'G^{(0)}(0,\mu;l,\mu')\Psi_0(l,\mu')\,d\mu',\qquad \mu<0.$$

(95)
$$0=\mu_0G^{(0)}(l,\mu;0,\mu_0)+\int_{-1}^{0}\mu'G^{(0)}(l,\mu;0,\mu')\Psi_0(0,\mu')\,d\mu' -\int_0^1\mu'G^{(0)}(l-\varepsilon,\mu;l,\mu')\Psi_0(l,\mu')\,d\mu',\qquad \mu<0,$$

and

(96)
$$\Psi_0(l,\mu)=\mu_0G^{(0)}(l,\mu;0,\mu_0)+\int_{-1}^{0}\mu'G^{(0)}(l,\mu;0,\mu')\Psi_0(0,\mu')\,d\mu' -\int_0^1\mu'G^{(0)}(l-\varepsilon,\mu;l,\mu')\Psi_0(l,\mu')\,d\mu',\qquad \mu>0.$$

'Clearly these equations are quite complicated. However, they are well suited for approximation procedure in limiting cases.

(i) *Thick slabs.* In first approximation we take $\Psi_0(0,\mu')$ to be that obtained for the half-plane problem, i.e. equations (74) and (76). It will be apparent that with this choice equations (93) and (94) are satisfied up to terms of order $\exp(-2l/\nu_0)$. With this approximation the terms in equations (95) and (96) involving $\Psi_0(0,\mu')$ become known inhomogeneous terms. Indeed,

(97)
$$\mu_0G^{(0)}(l,\mu;0,\mu_0)+\int_{-1}^{0}\mu'G^{(0)}(l,\mu;0,\mu')\Psi_0(0,\mu')\,d\mu'=J$$

is just 2π times the ϕ symmetrical part of equation (80) evaluated at $x=l$. Since we are interested in the case of large l, it suffices to use only the asymptotic part, i.e.

(98)
$$J\sim A\phi_{0+}(\mu),$$

where

(99)
$$A=\frac{\mu_0}{\nu_0+\mu_0}\,\frac{2\nu_0}{X(-\mu_0)}\,\frac{\phi_{0+}(\mu_0)X(-\nu_0)}{N_{0+}}\exp(-l/\nu_0).$$

Then equations (95) and (96) become

(100)
$$A\phi_{0+}(\mu)=\int_0^1\mu'G^{(0)}(l-\varepsilon,\mu;l,\mu')\Psi_0(l,\mu')\,d\mu',\qquad \mu<0,$$

and

$$\Psi_0(l, \mu) = A\phi_{0+}(\mu) - \int_0^1 \mu' G^{(0)}(l - \varepsilon, \mu; l, \mu')\Psi_0(l, \mu')d\mu'. \tag{101}$$

We remark that this is just the classical Milne problem (cf. [1]). The solution is quite similar to that of the half-plane albedo problem. In essence, equation (100) is a singular integral equation for the quantities

$$T_0 = -\frac{1}{N_{0-}}\int_0^1 \mu' \phi_{0-}(\mu')\Psi_0(l, \mu')\, d\mu',$$

and

$$T(\nu) = -\frac{1}{N(\nu)}\int_0^1 \mu' \phi_\nu(\mu')\Psi_0(l, \mu')\, d\mu', \qquad -1 \leqq \nu \leqq 0. \tag{102}$$

Given these, we have $\Psi_0(l, \mu)$ from equation (101). The result is

$$\Psi_0(l, \mu) = \frac{AX(-\nu_0)}{X(-\mu)} \frac{c\nu_0^2}{\nu_0^2 - \mu^2}. \tag{103}$$

Corrections to our first approximation for $\Psi_0(0, \mu)$ can now be readily found. We again only have to solve the Milne problem with inhomogeneous term determined by equation (103).

(ii) *Thin slabs.* Physically we might expect that this limiting case is even simpler. That this is indeed so is shown by the following elementary calculation. For $l \ll 1$ we take

$$\Psi_0(0, \mu') = 0 + \bar{\Psi}(0, \mu'), \quad \text{and} \quad \Psi_0(l, \mu') = \delta(\mu' - \mu_0) + \bar{\Psi}(l, \mu'), \tag{104}$$

where $\bar{\Psi}$ is to be of order l.[3] For the kernels occurring in equations (93)–(96) we note that

$$\begin{aligned} G^{(0)}(0, \mu; l, \mu') - G_-^{(0)}(\mu, \mu') &= O(l), \\ G^{(0)}(l, \mu; 0, \mu') - G_+^{(0)}(\mu, \mu') &= O(l). \end{aligned} \tag{105}$$

Then, for example, if we keep terms only up to order l in equation (93) we obtain

$$\begin{aligned} \delta(\mu - \mu_0) = {} & \mu_0 G_+^{(0)}(\mu, \mu_0) + \int_{-1}^0 \mu' G_+^{(0)}(\mu, \mu')\bar{\Psi}(0, \mu')\, d\mu' \\ & - \int_0^1 \mu' G_-^{(0)}(\mu, \mu')\{\delta(\mu' - \mu_0) + \bar{\Psi}(l, \mu')\}\, d\mu' \\ & - \int_0^1 \mu'\{G^{(0)}(0, \mu; l, \mu') - G_-^{(0)}(\mu, \mu')\}\delta(\mu' - \mu_0)\, d\mu', \qquad \mu > 0 \end{aligned} \tag{106}$$

[3] Here we are restricting ourselves to the situation where $l \ll \mu_0$.

Using the identity $\mu_0\{G_+ - G_-\} = \delta(\mu - \mu_0)$ reduces this to the statement that

$$\mu_0\{G^{(0)}(0, \mu; l, \mu') - G_-^{(0)}(\mu, \mu')\} = \int_{-1}^{0} \mu' G_+^{(0)}(\mu, \mu')\overline{\Psi}(0, \mu')\, d\mu' - \int_{0}^{1} \mu' G_-^{(0)}(\mu, \mu')\overline{\Psi}(l, \mu')\, d\mu', \qquad \mu > 0. \tag{107}$$

Similarly treating equation (96), we obtain

$$\overline{\Psi}(l, \mu) = \mu_0\{G^{(0)}(l, \mu; 0, \mu_0) - G_-^{(0)}(\mu, \mu_0)\} + \int_{-1}^{0} \mu' G_+^{(0)}(\mu, \mu')\overline{\Psi}(0, \mu')\, d\mu' - \int_{0}^{1} \mu' G_-^{(0)}(\mu, \mu')\overline{\Psi}(l, \mu')\, d\mu', \qquad \mu > 0. \tag{108}$$

Subtracting equation (107) from (108) then yields

$$\overline{\Psi}(l, \mu) = \mu_0\{G^{(0)}(l, \mu; 0, \mu_0) - G_+^{(0)}(\mu, \mu_0)\} + \mu_0\{G^{(0)}(0, \mu; l, \mu_0) - G_-^{(0)}(\mu, \mu_0)\}, \qquad \mu > 0. \tag{109}$$

In completely analogous fashion we find using equations (94) and (95) that

$$\Psi(0, \mu) = -\mu_0\{G^{(0)}(l, \mu; 0, \mu_0) - G_+^{(0)}(\mu, \mu_0)\} - \mu_0\{G^{(0)}(0, \mu; l, \mu_0) - G_-^{(0)}(\mu, \mu_0)\}, \qquad \mu < 0. \tag{110}$$

E. *Adjacent half-spaces.* It may be noted that in the examples discussed so far there has been a very natural breakup of our integral equation for the surface distribution into two: one holds for *incident* directions, the other for emergent distributions. The roles these two equations played were quite asymmetrical. Thus, the first equation was the significant one. It determined the expansion coefficients (Γ, T). The second equation essentially served only as an identity that helped simplify the form of the solution. To see that this is not what happens in general we consider the following simple example. Consider two semi-infinite media bounded by the plane $x = 0$. In both of these the transport equations are the same except that $c = c_1$ for $x > 0$ and $c = c_2$ for $x < 0$. Further, we assume plane symmetry and ϕ independence. Let the sources be present only in the right-hand half-space. Then, in region "1" we have the representation

$$\Psi(x, \mu) = \int_{-1}^{1}\int_{0}^{\infty} G^{(1)}(x, \mu; x', \mu')Q(x', \mu')\, dx'\, d\mu' + \int_{-1}^{1} G^{(1)}(x, \mu; 0, \mu')\mu'\Psi(0, \mu')\, d\mu', \qquad x > 0, \tag{111}$$

while in region "2,"

$$\Psi(x, \mu) = -\int_{-1}^{1} G^{(2)}(x, \mu; 0, \mu')\mu'\Psi(0, \mu')\, d\mu', \qquad x < 0. \tag{112}$$

(Here the $G^{(1,2)}$ are our previous $G^{(0)}$'s with, however, $c = c_1$ or c_2, respectively.)

If, further, the sources Q are far to the right of the origin, the contribution to equation (111) will be $-A\phi_{0-}(\mu)e^{-x/\nu_0}$ where A is some constant. Passing now to the limit $x \to 0$ we find that equations (111) and (112) become the integral equations

$$\Psi(0, \mu) = -A\phi_{0-}(\mu) + \int_{-1}^{1} G_+^{(1)}(\mu, \mu')\mu'\Psi(0, \mu')\, d\mu', \tag{113}$$

and

$$\Psi(0, \mu) = -A\phi_{0-}(\mu) - \int_{-1}^{1} G_+^{(2)}(\mu, \mu')\mu'\Psi(0, \mu')\, d\mu'. \tag{114}$$

Subtracting, this becomes

$$A\phi_{0-}(\mu) = \int_{-1}^{1} K_1(\mu, \mu')\mu'\Psi(0, \mu')\, d\mu' + \int_{-1}^{1} K_2(\mu, \mu')\mu'\Psi(0, \mu')\, d\mu', \tag{115}$$

where

$$K = \int_{\nu>0} \frac{\phi_\nu^{(1)}(\mu)\phi_\nu^{(1)}(\mu')\, d\nu}{N^{(1)}(\nu)} \quad \text{and} \quad K_2 = \int_{\nu<0} \frac{\phi_\nu^{(2)}(\mu)\phi_\nu^{(2)}(\mu')\, d\nu}{N^{(2)}(\nu)}. \tag{116}$$

To simplify equation (115) we define

$$\begin{aligned} \Gamma(\nu) &= \frac{1}{N^{(1)}(\nu)} \int_{-1}^{1} \mu'\phi_\nu^{(1)}(\mu')\Psi(0, \mu')\, d\mu', \qquad \nu > 0, \\ &= \frac{1}{N^{(2)}(\nu)} \int_{-1}^{1} \mu'\phi_\nu^{(2)}(\mu')\Psi(0, \mu')\, d\mu', \qquad \nu < 0. \end{aligned} \tag{117}$$

Then equation (115) becomes

$$A\phi_{0-}(\mu) = \int_{-1}^{1} \phi_\nu(\mu)\Gamma(\nu)\, d\nu, \qquad -1 \leqq \mu \leqq 1. \tag{118}$$

Here

$$\begin{aligned} \phi_\nu(\mu) &= \phi_\nu^{(1)}(\mu), \qquad \nu > 0, \\ &= \phi_\nu^{(2)}(\mu), \qquad \nu < 0. \end{aligned} \tag{119}$$

From equation (118) we can readily find $\Gamma(\nu)$ by standard techniques. (See, for example, [**1**].) We note that on inserting the normal mode expansions for $G^{(1)}$ and $G^{(2)}$ in equations (111) and (112) that knowing $\Gamma(\nu)$ is all that we need in order to know Ψ everywhere. On the other hand, there is no analog of the *regular* integral equation (for example, equations (69)) which we used to simplify the expressions for the emergent distribution after having found $\Gamma(\nu)$.

IV. **Plasma problems.** In §II we restricted ourselves to local kernels $f(\boldsymbol{v}, \boldsymbol{v}'; \boldsymbol{r})$. To see how the approach can be applied more generally, let us consider the following simplified linear plasma equations [**2**], [**3**]:

$$\frac{\partial \Psi}{\partial t}(\boldsymbol{r}, \boldsymbol{v}, t) + \boldsymbol{v} \cdot \nabla\Psi = \frac{en}{m} \boldsymbol{E} \cdot \nabla_v \Psi_0, \tag{120a}$$

$$\nabla \cdot \boldsymbol{E} = -4\pi e \int \Psi\, d^3v. \tag{121}$$

(Here Ψ_0 is given and e, n, m are constants.) By Fourier transforming with respect to time, we reduce equation (120a) to

(120b) $$\boldsymbol{v}\cdot\nabla\Psi - i\omega\Psi = (en/m)\boldsymbol{E}\cdot\nabla_v\Psi_0.$$

Our problem is to solve equations (120b) and (121) in a region V. Appropriate conditions on Ψ and $\boldsymbol{E}$ (or the corresponding potential) are presumed given on the founding surface(s) S.

As in §II it is convenient to generalize by adding to the right-hand side of equation (120b) a source term $Q(\boldsymbol{r}, \boldsymbol{v})$. Further it is useful to introduce an adjoint set of equations of the form

(122) $$\boldsymbol{v}\cdot\nabla\tilde{\Psi}(\boldsymbol{r}, \boldsymbol{v}) - i\omega\tilde{\Psi} = (en/m)\tilde{E} + \tilde{Q}(\boldsymbol{r}, v),$$

(123) $$\nabla\cdot\tilde{E} = -4\pi e\int\tilde{\Psi}\nabla_v\Psi_0\, d^3v.$$

(*Note:* In contrast to $\boldsymbol{E}$, $\tilde{E}$ is a *scalar*. The essential point here is that $\boldsymbol{E}$ occurs only in a scalar term in equation (120b).)

By simple manipulation of equations (120)–(123), we readily deduce the identity

(124) $$\{\Psi\iint_V(\boldsymbol{r}, \boldsymbol{v})\tilde{Q}(\boldsymbol{r}, -\boldsymbol{v}) - \tilde{\Psi}(\boldsymbol{r}, -\boldsymbol{v})Q(\boldsymbol{r}, \boldsymbol{v})\}\, d^3r\, d^3v$$
$$= \iint_S \boldsymbol{n}_i\cdot\boldsymbol{v}\tilde{\Psi}(\boldsymbol{r}, -\boldsymbol{v})\Psi(\boldsymbol{r}, \boldsymbol{v})\, dS\, d^3v - \frac{n}{4\pi m}\int_S(\boldsymbol{n}_i\cdot\boldsymbol{E})\tilde{E}\, dS.$$

One application of equation (124) is the following:

Let

(125) $$Q(\boldsymbol{r}, \boldsymbol{v}) = 0, \qquad \tilde{Q}(\boldsymbol{r}, \boldsymbol{v}) = \delta(\boldsymbol{r}, -\boldsymbol{r}_0)\delta(\boldsymbol{v} - \boldsymbol{v}_0),$$

i.e.

(126) $$\tilde{\Psi}(\boldsymbol{r}, \boldsymbol{v}) = \tilde{G}(\boldsymbol{r}, \boldsymbol{v}; \boldsymbol{r}_0, \boldsymbol{v}_0), \qquad \tilde{E}(\boldsymbol{r}) = \tilde{E}(\boldsymbol{r}; \boldsymbol{r}_0, \boldsymbol{v}_0).$$

Then we obtain

(127a) $$\Psi(\boldsymbol{r}_0, -\boldsymbol{v}_0) = \iint_S \boldsymbol{n}_i(\boldsymbol{r}'_S)\cdot\boldsymbol{v}'\tilde{G}(\boldsymbol{r}'_S, -\boldsymbol{v}'; \boldsymbol{r}_0, \boldsymbol{v}_0)\Psi(\boldsymbol{r}'_S, \boldsymbol{v}')\, dS'\, d^3v'$$
$$- \frac{n}{4\pi m}\int_S[\boldsymbol{n}_i(\boldsymbol{r}'_S)\cdot\boldsymbol{E}(\boldsymbol{r}'_S)]\tilde{E}(\boldsymbol{r}'_S; \boldsymbol{r}_0, \boldsymbol{v}_0)\, dS',$$

or, changing names,

(127b) $$\Psi(\boldsymbol{r}, \boldsymbol{v}) = \iint_S \boldsymbol{n}_i(\boldsymbol{r}'_S)\cdot\boldsymbol{v}'\tilde{G}(\boldsymbol{r}'_S, -\boldsymbol{v}'; \boldsymbol{r}, -\boldsymbol{v})\Psi(\boldsymbol{r}'_S, \boldsymbol{v}')\, dS'\, d^3v'$$
$$- \frac{n}{4\pi m}\int_S \boldsymbol{n}_i(\boldsymbol{r}_S)\cdot\boldsymbol{E}(\boldsymbol{r}'_S)\tilde{E}(\boldsymbol{r}'_S; \boldsymbol{r}, -\boldsymbol{v})\, dS'.$$

We remark that

(i) This is an integral representation for Ψ in terms of its boundary values—and those of the normal components of $\boldsymbol{E}$ on the boundary.

(ii) Passing to the limit $\boldsymbol{r} \to \boldsymbol{r}_S$, we obtain an integral equation for the needed boundary values of Ψ. In case the boundary conditions are *given* [**2**], [**3**] in terms of $\boldsymbol{n}_i \cdot \boldsymbol{E}$, we are finished with our formulation. If, however, the boundary conditions are given, for example, in terms of the potentials on S, we must supplement our integral equation with the corresponding integral equations of electrostatics (in the general case those of electrodynamics) for the normal components of $\boldsymbol{E}$. This can, of course, be quite a complication but one that should be expected. We are here combining transport theory with electrodynamics. As a bare minimum, we must expect the complications arising in the boundary value problems of the latter field.

(iii) In some special cases we can again eliminate much of the adjoint solution by a reciprocity-like principle. Thus, let the solution of our original equations with source $Q(\boldsymbol{r}, \boldsymbol{v}) = \delta(\boldsymbol{r} - \boldsymbol{r}_1)\delta(\boldsymbol{v} - \boldsymbol{v}_1)$ be $\tilde{G}(\boldsymbol{r}, \boldsymbol{v}; \boldsymbol{r}_1, \boldsymbol{v}_1)$. Now suppose it is possible to choose G, $\tilde{G}$, $\boldsymbol{E}$, $\tilde{\boldsymbol{E}}$ such that these all vanish at infinity. The identity equation (124) with V all of space then tells us that

$$G(\boldsymbol{r}_0, \boldsymbol{v}_0; \boldsymbol{r}_1, \boldsymbol{v}_1) = \tilde{G}(\boldsymbol{r}_1, -\boldsymbol{v}_1, \boldsymbol{r}_0, -\boldsymbol{v}_0). \tag{128}$$

In this case the integral representation of equation (127b) becomes

$$\begin{aligned} \Psi(\boldsymbol{r}, \boldsymbol{v}) &= \iint_S \boldsymbol{n}_i(\boldsymbol{r}'_S) \cdot \boldsymbol{v}' G(\boldsymbol{r}, \boldsymbol{v}; \boldsymbol{r}'_S, \boldsymbol{v}') \Psi(\boldsymbol{r}'_S, \boldsymbol{v}') \, dS' \, d^3v' \\ &\quad - \frac{n}{4\pi m} \int_S \boldsymbol{n}_i(\boldsymbol{r}') \cdot \boldsymbol{E}(\boldsymbol{r}'_S) \quad (\boldsymbol{r}'_S; \boldsymbol{r}, -\boldsymbol{v}) \, dS'. \end{aligned} \tag{129}$$

Again passing to the limit $\boldsymbol{r} \to \boldsymbol{r}_S$ gives an integral equation for $\Psi(\boldsymbol{r}_S, \boldsymbol{v})$. In general this too must be supplemented by the appropriate equations for $\boldsymbol{n}_i \cdot \boldsymbol{E}$.

V. **Conclusion.** It is hoped that it has been shown that the boundary value problems of transport theory can be formulated in a fashion quite analogous to the classical boundary value problems. The resulting equations are integral equations on a five dimensional surface in the six dimensional phase space. In general these are quite complicated. However, the kernels themselves have some remarkably simple properties. The method seems to lead to the solutions of the soluble problems at least as quickly as any other. In particular, it seems well suited for approximation methods in various limiting situations.

References

1. K. M. Case and P. F. Zweifel, *Linear transport theory*, Addison-Wesley, Reading, Mass. 1967.

2. L. Landau, *On the vibrations of the electronic plasma*, J. Phys. U.S.S.R. **10** (1946), 25.

3. K. M. Case, "Singular eigenfunctions and plasma problems" in *Proceedings of Symposia in Applied Mathematics*, Vol. XVIII, Amer. Math. Soc., Providence, R.I., 1967.

University of Michigan,
Ann Arbor, Michigan

On the structure of isotropic transport operators in space[1]

Erwin H. Bareiss and Ibrahim K. Abu-Shumays

Abstract Based on the idea of separation of variables, a spectral theory for the three-dimensional, stationary isotropic transport operator is presented in a vector space of complex-valued Borel functions resulting in continuous sets of regular and generalized eigenfunctions. Because of the nonselfadjoint nature of this operator, the results could not be anticipated intuitively from the known decomposition of this operator in the special case of plane geometry. The results obtained indicate a promising new approach to the analytic solution of the linear transport equation in higher space dimensions. This work summarizes and extends a previous Argonne National Laboratory Report (ANL-6914). Examples are given for slab geometry with and without axial symmetry, spherical symmetry, and cylindrical symmetries.

I. **Introduction.** Ever since Wiener and Hopf [**22**] discovered a powerful method for the analytic solution of certain problems in radiative transfer, the search for more general, more powerful, and perhaps simplified methods has not ceased. In 1960, Case [**12**] adapted a method by Davison [**17**] and Van Kampen [**24**], based on dispersion theory, to the stationary isotropic transport operator in one space dimension with axial symmetry in the angular distribution of the solution space. The method has been extended to anisotropic scattering and multicell and multigroup problems [**7**], [**19**], [**30**], [**35**]–[**37**], [**41**]–[**44**], [**50**]. Mathematically, finding expansion modes of an operator is related to the spectral theory of the operator. Since the linear transport operator is the sum of several linear operators, different approaches to a spectral theory are possible and have been attempted [**1**], [**4**], [**5**], [**10**], [**18**], [**34**], [**39**], [**46**]–[**49**]. The introduction of the Dirac δ-function as in [**12**] to construct the singular modes has symbolic significance, since the nature of this δ-function in the solution of the transport equation is not obvious. However, it can be demonstrated [**4**] that this δ-function has its origin in the application of

[1] Work performed under the auspices of the United States Atomic Energy Commission.

Cauchy's second theorem when a variable moves from the inside of a closed contour across to the outside.

The use of elementary functions to construct solutions for more than one space dimension, if only for special geometries, was attempted in **[25]**, **[35]**, **[36]**, and **[38]**. Since often the continuous part of the spectrum was overlooked, at most asymptotically correct answers were found. On the following pages, a decomposition of the transport operator in three space dimensions is carried out. The general idea is that of **[12]** and **[17]**, for the senior author was unable to generalize his ideas in **[4]**. The reason is that the transport operator is not selfadjoint, and therefore no general spectral theory has existed for it, as of today. The results of the expansion theorem proved rather surprising and could not be anticipated by intuition. On the other hand, the known expansion theorem for plane geometry with its much simpler structure is a special case of the general theory. The method used here is suggested by the principle of separation of variables, which is widely used for the decomposition of linear partial differential equations of mathematical physics. The solution space for the transport equation (1.4) below is mathematically defined as a set of complex-valued Borel functions. To attempt completeness, it was necessary to add to the set of regular eigenfunctions, $\Phi^0_\Lambda(\Omega)$, and the corresponding regular expansion modes $\Psi^0_\Lambda(r, \Omega)$ over a complex continuum, a set of generalized eigenfunctions, $\Phi_{\Lambda,\Omega_0}(\Omega)$, and the corresponding generalized expansion modes $\Psi_{\Lambda,\Omega_0}(r, \Omega)$ over a complex continuum of even higher dimension.

The regular eigenfunctions are characterized by a complex vector, which we may call a regular (or proper) eigenvector, Λ. These eigenvectors, however, have to satisfy the condition $\Lambda \cdot \Lambda = \lambda_0^2$, where λ_0^2 assumes exactly one value for any given equation; therefore $\pm\lambda_0$ can be called eigenvalues. To obtain the generalized eigenfunctions, it is necessary to evaluate the improper integral which occurs in equation (2.13) below. Its regular form as it appears in (2.9) can be reduced to integrals with known analytic expressions, but its improper form and its interpretation requires careful analysis in order to formulate the decomposition theorem. Essentially, the "difficult" integrals can be considered Cauchy principal-value integrals of higher dimension. Although principal-value integrals have been considered extensively for problems in potential theory **[20]**, **[45]**, the integrals encountered here do not quite fit their pattern. What may make the following pages interesting is the fact that most results are not quite what intuition at first would suggest.

A basic result of these research efforts is the general solution to the transport equation (1.4) below in a sense analogous to that of the theory of differential equations. Special representations are obtained by imposing geometrical restrictions on the general solution. They enable the numerical analyst to construct exact test problems for checking the performance of general multidimensional transport codes, such as described in [3], [9], **[11]**, **[18]**, **[23]**, **[29]**, and **[33]**, and therefore can be most helpful in error analysis. The calculation of singular integrals is not difficult if properly executed **[6]**. There is little doubt that the results of the following pages can be generalized to anisotropic scattering, multiregion, and multigroup problems. A similar theory can also be conceived for the discrete ordinate approximations; their spectrum, modes, and general solutions can then be compared with the continuous case presented here to yield a measure of fitness for the approximate solutions.

Many papers on the spectral representation of the time-dependent equation give a general expansion in exponentials, although not much is said on the structure of the time-independent coefficient functions, which very often satisfy an equation of type (1.4). This gap is filled here.

We turn now to the description of the operator. Given is the transport equation for constant energy

$$\frac{\partial \Psi}{\partial t} + v\Omega \cdot \nabla\Psi + v\sigma\Psi = cv\sigma \int_{\{\Omega'\}} f(\Omega \cdot \Omega')\Psi(r, \Omega', t)\, d_2\Omega' + Q(r, \Omega, v, t), \tag{1.1}$$

where $\Psi = \Psi(r, \Omega, t)$ is the directional flux, a scalar function;
t is the time;
r is the position vector;
v is a constant velocity in the direction of the unit vector Ω;
σ is the total macroscopic cross section at velocity v;
c is the net number of particles produced per collision $(= (\sigma_s + \nu\sigma_f)/\sigma)$;
f is the scattering function;
Q is the production of particles by sources;

and

$\nabla \equiv$ grad, operates with respect to r only.

The scattering function is so normalized that[2]

$$\int f(\Omega \cdot \Omega')\, d_2\Omega' = 1. \tag{1.2}$$

We investigate the decomposition of equation (1.1) under the following simplifying assumptions:

(1.3)
(a) $f = 1/(4\pi)$; i.e. we have isotropic scattering;
(b) σ is constant;
(c) $Q(r, \Omega, v, t) = 0$;
(d) $\partial\Psi/\partial t \equiv 0$ (stationary case).

Let the mean free path, $1/\sigma$, be introduced as unit length. This is equivalent to setting $\sigma = 1$ in equation (1.1). Since $v \neq 0$, equation (1.1) reduces to

$$\Omega \cdot \nabla\Psi + \Psi - \frac{c}{4\pi} \int_{\{\Omega'\}} \Psi(r, \Omega')\, d_2\Omega' = 0. \tag{1.4}$$

[2] Subsequently, the operator $\int \cdot\, d_2\Omega'$ will mean integration over the entire unit sphere in the sense defined in §§III and IV.

If assumption (d) does not hold, one often lets $\Psi = e^{-\alpha t}\Psi''$, which leads to a stationary equation with

$$\sigma' = \sigma\beta, \qquad c' = c/\beta, \qquad \beta = 1 - \alpha/\sigma v.$$

When α is taken complex, σ' and c' are complex, and the results of this report need appropriate modification.

II. **Separation of variables and the characteristic equation.** We look for all solutions that have the form

$$\Psi_\Lambda(r, \Omega) = R_\Lambda(r)\Phi_\Lambda(\Omega). \tag{2.1}$$

Substitution of equation (2.1) into equation (1.4) yields

$$-\frac{\Omega \cdot \nabla R_\Lambda}{R_\Lambda} = 1 - \frac{c}{4\pi\Phi_\Lambda}\int \Phi_\Lambda(\Omega')\, d_2\Omega' \tag{2.2}$$

under the assumption $R_\Lambda \neq 0$, $\Phi_\Lambda \neq 0$. Since $\nabla R_\Lambda/R_\Lambda$ does not depend on Ω by definition (2.1) and does not depend on r by (2.2), it must be proportional to a parameter vector, $-\Lambda$, which will be in general a complex vector,

$$\nabla R_\Lambda/R_\Lambda = -\Lambda, \tag{2.3}$$

whence

$$R_\Lambda = \text{const} \cdot e^{-\Lambda \cdot r}. \tag{2.4}$$

Therefore, the right-hand side of equation (2.2) is equal to $\Omega \cdot \Lambda$, and equation (2.2) takes the form

$$(1 - \Omega \cdot \Lambda)\Phi_\Lambda - \frac{c}{4\pi}\int \Phi_\Lambda(\Omega')\, d_2\Omega' = 0. \tag{2.5}$$

We normalize Φ_Λ such that

$$\int \Phi_\Lambda(\Omega')\, d_2\Omega' = \kappa, \tag{2.6}$$

where κ is a conveniently chosen, but fixed normalization constant (e.g. $\kappa = 1$ or $4\pi/c$). Under the condition that

$$\Omega \cdot \Lambda \neq 1 \quad \text{for all } \Omega, \tag{2.7}$$

the solution of (2.5) satisfying equation (2.6) is denoted by Φ_Λ^0 and is

$$\Phi_\Lambda^0 = \frac{\kappa c}{4\pi}\frac{1}{1 - \Omega \cdot \Lambda}. \tag{2.8}$$

Substitution of equation (2.8) in equation (2.6) yields a condition for Λ, namely, the *characteristic equation*,

$$1 - \frac{c}{4\pi}\int \frac{d_2\Omega}{1 - \Omega \cdot \Lambda} = 0. \tag{2.9}$$

DEFINITION 2.1. The set of solutions of equation (2.9) satisfying equation (2.7) is denoted by $\{\Lambda^0\}$. An element $\Lambda \in \{\Lambda^0\}$ is called a *regular eigenvector*. The corresponding functions Φ_Λ^0 are called *regular eigenfunctions* of equation (2.5).

Hence, we can write for equation (2.1)

$$\Psi_\Lambda^0 = e^{-\Lambda\cdot r}\Phi_\Lambda^0(\Omega), \qquad \Lambda \in \{\Lambda^0\}, \tag{2.10}$$

and Φ_Λ^0 is given by equation (2.8). The function $\Psi_\Lambda^0(r, \Omega)$ has been so normalized that $\Psi_\Lambda^0(0, \Omega) = \Phi_\Lambda^0(\Omega)$.

Consider now the case where, for a given Λ, equation (2.7) is *not* satisfied. Then, at certain directions $\Omega = \Omega_0$, Φ_Λ in equation (2.8) is unbounded, and the meaning of the integral (2.6) must be redefined. Assuming this is done, we then obtain another characteristic equation with possible solutions for Λ, and a set of elementary solutions corresponding to equation (2.10), but unbounded in Ω. These functions are, however, a subset of an even wider class of elementary solutions of the general form (2.1), when we extend the class of admissible functions to be generalized functions. We extend our function space by the set of functions $\delta(\Omega - \Omega_0)$ having the following definition.

DEFINITION 2.2. Given any function $G(\Omega)$, which is continuous with respect to Ω, $|\Omega| = 1$, in a neighborhood $D(\Omega)$ of a given direction Ω_0, then $\delta(\Omega - \Omega_0)$ is defined by

$$\begin{aligned}\int_{D(\Omega)} G(\Omega)\delta(\Omega - \Omega_0)\, d_2\Omega &= G(\Omega_0) \quad \text{for} \quad \Omega_0 \in D(\Omega),\\ &= 0 \qquad\quad\ \text{for} \quad \Omega_0 \notin D(\Omega).\end{aligned}$$

Now, we introduce the *improper eigenfunctions* (*generalized eigenfunctions*) as

$$\Phi_{\Lambda,\Omega_0} = \frac{\kappa c}{4\pi}\frac{1}{1 - \Omega\cdot\Lambda} + \kappa G\delta(\Omega - \Omega_0), \tag{2.11}$$

where the *improper eigenvectors* Λ are such that

$$\Lambda \in \{\Lambda : 1 - \Omega_0\cdot\Lambda = 0\}, \tag{2.12}$$

and G is a function possibly of Λ and Ω. Substitution of equation (2.11) into equation (2.5) under consideration of equation (2.6), and integration with respect to Ω over the unit sphere, show that the Φ's of equation (2.11) are indeed weak solutions. The function G is defined by substituting equation (2.11) into equation (2.6). Hence

$$G = 1 - \frac{c}{4\pi}\int \frac{d_2\Omega}{1 - \Omega\cdot\Lambda} \tag{2.13}$$

and is a function of Λ only. An elementary solution corresponding to equation (2.11) is

$$\Psi_{\Lambda,\Omega_0} = e^{-\Lambda\cdot r}\Phi_{\Lambda,\Omega_0}, \tag{2.14}$$

where Φ_{Λ,Ω_0} is given by equation (2.11), and Λ is an element from the set (2.12). The function Ψ_{Λ,Ω_0} is so normalized that $\Psi_{\Lambda,\Omega_0}(0, \Omega) = \Phi_{\Lambda,\Omega_0}(\Omega)$.

It is seen from (2.5) and is readily verified from (2.6) to (2.9) and from (2.11) to (2.13) that if Λ is a (regular or improper) eigenvector, $-\Lambda$ is also an eigenvector with

$$\Phi^0_{-\Lambda}(\Omega) = \Phi^0_\Lambda(-\Omega), \qquad \Phi_{-\Lambda,-\Omega_0}(\Omega) = \Phi_{\Lambda,\Omega_0}(-\Omega),$$

and if c is real, $\bar{\Lambda}$ is also an eigenvector with

$$\Phi^0_{\bar{\Lambda}}(\Omega) = \overline{\Phi^0_\Lambda(\Omega)}, \qquad \Phi_{\bar{\Lambda},\Omega_0}(\Omega) = \overline{\Phi_{\Lambda,\Omega_0}(\Omega)}$$

as corresponding eigenfunctions.

Any linear combination of solutions of the forms of equations (2.10) and (2.14) are again solutions of the linear equation (1.4). Therefore the general solution of equation (1.4), which can be represented by a linear combination of the regular expansion modes $\Psi^0_\Lambda(r, \Omega)$ of equation (2.10) and the improper expansion modes $\Psi_{\Lambda,\Omega_0}(r, \Omega)$ of equation (2.14), is given by

$$(2.15)\quad \Psi(r, \Omega) = \int_{\{\Lambda^0\}} A^0(\Lambda)\Psi^0_\Lambda \, d_4\Lambda + \int_{\{\Omega_0\}} d_2\Omega_0 \int_{\{\Lambda:\Omega_0\cdot\Lambda=1\}} d_4\Lambda A(\Lambda, \Omega_0)\Psi_{\Lambda,\Omega_0},$$

where $\{\Lambda^0\}$ is given by equation (2.9), and $\{\Omega_0\}$ is the set of all real directions. The first and second terms on the right-hand side of equation (2.15) are given explicitly below by equations (5.12) and (6.9), respectively. Another representation is given in (7.19).

Conjecture. Every solution of the homogeneous equation (1.4) in the class of functions that are integrable[3] with respect to Ω and continuous in $|r| < \infty$ have a representation of the form of equation (2.15).

The following sections will be specific as to the meaning of integration with respect to Ω and Λ.

III. Integration of $(1 - \Omega \cdot \Lambda)^{-1}$.

A. *Geometrical considerations.* To evaluate equations (2.9) and (2.13), a specific meaning is required for

$$(3.1)\qquad I_\Lambda = \int \frac{d_2\Omega}{1 - \Omega\cdot\Lambda}.$$

This integral is invariant under the rotation group $\mathscr{R}$ of Ω. The complex vector Λ can be represented as the sum of a real vector $\Lambda_1 \in E_3$ and an imaginary vector $i\Lambda_2$ with $\Lambda_2 \in E_3$; that is,

$$(3.2)\qquad \Lambda = \Lambda_1 + i\Lambda_2. \quad \text{(Definition)}$$

The integral I_Λ is therefore only a function of

$$(3.3)\qquad \left.\begin{aligned} \lambda_1 &= |\Lambda_1|; \\ \lambda_2 &= |\Lambda_2|; \\ \gamma &= \cos^{-1}[(\Lambda_1\cdot\Lambda_2)/(\lambda_1\lambda_2)]. \end{aligned}\right\} \quad \text{(Definition)}$$

[3] See footnote 2.

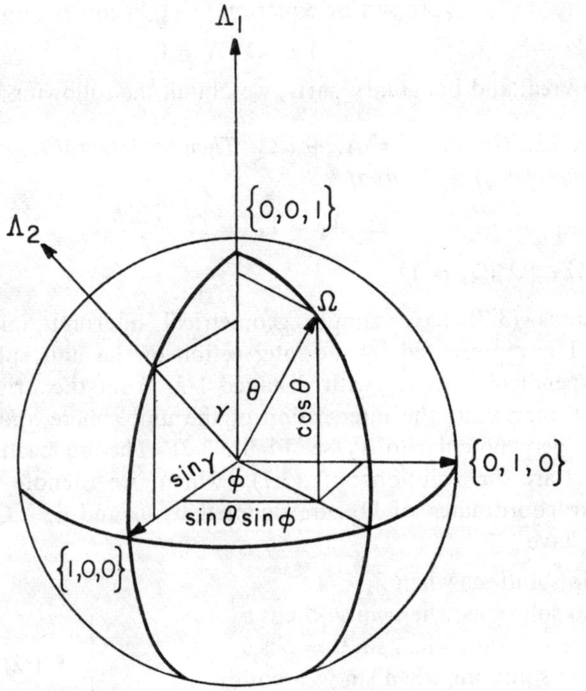

FIGURE 3.1. Reference system for the computation of $\int \frac{d_2\Omega}{1 - \Omega \cdot \Lambda}$.

It turns out to be convenient to choose Cartesian coordinates such that

$$\begin{aligned} \Lambda_1 &= \lambda_1\{0, 0, 1\}; \\ \Lambda_2 &= \lambda_2\{\sin \gamma, 0, \cos \gamma\}; \\ \Omega &= \{\cos \phi \sin \theta, \sin \phi \sin \theta, \cos \theta\}. \end{aligned} \tag{3.4}$$

This means that Λ_1 lies on the z-axis, Λ_2 lies in the x, z-plane, and Ω is a point on the unit sphere, with θ and ϕ representing spherical coordinates as illustrated in Figure 3.1.

We define[4] the inner product $\Omega \cdot \Lambda = (\Omega, \Lambda)$ to be

$$\begin{aligned} \Omega \cdot \Lambda &= \Omega \cdot \Lambda_1 + i\Omega \cdot \Lambda_2; \\ \Omega \cdot \Lambda_1 &= \lambda_1 \cos \theta; \\ \Omega \cdot \Lambda_2 &= \lambda_2(\cos \phi \sin \theta \sin \gamma + \cos \theta \cos \gamma). \end{aligned} \tag{3.5}$$

[4] Note that this definition differs from the usual definition in complex inner-product spaces, where $\Omega \cdot \Lambda = \Omega \cdot \Lambda_1 - i\Omega \cdot \Lambda_2$.

The integral I_Λ, expressed by equation (3.1), becomes improper when

$$1 - \Omega \cdot \Lambda = 0. \tag{3.6}$$

Separating real and imaginary parts, we obtain the following lemma:

LEMMA 3.1. *Given:* $\Lambda = \Lambda_1 + i\Lambda_2$. *Then the integral* I_Λ, *expressed by equation* (3.1), *is improper if and only if*

$$\Omega \cdot \Lambda_1 = 1, \qquad \Omega \cdot \Lambda_2 = 0, \tag{3.7}$$

for some $\Omega \in \{\Omega : |\Omega| = 1\}$.

Equations (3.7) have simple geometrical interpretations. The condition $\Omega \cdot \Lambda_1 = 1$ is represented by the intersection of the unit sphere $|\Omega| = 1$ and a plane perpendicular to Λ_1 with distance $1/\lambda_1$ from the origin. The condition $\Omega \cdot \Lambda_2 = 0$ represents the intersection of the unit sphere and the plane through the origin, perpendicular to Λ_2 (see Figure 3.2). The intersections, if any, of these two circles are the solutions of (3.7), which we denote by Ω_0^+ and Ω_0^-. The polar coordinates of $\Omega_0^\pm$ are denoted by θ_0 and $\phi_0^\pm$. Obviously, $\cos\theta_0 = 1/\lambda_1$. We have

$$\begin{array}{l} \text{no solutions when } \lambda_1 < 1 \\ \left.\begin{array}{l} \text{no solutions when } \sin\gamma < \cos\theta_0 \\ \text{one solution when } \sin\gamma = \cos\theta_0 \\ \text{two solutions when } \sin\gamma > \cos\theta_0 \\ \text{infinitely many solutions when } \lambda_2 = 0, \cos\theta_0 \end{array}\right\} = 1/\lambda_1, \quad \lambda_1 \geqq 1. \end{array} \tag{3.8}$$

The last line of (3.8) is a consequence of $\Omega \cdot \Lambda_1 = 1$ of (3.7) only, since for $\lambda_2 = 0$, $\Omega \cdot \Lambda_2 = 0$ is always satisfied. We conclude

COROLLARY 3.1.[5] *The integral* I_Λ *of* (3.1) *is proper when* $\lambda_1 < 1$ *or* $\lambda_1 \sin\gamma < 1$ *for* $\lambda_2 \neq 0$. *It is improper, when* $\lambda_1 \geqq 1$ *for* $\lambda_2 = 0$ *or* $\lambda_1 \sin\gamma \geqq 1$ *for* $\lambda_2 \neq 0$.

From Figure 3.2, (p. 45) we can also deduce

$$\cos\theta_0 = 1/\lambda_1, \qquad \cos(\pm\phi_0) = -\cot\theta_0 \cot\gamma, \tag{3.9}$$

a useful relation to determine Ω_0.

We can now proceed with the actual evaluation of I_Λ, expressed by equation (3.1). From equation (3.5), we conclude that

$$I_\Lambda = \int \frac{d_2\Omega}{1 - \Omega \cdot \Lambda} = \int_0^\pi \sin\theta \, d\theta \int_{-\pi}^{\pi} \frac{d\phi}{\alpha + \beta\cos\phi}, \tag{3.10}$$

where

$$\begin{aligned} \alpha &= 1 - \lambda_1 \cos\theta - i\lambda_2 \cos\gamma \cos\theta; \\ \beta &= -i\lambda_2 \sin\gamma \sin\theta. \end{aligned} \tag{3.11}$$

[5] Corollary 3.1 can also be deduced from Figure 6.2 by noting, for example, that Λ_2, for $\Lambda_2 \neq 0$, lies on a cone of aperture γ with Λ_1 as axis. The intersections of this cone with the plane $\Omega_0 \cdot \Lambda_2 = 0$, if any, are the desired solutions. An algebraic proof of Corollary 3.1 is given in **[5]**.

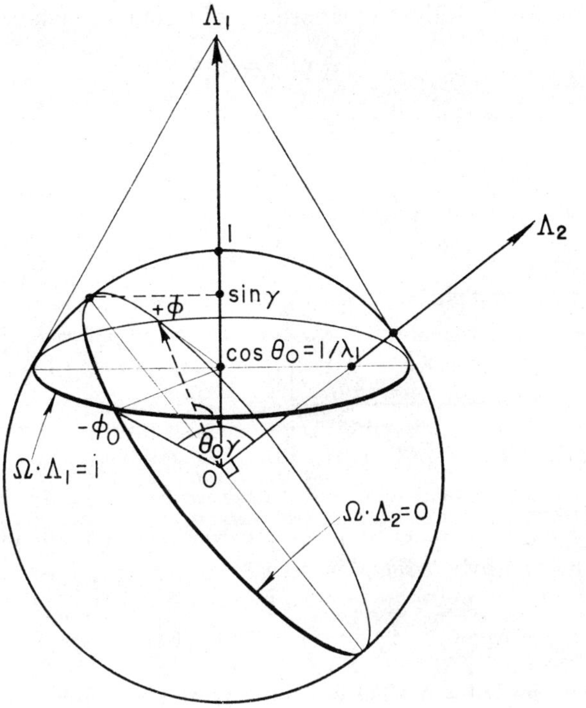

FIGURE 3.2. Geometric solution of $1 - \Omega_0 \cdot \Lambda = 0$; $\Omega_0^{\pm} = (\theta_0, \pm\phi_0)$.

First we evaluate

$$I_\theta = \int_{-\pi}^{+\pi} \frac{d\phi}{\alpha + \beta \cos \phi} \tag{3.12}$$

then we evaluate

$$I_\Lambda = \int_0^\pi I_\theta \sin \theta \, d\theta. \tag{3.13}$$

B. *Evaluation of* I_θ. If $\beta = 0$, then $I_\theta = 2\pi/\alpha$.

Assume $\beta \neq 0$, and introduce the transformation

$$z = e^{i\phi}, \qquad dz = iz\, d\phi. \tag{3.14}$$

With $2 \cos \phi = z + z^{-1}$, the right side of (3.12) becomes

$$I_\theta = \frac{2}{i} \oint \frac{dz}{\beta z^2 + 2\alpha z + \beta} = -\frac{2i}{\beta} \oint \frac{dz}{(z - z_1)(z - z_2)}, \tag{3.15}$$

where the integral extends over the unit circle. A convenient form for the zeros z_1 and z_2 of the quadratic $z^2 + 2(\alpha/\beta)z + 1$ is

$$z_{1,2} = -\frac{\alpha}{\beta} \pm \frac{\alpha - \beta}{\beta}\left[+ \left(\frac{\alpha + \beta}{\alpha - \beta}\right)^{1/2}\right]. \tag{3.16}$$

We define the "positive" square root z of a complex number w in the usual way,

$$+w^{1/2} = z \Rightarrow \operatorname{Re} z \geqq 0 \tag{3.17}$$

that is the w-plane is cut along the negative real axis so that

$$\operatorname{Re} z = 0 \Rightarrow \operatorname{Im} z > 0.$$

The "negative" square root of w is defined by

$$-w^{1/2} = -(+w^{1/2}).$$

It is obvious that $z_1 z_2 = 1$. Now,

$$1 \geqq \left| \frac{1 - \left[+\left(\frac{\alpha+\beta}{\alpha-\beta}\right)^{1/2}\right]}{1 + \left[+\left(\frac{\alpha+\beta}{\alpha-\beta}\right)^{1/2}\right]} \right| \equiv \left| -\frac{\alpha - (\alpha-\beta)\left[+\left(\frac{\alpha+\beta}{\alpha-\beta}\right)^{1/2}\right]}{\beta} \right| = |z_1|. \tag{3.18}$$

Thus, $|z_1| \leqq 1$ and $|z_2| \geqq 1$.

First, consider $|z_1| < 1$. Since z_1 is the only singularity in the unit circle, (3.15) yields, by *Cauchy's Residue Theorem*,[6]

$$I_\theta = 2\pi \Big/ (\alpha - \beta)\left[+\left(\frac{\alpha+\beta}{\alpha-\beta}\right)^{1/2}\right]. \tag{3.19}$$

Next consider $|z_1| = 1$. The equality sign in (3.18) holds only when

$$+[(\alpha+\beta)/(\alpha-\beta)]^{1/2}$$

is pure imaginary, implying $(\alpha/\beta) \in [-1, 1]$, and $|z_1| = |z_2| = 1$. In this case, the two singularities lie on the unit circle, and we define the integral of (3.15) by its *Cauchy's Principal Value.* If $-1 < \alpha/\beta < 1$, this is equivalent to deforming the contour of the integral in (3.15) slightly to bring the two singularities inside the contour, and taking half the sum of the residues. Since $\operatorname{Res} z_1 = -\operatorname{Res} z_2$, this sum is zero; thus,

$$I_\theta = 0. \tag{3.20}$$

If $\alpha/\beta = \pm 1$, we have $z_1 = z_2 = \mp 1$ in (3.15), and consequently I_θ does not exist as a Cauchy principal value. This result can also be obtained by direct inspection of the integrand in (3.12).

We express the condition $-1 < \alpha/\beta < 1$ in terms of the original parameters. From (3.11), it follows that α/β is real if and only if

$$1 - \lambda_1 \cos\theta = 0. \tag{3.21}$$

Furthermore, $-1 < \alpha/\beta < 1$ if and only if (3.21) holds and

$$-1 < \cos\gamma \cos\theta / \sin\gamma \sin\theta < 1. \tag{3.22}$$

---

[6] This part of our integral is known in the literature as $I_0 = 2\pi/(\alpha^2 - \beta^2)^{1/2}$. See, e.g., **[21]**.

Squaring, rearranging terms, and considering that $\cos\theta = 1/\lambda_1$ by (3.21), yields $\lambda_1^2 \sin^2\gamma > 1$. Since $\sin\gamma \geqq 0$, (3.22) reduces to

$$\lambda_1 \sin\gamma > 1. \tag{3.23}$$

For $\alpha/\beta = \pm 1$, the inequality sign in (3.23) is replaced by an equality sign.

In summary, combining (3.19) to (3.23) yields

$$\begin{aligned} I_\theta = \int_{-\pi}^{+\pi} \frac{d\phi}{\alpha + \beta\cos\phi} &= 0 \quad \text{for} \quad \lambda_1 \cos\theta = 1 \quad \text{and} \quad \lambda_1 \sin\gamma > 1, \\ &= 2\pi \Big/ (\alpha - \beta)\left[+\left(\frac{\alpha+\beta}{\alpha-\beta}\right)^{1/2}\right] \quad \text{otherwise,} \end{aligned} \tag{3.24}$$

where α and β are defined by (3.11). Note that this representation includes the case $\beta = 0$, for which $I_\theta = 2\pi/\alpha$, and the case $\alpha/\beta = \pm 1$, or equivalently $\lambda_1 \cos\theta = 1$, $\lambda_1 \sin\gamma = 1$, for which $I_\theta \to \infty$ as $\alpha \to \pm\beta$. Equation (3.24) was derived in **[5]**, using a different transformation.

IV. **Further integration of** $(1 - \Omega \cdot \Lambda)^{-1}$.

A. *The function* $(Z(\mu))^{1/2}$. In §III we gave a closed expression for I_θ. We must now consider I_θ as a function of θ, and determine I_Λ as indicated in equation (3.13). It will be convenient below to consider I_θ as a function of $\cos\theta$. Therefore we introduce the notation

$$\begin{aligned} &\mu \equiv \cos\theta; \\ &\mu_0 \equiv \cos\theta_0 \equiv 1/\lambda_1, \quad \text{for} \quad \lambda_1 \geqq 1; \\ &I_\mu(\mu) \equiv I_\theta(\theta). \end{aligned} \tag{4.1}$$

To set the proper background for the integration of I_μ, we discuss first the function

$$(Z(\mu))^{1/2} \equiv (\alpha - \beta)\left[+\left(\frac{\alpha+\beta}{\alpha-\beta}\right)^{1/2}\right]. \tag{4.2}$$

From equations (3.11) and (4.1), we have

$$\begin{aligned} \frac{\alpha+\beta}{\alpha-\beta} &= \frac{(\alpha+\beta)(\bar\alpha - \bar\beta)}{|\alpha-\beta|^2} \\ &= \frac{(1-\lambda_1\mu)^2 + \lambda_2^2(\mu^2 - \sin^2\gamma) - 2i\lambda_2(1-\lambda_1\mu)\sin\gamma\sin\theta}{(1-\lambda_1\mu)^2 + \lambda_2^2\cos^2(\gamma+\theta)} \end{aligned} \tag{4.3}$$

as a function of μ. The definition of its square root was given in (3.17). The behavior of the positive square root of equation (4.3) as a function of μ is as follows:

(a) If $\lambda_1 < 1$, it follows immediately from equation (4.3) that

$$\operatorname{Im} \frac{\alpha+\beta}{\alpha-\beta} < 0, \tag{4.4}$$

and hence $[+((\alpha+\beta)/(\alpha-\beta))^{1/2}]$ is continuous in the interval $-1 \leqq \mu \leqq +1$.

(b) If $\lambda_1 \geqq 1$, it follows from equation (4.3) that, respectively,

$$\text{(4.5)} \qquad \operatorname{Im} \frac{\alpha+\beta}{\alpha-\beta} \lesseqqgtr 0 \quad \text{for} \quad \mu \lesseqqgtr \mu_0.$$

At $\mu = \mu_0$,

$$\text{(4.6)} \qquad \operatorname{Re} \frac{\alpha+\beta}{\alpha-\beta} \gtreqqless 0 \quad \text{for} \quad \lambda_1 \sin\gamma \lesseqqgtr 1.$$

We set

$$\text{(4.7)} \qquad \rho = \left| \left(\frac{\mu_0^2 - \sin^2\gamma}{\cos^2(\gamma+\theta)} \right)^{1/2} \right|.$$

As μ approaches μ_0 from both sides, we have in the limit

$$\text{(4.8)} \qquad \begin{aligned} \left[+ \left(\frac{\alpha+\beta}{\alpha-\beta} \right)^{1/2} \right] &= \rho \qquad \text{for} \quad \mu = \mu_0 \pm 0 \quad \text{and} \quad \lambda_1 \sin\gamma < 1; \\ &= i\rho \qquad \left. \begin{aligned} &\text{for} \quad \mu = \mu_0 + 0 \\ = -i\rho \quad &\text{for} \quad \mu = \mu_0 - 0 \end{aligned} \right\} \quad \text{and} \quad \lambda_1 \sin\gamma > 1. \end{aligned}$$

The square root is $\rho = 0$ for $\mu = \mu_0 \pm 0$ and $\lambda_1 \sin\gamma = 1$. Hence, (4.8) is continuous in the interval $-1 \leqq \mu \leqq 1$ for $\lambda_1 \sin\gamma \leqq 1$, but has a discontinuity at $\mu = \mu_0$ in the case $\lambda_1 \sin\gamma > 1$.

Since $(\alpha - \beta)$ is a continuous function in μ, we can conclude that $(Z(\mu))^{1/2}$ is continuous in $-1 \leqq \mu \leqq 1$ for $\{\lambda_1 < 1\}$, and $\{\lambda_1 \geqq 1, \lambda_1 \sin\gamma < 1\}$, but has a discontinuity at $\mu = \mu_0$ for $\{\lambda_1 > 1, \lambda_1 \sin\gamma > 1\}$. In the case of $\lambda_1 \sin\gamma = 1$, $(Z(\mu_0))^{1/2} = 0$, and $(Z(\mu))^{1/2}$ is therefore also continuous.

We are now in a position to attribute signs to $(Z(\mu))^{1/2}$.

The sign of the square root of $Z(\mu)$ is determined by substituting $\mu = \pm 1$ in equation (4.2). We obtain immediately from equation (4.3), observing that $\mu = \pm 1$ implies $\sin\theta = 0$,

$$\text{(4.9)} \qquad (\alpha+\beta)/(\alpha-\beta) = 1, \qquad \text{for } \mu = \pm 1;$$

and from (3.11)

$$\text{(4.10)} \qquad \alpha - \beta = (1 \mp \lambda_1) \mp i\lambda_2 \cos\gamma, \qquad \text{for } \mu = \pm 1.$$

Hence, by the discussion above, we have for $\lambda_1 < 1$,

$$\text{(4.11)} \qquad Z^{1/2} = +Z^{1/2} \qquad \text{for } -1 \leqq \mu \leqq 1;$$

and for $\lambda_1 \geqq 1$,

$$\text{(4.12)} \qquad \begin{aligned} Z^{1/2} &= +Z^{1/2} \qquad \text{for } -1 \leqq \mu < \mu_0, \\ &= -Z^{1/2} \qquad \text{for } \mu_0 < \mu \leqq 1, \end{aligned}$$

$$\text{(4.12a)} \qquad +(Z(\mu_0 - 0))^{1/2} = -(Z(\mu_0 + 0))^{1/2} \qquad \text{for } \lambda_1 \sin\gamma < 1,$$

$$\text{(4.12b)} \qquad +(Z(\mu_0 - 0))^{1/2} = -[-(Z(\mu_0 + 0))^{1/2}] \qquad \text{for } \lambda_1 \sin\gamma > 1,$$

$$\text{(4.12c)} \qquad (Z(\mu_0 \pm 0))^{1/2} = 0 \qquad \text{for } \lambda_1 \sin\gamma = 1;$$

where the last three equations are a consequence of equation (4.8) and the subsequent remarks.

We are now ready to integrate I_μ with respect to μ.

B. *Integration of* I_μ. Comparison of equations (3.24), (4.1), and (4.2) shows that

$$(4.13)\quad I_\mu = 2\pi/(Z(\mu))^{1/2} \qquad \text{for } \lambda_1\mu \neq 1;\ \lambda_1\mu = 1 \ \text{ and } \ \lambda_1 \sin\gamma \leqq 1.$$

For $\lambda_1 \sin\gamma = 1$ and $\lambda_1\mu = 1$, I_μ is unbounded. For $\lambda_1 \sin\gamma > 1$ and $\lambda_1\mu = 1$, I_μ is zero. If we use Riemannian integration, we have, under consideration of equations (4.11), (4.12), and (3.24), two cases, which can be integrated, (a) where $Z^{1/2}$ is continuous, and (b) where $Z^{1/2}$ has a discontinuity:

(a) $\lambda_1 < 1$, or $\lambda_1 \geqq 1$ and $\lambda_1 \sin\gamma < 1$:

$$(4.14)\qquad I_\Lambda = 2\pi \int_{-1}^{+1} \frac{d\mu}{(Z(\mu))^{1/2}}.$$

(b) $\lambda_1 \geqq 1$, $\lambda_1 \sin\gamma > 1$:

$$(4.15)\qquad I_\Lambda = 2\pi \left\{ \int_{-1}^{\mu_0 - 0} \frac{d\mu}{+(Z(\mu))^{1/2}} + \int_{\mu_0+0}^{1} \frac{d\mu}{-(Z(\mu))^{1/2}} \right\}.$$

The situation $\lambda_1 \geqq 1$, $\lambda_1 \sin\gamma = 1$, leads to an improper integral, which must be obtained by a limiting process. This process is carried through in §C(c) below.

To obtain an analytic expression for the desired integral, we observe from equation (4.2) that

$$(4.16)\qquad Z = \alpha^2 - \beta^2.$$

From (3.11) and (4.1), it follows then that

$$(4.17)\qquad Z(\mu) = a\mu^2 - 2b\mu + c,$$

where

$$(4.18)\qquad \begin{aligned} a &= \lambda_1^2 - \lambda_2^2 + 2i\lambda_1\lambda_2 \cos\gamma, \\ b &= \lambda_1 + i\lambda_2 \cos\gamma, \\ c &= 1 + \lambda_2^2 \sin^2\gamma. \end{aligned}$$

The indefinite integral of $Z(\mu)^{-1/2}$ is (as can be verified)

$$(4.19\text{a})\qquad \int \frac{d\mu}{Z^{1/2}} = \pm \frac{1}{a^{1/2}} \log(a\mu - b \pm a^{1/2}Z^{1/2}) + \text{const}, \quad \text{for} \quad b^2 \neq ac$$

$$(4.19\text{b})\qquad = \frac{1}{a^{1/2}} \log(a\mu - b) + \text{const}, \qquad \text{for} \quad b^2 = ac.$$

Equation (4.19b) is contained in (4.19a), provided the sign of $a^{1/2}$ is chosen so that

$$(4.19\text{c})\qquad Z^{1/2} = a^{1/2}(\mu - b/a).$$

Since I_Λ of (3.10) can be directly evaluated for $b^2 = ac$ without reference to (4.19b), we treat this case first to have it available for comparison with the limit cases of the general integrals (4.14) and (4.15).

Case $b^2 = ac$. Using (4.18), the condition $b^2 = ac$ becomes

$$\text{(4.20a)} \qquad \lambda_2^2 \sin^2 \gamma(\lambda_1^2 - \lambda_2^2 - 1) = 0, \quad \text{for} \quad \operatorname{Re} b^2 = \operatorname{Re} ac$$

$$\text{(4.20b)} \qquad \lambda_1\lambda_2^3 \cos \gamma \sin^2 \gamma = 0, \quad \text{for} \quad \operatorname{Im} b^2 = \operatorname{Im} ac.$$

Conditions (4.20a) and (4.20b) are simultaneously satisfied if at least one of the following equations is true:

(4.21a) (i) $\lambda_2 = 0$;

(4.21b) (ii) $\sin \gamma = 0$;

(4.21c) (iii) $\cos \gamma = 0$ and $\lambda_1^2 - \lambda_2^2 = 1$.

(i) If we set $\lambda_2 = 0$, then from (3.10) and (3.11),

$$\text{(4.22)} \qquad I_\Lambda = 2\pi \int_{-1}^{+1} \frac{d\mu}{1 - \lambda_1\mu} = \frac{2\pi}{\lambda_1} \log \left| \frac{1 + \lambda_1}{1 - \lambda_1} \right|, \quad \text{for} \quad \lambda_2 = 0$$

according to results obtained in the case of plane geometry. The Cauchy principal value was used to obtain the result for $\lambda_1 > 1$.

(ii) If we set $\sin \gamma = 0$, it follows from (3.11) that

$$\text{(4.23)} \quad I_\Lambda = 2\pi \int_{-1}^{+1} \frac{d\mu}{1 - (\lambda_1 \pm i\lambda_2)\mu} = \frac{2\pi}{\lambda_1 \pm i\lambda_2} \log \frac{1 + \lambda_1 \pm i\lambda_2}{1 - \lambda_1 \mp i\lambda_2}, \qquad \text{for} \quad \sin \gamma = 0,$$

where the upper sign corresponds to the value $\cos \gamma = +1$ and the lower sign corresponds to $\cos \gamma = -1$. It will be shown that for the proper integral, $\sin \gamma = 0$ implies $\lambda_1\lambda_2 = 0$ [see equation (5.5)], and that for the improper integral, (4.23) has meaning only as a limit case [$\lambda_1 \to \infty$, implying $I_\Lambda = 0$; see equation (4.41)].

(iii) If we set $\cos \gamma = 0$ ($\sin \gamma = 1$) and $\lambda_1^2 - \lambda_2^2 = 1$, then $\lambda_1 \sin \gamma \geqq 1$. From (4.16) to (4.18), we obtain $Z = \alpha^2 - \beta^2 = (\mu - \lambda_1)^2$, and consequently from (4.12), (4.12b), and (4.15)

$$\begin{aligned} Z^{1/2} &= +Z^{1/2} = \lambda_1 - \mu \qquad \text{for } -1 \leqq \mu < \mu_0; \\ &= -Z^{1/2} = \mu - \lambda_1 \qquad \text{for } \mu_0 < \mu \leqq 1. \end{aligned}$$

Thus

$$\begin{aligned} I_\Lambda &= 2\pi \int_{-1}^{\mu_0 - 0} \frac{d\mu}{+Z^{1/2}} + 2\pi \int_{\mu_0+0}^{1} \frac{d\mu}{-Z^{1/2}} = 2\pi \int_{-1}^{1/\lambda_1 - 0} \frac{d\mu}{\lambda_1 - \mu} + 2\pi \int_{1/\lambda_1 + 0}^{1} \frac{d\mu}{\mu - \lambda_1} \\ &= 2\pi \log \frac{\lambda_1 + 1}{\lambda_1 - 1/\lambda_1} + 2\pi \log \frac{\lambda_1 - 1}{\lambda_1 - 1/\lambda_1} = 2\pi \log \frac{\lambda_1^2}{\lambda_1^2 - 1} = 2\pi \log \frac{\lambda_1^2}{\lambda_2^2}. \end{aligned}$$

Thus,

$$\text{(4.24)} \qquad I_\Lambda = 4\pi \log \frac{\lambda_1}{\lambda_2}, \quad \text{for} \quad \cos \gamma = 0 \quad \text{and} \quad \lambda_1^2 - \lambda_2^2 = 1.$$

We return now to the general case.

Case $b^2 \neq ac$. As explained at the beginning of this section, we distinguish only two cases, (a) and (b) corresponding to (4.14) and (4.15), respectively:

(a) To obtain the integral (4.14), we must evaluate $Z^{1/2}$ at $\mu = \pm 1$ under the conditions (4.11) and (4.12). This can be done using equations (4.17) and (4.18). Then

$$Z(\pm 1) = a \mp 2b + c = (1 \mp b)^2.$$

Hence, by equations (4.11) and (4.12),

$$Z^{1/2}(\pm 1) = 1 \mp b, \tag{4.25}$$

and by (4.19a), we obtain for (4.14)

$$\begin{aligned} \int_{-1}^{+1} \frac{d\mu}{Z^{1/2}} &= \frac{1}{a^{1/2}} \log \frac{a - b + a^{1/2}(1 - b)}{-a - b + a^{1/2}(1 + b)} \\ &= \frac{1}{a^{1/2}} \log \frac{(1 + a^{1/2})(a^{1/2} - b)}{(1 - a^{1/2})(a^{1/2} - b)}. \end{aligned} \tag{4.26}$$

We assign the negative sign to $a^{1/2}$. In this case, even for $\lambda_2 \to 0$, $(a^{1/2} - b) \neq 0$ and can be canceled.

Now we define the inner product $\Lambda^2 \equiv \Lambda \cdot \Lambda$ to be, for $\Lambda = \Lambda_1 + i\Lambda_2$,

$$\begin{aligned} \Lambda^2 &\equiv (\Lambda_1 + i\Lambda_2) \cdot (\Lambda_1 + i\Lambda_2) \\ &\equiv \Lambda_1^2 - \Lambda_2^2 + 2i\Lambda_1 \cdot \Lambda_2 \\ &= \lambda_1^2 - \lambda_2^2 + 2i\lambda_1\lambda_2 \cos\gamma. \end{aligned} \tag{4.27}$$

Comparing this expression with (4.18), we see that

$$\Lambda^2 = a. \tag{4.28}$$

We substitute[7]

$$a^{1/2} = -a^{1/2} = -(+a^{1/2}) = \sqrt{\Lambda^2} \tag{4.29}$$

in equation (4.26), simplify, and multiply by 2π to conform with (4.14). This yields the desired explicit expression for the proper integral in the characteristic equation (2.9),

$$I_\Lambda = \int \frac{d_2\Omega}{1 - \Omega \cdot \Lambda} = \int_{-1}^{+1} I_\mu \, d\mu = \frac{2\pi}{\sqrt{\Lambda^2}} \log \frac{1 + \sqrt{\Lambda^2}}{1 - \sqrt{\Lambda^2}} = \frac{4\pi}{\sqrt{\Lambda^2}} \tanh^{-1} \sqrt{\Lambda^2} \tag{4.30}$$

$$\text{for } \{\lambda_1 < 1\} \text{ and } \{\lambda_1 \geq 1, \lambda_1 \sin\gamma < 1\},$$

where Λ^2 is defined by (4.27), and the Λ^2-plane is cut along the negative axis. Note that $\sqrt{\Lambda^2}$ in the final form (4.30) can be either $\sqrt[-]{\Lambda^2}$ or $\sqrt[+]{\Lambda^2}$.

(b) To obtain the integral (4.15), we must evaluate $Z^{1/2}$ at $\mu = \pm 1$ and

[7] Note that $\sqrt{\Lambda^2}$ is a complex scalar and not the vector Λ.

$\mu = \mu_0 \pm 0$. We obtain, from (4.1), (4.17), (4.18), and (4.12b)

$$(4.31) \qquad (Z(\mu_0 \pm 0))^{1/2} = \pm\lambda_2[+(\sin^2\gamma - \mu_0^2)^{1/2}],$$

while

$$(4.25) \qquad (Z(\pm 1))^{1/2} = 1 \mp b$$

as before. Therefore, by equations (4.18) and (4.28),

(4.32)

$$\int_{-1}^{\mu_0-0} \frac{d\mu}{+Z^{1/2}} + \int_{\mu_0+0}^{+1} \frac{d\mu}{-Z^{1/2}}$$

$$= \frac{1}{a^{1/2}} \log \frac{a\mu_0 - b + a^{1/2}\lambda_2(\sin^2\gamma - \mu_0^2)^{1/2}}{-a - b + a^{1/2}(1+b)}$$

$$+ \frac{1}{a^{1/2}} \log \frac{a - b + a^{1/2}(1-b)}{a\mu_0 - b - a^{1/2}\lambda_2(\sin^2\gamma - \mu_0^2)^{1/2}}$$

$$= \frac{1}{a^{1/2}} \log \left[\frac{(1+a^{1/2})(a^{1/2}-b)}{(1-a^{1/2})(a^{1/2}-b)} \cdot \frac{\lambda_2[-\lambda_2/\lambda_1 + i\cos\gamma + (1/\lambda_1)a^{1/2}(\lambda_1^2\sin^2\gamma - 1)^{1/2}]}{\lambda_2[-\lambda_2/\lambda_1 + i\cos\gamma - (1/\lambda_1)a^{1/2}(\lambda_1^2\sin^2\gamma - 1)^{1/2}]} \right]$$

$$= \frac{1}{\sqrt{\Lambda^2}} \log \left[\frac{1+\sqrt{\Lambda^2}}{1-\sqrt{\Lambda^2}} \cdot \frac{-\lambda_2 + i\lambda_1\cos\gamma + (\sqrt{\Lambda^2})(\lambda_1^2\sin^2\gamma - 1)^{1/2}}{-\lambda_2 + i\lambda_1\cos\gamma - (\sqrt{\Lambda^2})(\lambda_1^2\sin^2\gamma - 1)^{1/2}} \right].$$

Substituting the last line of (4.32) into (4.15) yields the desired explicit expression for the improper integral in (2.13),

$$I_\Lambda = \int \frac{d_2\Omega}{1 - \Omega\cdot\Lambda} = \int_{-1}^{+1} I_\mu \, d\mu$$

$$(4.33) \qquad = \frac{2\pi}{\sqrt{\Lambda^2}} \log \left[\frac{\sqrt{\Lambda^2}+1}{\sqrt{\Lambda^2}-1} \cdot \frac{(\sqrt{\Lambda^2})(\lambda_1^2\sin^2\gamma - 1)^{1/2} - \lambda_2 + i\lambda_1\cos\gamma}{(\sqrt{\Lambda^2})(\lambda_1^2\sin^2\gamma - 1)^{1/2} + \lambda_2 - i\lambda_1\cos\gamma} \right]$$

$$\text{for} \quad \{\lambda_1 \geq 1, \text{ and } \lambda_1 \sin\gamma > 1\},$$

where Λ^2 is defined by (4.28), and the Λ^2-plane is cut along the negative axis.

As in (4.30), we see from the final equation (4.33) that the sign of $a^{1/2}$ is of no practical interest; i.e. replacing $\sqrt{\Lambda^2}$ by $-\sqrt{\Lambda^2}$ leaves this equation unchanged.

We note that

$$(4.34) \qquad -\lambda_2 + i\lambda_1\cos\gamma = -(\lambda_1^2\sin^2\gamma - a)^{1/2}.$$

Use of equation (4.34) would give equation (4.33) a more symmetric appearance, but no additional insight.

C. *Limit cases.* In this section we consider all limit cases of interest. First we see (a) how the special cases, when $b^2 = ac$, are related. Then, (b) we consider limits of the proper integral (4.30) and its relation to the special cases, and finally, (c) we study limits of the improper integral (4.33) and its relation to (4.30), the special cases and their limits.

(a) If we let λ_2 go to zero in equation (4.23), we have

$$(4.35)\quad \lim_{\lambda_2\to 0} I_\Lambda = \frac{2\pi}{\lambda_1}\log\left|\frac{1+\lambda_1}{1-\lambda_1}\right| + \begin{cases} 0 & \text{for } \lambda_1 < 1, \\ i\pi & \text{for } \cos\gamma = 1 \\ -i\pi & \text{for } \cos\gamma = -1 \end{cases} \quad \text{and } \lambda_1 > 1.$$

Hence, I_Λ in equation (4.22) is the arithmetic mean of I_Λ in equation (4.35).

(b) Now we investigate equation (4.30) and obtain, by (4.27),

$$(4.36)\quad \begin{aligned} &\lim_{\lambda_2\to 0}\frac{2\pi}{\sqrt{\Lambda^2}}\log\frac{1+\sqrt{\Lambda^2}}{1-\sqrt{\Lambda^2}} \\ &\quad = \frac{2\pi}{\lambda_1}\log\left|\frac{1+\lambda_1}{1-\lambda_1}\right| + \begin{cases} 0 & \text{for } \lambda_1 < 1, \\ i\pi & \text{for } \cos\gamma > 0 \\ -i\pi & \text{for } \cos\gamma < 0 \end{cases} \quad \text{and } \lambda_1 > 1. \end{aligned}$$

Therefore, equation (4.35) is a special case of (4.36) since the case $\sin\gamma = 0$ corresponds exactly to equation (4.35).

Next we consider the case $\cos\gamma = 0$ in (4.30). This case occurs in §V in connection with the regular eigenfunctions. Note that $\cos\gamma = 0$ for the proper eigenvectors Λ^0 implies $\lambda_1 \sin\gamma = \lambda_1 < 1$; therefore (4.30) reduces to

$$(4.37)\quad I_\Lambda = \frac{2\pi}{(\lambda_1^2-\lambda_2^2)^{1/2}}\log\frac{1+(\lambda_1^2-\lambda_2^2)^{1/2}}{1-(\lambda_1^2-\lambda_2^2)^{1/2}} \quad \text{for } \cos\gamma = 0,\ \lambda_1 < 1.$$

As $\lambda_2 \to 0$, equation (4.37) reduces to equations (4.36), (4.35), and (4.22) for $\lambda_1 < 1$.

For the sake of completeness, we mention that the limit $\lambda_1 \to 0$ does not cause any difficulties, since in $\lim_{\lambda_1\to 0}\sqrt{\Lambda^2} = \pm i\lambda_2$ either sign may be used in equation (4.30).

(c) In a similar manner, we investigate equation (4.33), where $\lambda_1 \geqq 1$ always. We have

$$(4.38)\quad \lim_{\lambda_2\to 0} I_\Lambda = \frac{2\pi}{\lambda_1}\log\left(\frac{\lambda_1+1}{\lambda_1-1}\,\frac{(\lambda_1^2\sin^2\gamma-1)^{1/2}+i\cos\gamma}{(\lambda_1^2\sin^2\gamma-1)^{1/2}-i\cos\gamma}\right).$$

The limit depends on the angle γ. For $\cos\gamma = 0$, equation (4.38) reduces to equation (4.22) for $\lambda_1 \geqq 1$. One can also consider I_Λ in equation (4.22) as the mean value of I_Λ in equation (4.33) with respect to γ.

As $(\lambda_1\sin\gamma - 1) \to 0+$,

$$(4.39)\quad I_\Lambda = \frac{2\pi}{\sqrt{\Lambda^2}}\log\frac{1+\sqrt{\Lambda^2}}{1-\sqrt{\Lambda^2}}.$$

Hence equation (4.33) changes continuously to equation (4.30).

From equations (4.38) and (4.39), we conclude

$$(4.40)\quad \lim_{\lambda_2\to 0}\ \lim_{\lambda_1\sin\gamma\to 1+} I_\Lambda = \lim_{\lambda_1\sin\gamma\to 1+}\ \lim_{\lambda_2\to 0} I_\Lambda,$$

and the result coincides with equation (4.36) for $\lambda_1 > 1$.

If $\gamma \to 0$, it follows from $\lambda_1 \sin \gamma > 1$ that $\lambda_1 \to \infty$. In this case, from equation (4.33),

$$\lim_{\lambda_1 \to \infty} I_\Lambda \to 0, \qquad (\gamma \to 0). \tag{4.41}$$

Similarly, $\lim_{\lambda_2 \to \infty} I_\Lambda \to 0$.

Another type of limit case for (4.33) is $\sqrt{\Lambda^2} \to 1$ with $\lambda_2 \neq 0$. Let $\lambda_1^2 - \lambda_2^2 = 1 + \varepsilon$ and $\cos \gamma = \eta$, where ε and $\eta \to \pm 0$. Then

$$\Lambda^2 = 1 + \varepsilon + 2i\lambda_1\lambda_2\eta,$$

$$\sqrt{\Lambda^2} = 1 + \varepsilon/2 + i\lambda_1\lambda_2\eta + O(\varepsilon^2, \varepsilon\eta, \eta^2),$$

and

$$(\lambda_1^2 \sin^2 \gamma - 1)^{1/2} = \lambda_2 + \varepsilon/2\lambda_2 + O(\varepsilon^2, \eta^2).$$

Inserting these expressions in (4.33), rearranging terms, and simplifying yields

$$\lim_{\varepsilon \to 0} \lim_{\eta \to 0} I_\Lambda = \lim_{\eta \to 0} \lim_{\varepsilon \to 0} I_\Lambda = 4\pi \log \frac{\lambda_1}{\lambda_2} \tag{4.42}$$

in agreement with the special case (4.24). For $\lambda_2 \to 0$, (4.42) is unbounded.

We note that in (4.30) the limit case $\sqrt{\Lambda^2} \to 1$ corresponds to $\lambda_1^2 - \lambda_2^2 = 1 - \varepsilon$ ($\varepsilon \downarrow 0$) and $\cos \gamma \to 0$. This requires $\lambda_2 \to 0$, $\lambda_1 \to 1$. We conclude then from (4.37) that I_Λ of (4.30) is always unbounded for $\Lambda^2 = 1$.

V. **Solution of the characteristic equation and the regular eigenfunctions.** In equations (2.7) to (2.9), we defined the regular eigenfunctions and the characteristic equation. In Corollary 3.1, we asserted that the integral I_Λ is proper if and only if

$$\lambda_1 < 1; \quad \text{or} \quad \lambda_1 \sin \gamma < 1 \quad \text{and} \quad \lambda_2 \neq 0. \tag{5.1}$$

We shall prove

LEMMA 5.1. *Regular eigenfunctions of the form* (2.8) *exist only if* $\lambda_1 < 1$.

In §IV we showed that the integral I_Λ, appearing in the characteristic equation (2.9) and satisfying conditions (5.1), is given by equation (4.30). Therefore, we can write the following expression for the *characteristic equation*:

$$1 - (c/z) \tanh^{-1} z = 0; \qquad \sqrt{\Lambda^2} = z,\ \Lambda \in \{\Lambda^0\}. \tag{5.2}$$

This is the same equation in z as was found for the characteristic equation in plane geometry, and has two solutions, λ_0 and $-\lambda_0$, for given c as follows:

$$\begin{aligned} &\text{(a)} \quad 0 < c < 1, \quad \lambda_0 \text{ is real, and} \quad \lambda_0^2 < 1; \\ &\text{(b)} \quad c = 1, \lambda_0 = 0, \quad \text{and} \quad \lambda_0^2 = 0; \\ &\text{(c)} \quad 1 < c < \infty, \quad \lambda_0 \text{ is imaginary, and} \quad \lambda_0^2 < 0. \end{aligned} \tag{5.3}$$

Hence, λ_0 can be considered known for given c, and thus

$$\{\Lambda^0\} = \{\Lambda : \Lambda^2 = \lambda_0^2, \forall \Omega \to \Omega \cdot \Lambda \neq 1\}. \quad \text{(Definition)} \tag{5.4}$$

This means that every vector Λ that satisfies $\Lambda^2 = \lambda_0^2$ and $\Omega \cdot \Lambda \neq 1$ for each Ω is an element of $\{\Lambda^0\}$. Since λ_0^2 is real, we have, by (4.27), the condition

$$\lambda_1^2 - \lambda_2^2 = \lambda_0^2, \qquad \lambda_1\lambda_2 \cos\gamma = 0, \tag{5.5}$$

for $\Lambda^2 = \lambda_0^2$. The condition $\Omega \cdot \Lambda \neq 1$ for each Ω, yielded the conditions (5.1). These conditions impose only restrictions on the magnitudes λ_1 and λ_2 of the real and imaginary components Λ_1 and Λ_2, respectively, of $\Lambda = \Lambda_1 + i\Lambda_2$ and the mutual angle γ in E_3. From equation (5.5), if $\lambda_1\lambda_2 \neq 0$, then $\cos\gamma = 0$. Therefore, in (5.1), $\sin\gamma = 1$ and for $\lambda_2 \neq 0$ necessarily

$$\lambda_1 < 1. \tag{5.6}$$

This proves Lemma 5.1.

We treat the three cases of (5.3) separately in the following paragraphs. They contain the cases $\lambda_1\lambda_2 = 0$ and $\lambda_1\lambda_2 \neq 0$ but $\cos\gamma = 0$.

(a) $0 < c < 1$.

From equation (5.5), we obtain

$$\lambda_1 = (\lambda_2^2 + \lambda_0^2)^{1/2} \tag{5.7a}$$

and from (5.3) and (5.6), we obtain

$$\lambda_0 \leq \lambda_1 < 1; \qquad 0 \leq \lambda_2 < (1 - \lambda_0^2)^{1/2}. \tag{5.7b}$$

(b) $c = 1$.

By the same equations as used in (a), we obtain

$$\lambda_1 = \lambda_2; \tag{5.8a}$$

$$\begin{aligned} 0 &\leqq \lambda_1 < 1; \\ 0 &\leqq \lambda_2 < 1. \end{aligned} \tag{5.8b}$$

(c) $1 < c < \infty$.

Again, by the same equations as used in (a), we obtain

$$\lambda_1 = (\lambda_2^2 - |\lambda_0|^2)^{1/2}; \tag{5.9a}$$

$$0 \leq \lambda_1 < 1; \qquad |\lambda_0| \leq \lambda_2 < (1 + |\lambda_0|^2)^{1/2}. \tag{5.9b}$$

Because $\lambda_0^2 < 0$, we can write equation (5.9a) as

$$\lambda_1 = (\lambda_2^2 + \lambda_0^2)^{1/2} \tag{5.9c}$$

to conform with equation (5.7a).

The relationship between λ_1 and λ_2 is illustrated in Figure 5.1 (p. 56).

We mentioned at the beginning of §III that I_Λ is invariant under the rotation group $\mathscr{R}$ on Ω. This is also true of the characteristic equation. Hence, if $\Lambda \in \{\Lambda^0\}$, then $\mathscr{R}\Lambda \subset \{\Lambda^0\}$ also. Thus, if $\Phi_\Lambda^0(\Omega)$ is an eigenfunction, so is any element of $\{\Phi_{\mathscr{R}\Lambda}^0(\Omega)\}$. The range of $\{\Lambda^0\}$ is indicated in Figures 5.2 and 5.3 (p. 57) for the cases $c < 1$ and $c > 1$, respectively, for the projections of $\{\Lambda^0\}$ into E_3.

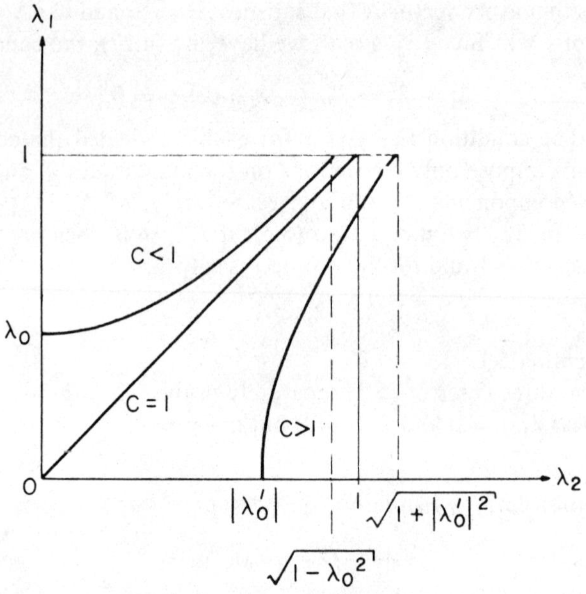

FIGURE 5.1. Relationship between λ_1 and λ_2 for $c < 1$, $c = 1$, and $c > 1$.

The functions $\Psi(r, \Omega)$ that can be represented by the regular eigenfunctions (2.8) are given by the first integral in equation (2.15). We select a convenient reference direction from $\{\Lambda^0\}$, say $w^0 + iu^0$, such that, by equations (5.7a) and (5.9c),

$$\Lambda^0 = \Lambda_1^0 + i\Lambda_2^0 = \sqrt{\lambda_2^2 + \lambda_0^2}w^0 + i\lambda_2 u^0, \tag{5.10}$$

where

$$w^0 = \Lambda_1^0/\lambda_1 \quad \text{and} \quad u^0 = \Lambda_2^0/\lambda_2. \tag{5.11}$$

Note that $w^0 \cdot u^0 = 0$ since $\cos\gamma = 0$. Let an element of $\mathscr{R}$ be denoted by R. Then

$$\int_{\{\Lambda^0\}} A^0(\Lambda)\Psi_\Lambda^0\, d_4\Lambda = \int_{\lambda_2=\begin{cases}0 \text{ for } c\le 1\\ |\lambda_0| \text{ for } c\ge 1\end{cases}}^{\lambda_2=(1-\lambda_0^2)^{1/2}} d\lambda_2 \int_{\mathscr{R}} d_3R A^0(R\Lambda^0)\Psi_{R\Lambda^0}(r,\Omega). \tag{5.12}$$

The order of integration on the right-hand side of this equation may be interchanged.

For applications, it may be convenient to choose for the reference direction an orthonormal system; e.g.

$$u^0 = \{1, 0, 0\}; \qquad v^0 = \{0, 1, 0\}; \qquad w^0 = \{0, 0, 1\}. \tag{5.13}$$

Then Λ_1^0 is in the "z-direction," and Λ_2^0 in the "x-direction," and an element R is given by the transformation matrix that transforms the orthonormal system (5.13) into another orthonormal system (u, v, w); i.e. R is an orthogonal transformation. The reduction of $\int A(\Lambda^0)\Psi^0_{R\Lambda^0}(r,\Omega)\, d_3R$ to a triple integral of the Riemannian type will depend on the geometric properties of the problem at hand.

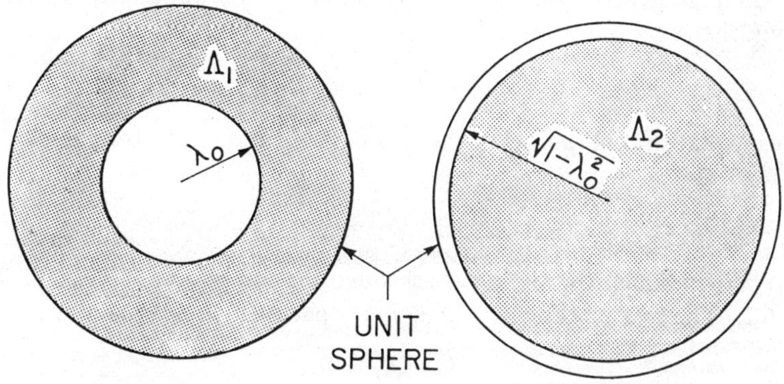

FIGURE 5.2. Range of $\Lambda_1 + i\Lambda_2 = \Lambda \in \{\Lambda^0\}$ for $c < 1$.

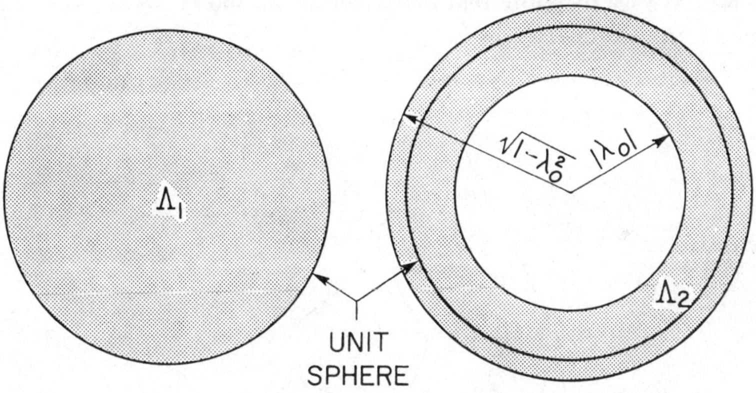

FIGURE 5.3. Range of $\Lambda_1 + i\Lambda_2 = \Lambda \in \{\Lambda^0\}$ for $c > 1$.

VI. **Generalized eigenfunctions and the general solution of the homogeneous equation.** In equations (2.11) and (2.14), we defined the generalized eigenfunction and the generalized expansion modes, respectively. In §IV, parts B and C, we determined the integral term in equation (2.13) to be I_Λ, as given by equation (4.33) or its limit cases. As has been anticipated, G is a function of only λ_1, λ_2, and γ, and does not depend on the particular direction of Ω_0 or the complex vector, Λ. It depends only on the relative position of the projections of the real and imaginary parts of Λ into E_3. Hence,

$$G = G(\Lambda) = G(\lambda_1, \lambda_2, \gamma) = G(|\Lambda_1|, |\Lambda_2|, \Lambda_1 \cdot \Lambda_2), \tag{6.1}$$

or explicitly

$$G = 1 - (c/4\pi)I_\Lambda, \tag{6.2}$$

where I_Λ is given by equation (4.33) for $\lambda_1 \geq 1$ and $\lambda_1 \sin\gamma \geq 1$ [and by equation (4.22) for $\lambda_2 \equiv 0$ and $\lambda_1 \geq 1$].

To obtain all generalized eigenfunctions that are unbounded at Ω_0, we observe that, by (2.12),

$$1 - \Omega_0 \cdot \Lambda = 0. \tag{6.3}$$

This means that the end points of the real part of the complex vector, Λ, lie in the plane

$$\Omega_0 \cdot \Lambda_1 = 1, \tag{6.4}$$

which is perpendicular to, and passes through the end point of Ω_0, as illustrated in Figure 6.1. The end points of the imaginary part of Λ lie in the plane

$$\Omega_0 \cdot \Lambda_2 = 0, \tag{6.5}$$

which is perpendicular to Ω_0 and passes through the origin, as illustrated also in Figure 6.1. We see therefore that these vectors, Λ, satisfy the conditions given in Lemma 3.1 and Corollary 3.1.

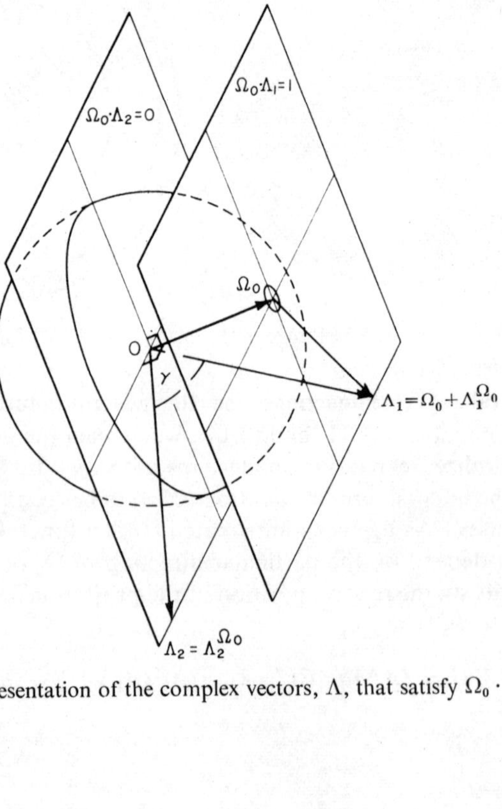

FIGURE 6.1. Representation of the complex vectors, Λ, that satisfy $\Omega_0 \cdot \Lambda = 1$.

DEFINITION 6.1. All vectors lying in the real plane $\Omega_0 \cdot \Lambda_2 = 0$ shall be denoted by $\{\Lambda^{\Omega_0}\}$.

With this definition, every vector Λ_1 satisfying equation (6.5) is defined by

$$\Lambda_2 = \Lambda_2^{\Omega_0}, \tag{6.6}$$

and every vector Λ_1 satisfying (6.4) is defined by

$$\Lambda_1 = \Omega_0 + \Lambda_1^{\Omega_0}, \tag{6.7}$$

since the two planes (6.4) and (6.5) are parallel (see Figure 6.2).

Therefore, the second term on the right-hand side of equation (2.15) has the representation

(6.8)

$$\int_{\{\Omega_0\}} d_2\Omega_0 \int_{\{\Lambda:\Omega_0\cdot\Lambda=1\}} d_4\Lambda A(\Lambda, \Omega_0)\Psi_{\Lambda,\Omega_0}(r, \Omega)$$
$$= \int_{\{\Omega_0\}} d_2\Omega_0 \int_{\{\Lambda^{\Omega_0}\}} d_2\Lambda_1^{\Omega_0} \int_{\{\Lambda^{\Omega_0}\}} d_2\Lambda_2^{\Omega_0} A(\Omega_0 + \Lambda_1^{\Omega_0} + i\Lambda_2^{\Omega_0}, \Omega_0)\Psi_{\Lambda,\Omega_0}(r, \Omega)$$

(6.9)

$$= \kappa \int_{\{\Lambda^{\Omega}\}} d_2\Lambda_1^{\Omega} \int_{\{\Lambda^{\Omega}\}} d_2\Lambda_2^{\Omega} A(\Lambda, \Omega)G(\Lambda) \exp[-(r, \Omega + \Lambda_1^{\Omega} + i\Lambda_2^{\Omega})]$$
$$+ \kappa \int_{\{\Omega_0\}} d_2\Omega_0 \int_{\{\Lambda^{\Omega_0}\}} d_2\Lambda_1^{\Omega_0} \int_{\{\Lambda^{\Omega_0}\}} d_2\Lambda_2^{\Omega_0} \frac{\exp[-(r, \Omega_0 + \Lambda_1^{\Omega_0} + i\Lambda_2^{\Omega_0})]}{1 - (\Omega, \Omega_0 + \Lambda_1^{\Omega_0} + i\Lambda_2^{\Omega_0})} A(\Lambda, \Omega_0),$$

where $\Lambda = \Omega + \Lambda_1^{\Omega} + i\Lambda_2^{\Omega}$ in $A(\Lambda, \Omega)$ and $G(\Lambda)$ of the first part of (6.9). G is reduced further by equation (6.1) [or equation (6.12) below].

The first term in (6.9) was obtained from (6.8) by interchanging Ω and Ω_0 in Definition 2.2 and corresponds to the second term of Φ_{Λ,Ω_0} in equation (2.11).

If the integration over Ω_0 is carried out in polar coordinates, we can replace Definition 2.2 by

$$\delta(\Omega - \Omega_0) = \delta(\mu - \mu_0)\delta(\phi - \phi_0) \tag{6.10}$$

and thus reduce the two-dimensional δ-function to the product of two one-dimensional δ-functions in a natural way.

In both terms of (6.9), the order of integration with respect to $\Lambda_1^{\Omega_0}$ and $\Lambda_2^{\Omega_0}$ can be interchanged. The actual integration in the $\{\Lambda^{\Omega_0}\}$-plane can be carried out in any way convenient for the specific problem at hand, e.g. in two-dimensional polar coordinates or two-dimensional Cartesian coordinates.

Further simplifications are possible, e.g. as follows: Let us denote the angle between $\Lambda_1^{\Omega_0}$ and $\Lambda_2^{\Omega_0}$ by γ^{Ω_0}, as indicated in Figure 6.2. Now

$$\begin{aligned} \lambda_1 &= (1 + (\lambda_1^{\Omega_0})^2)^{1/2} \\ \Lambda_1 \cdot \Lambda_2 &= (\Lambda_1^{\Omega_0} + \Omega_0, \Lambda_2^{\Omega_0}) \\ &= \lambda_1\lambda_2 \cos\gamma = \lambda_1^{\Omega_0}\lambda_2^{\Omega_0} \cos\gamma^{\Omega_0}. \end{aligned} \tag{6.11}$$

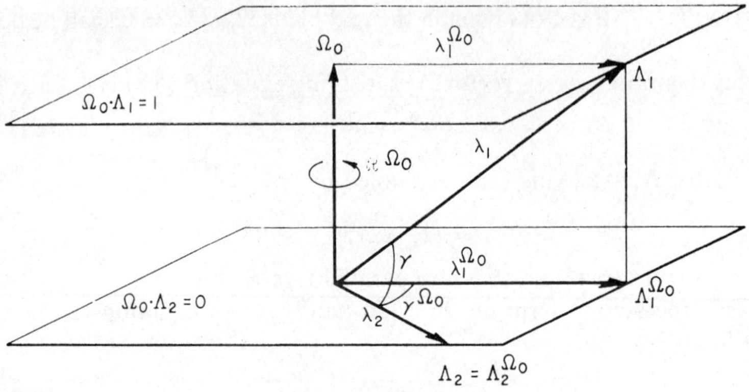

FIGURE 6.2. Example of specific coordinates for integration of (6.9).

Hence, λ_1, λ_2, and γ are determined by $\lambda_1^{\Omega_0}$, $\lambda_2^{\Omega_0}$, and γ^{Ω_0}, respectively, and we can write

$$G = G(\Lambda) = G(\lambda_1^{\Omega_0}, \lambda_2^{\Omega_0}, \gamma^{\Omega_0}). \tag{6.12}$$

Thus the first integral in (6.9) can be represented by

$$\begin{aligned} &\int_{\{\Lambda\Omega\}} d_2\Lambda_1^\Omega \int_{\{\Lambda\Omega\}} d_2\Lambda_2^\Omega A(\Lambda,\Omega)G(\Lambda)e^{-(r,\Lambda)} \\ &\quad = \int_0^\infty d\lambda_1^\Omega \int_0^\infty d\lambda_2^\Omega \int_0^{2\pi} d\gamma^\Omega G \int_{\mathscr{R}\Omega} dR^\Omega\, A(R^\Omega\Lambda^I, \Omega)e^{-(r,R^\Omega\Lambda^I)} \end{aligned} \tag{6.13}$$

where R^Ω is an element of the *plane* rotation group $\mathscr{R}^\Omega$, and Λ^I is the reference vector for the rotation such that $\Omega \cdot \Lambda^I = 1$, and $R^\Omega\Lambda^I = \Lambda$.

Hence the most general solution that can be obtained for equation (1.4) by superposition of solutions (2.1) is equation (2.15), where the integrals appearing in that equation are explicitly given by equations (5.12) and (6.9). In manipulating these expressions, one should bear in mind that $G(\Lambda)$ *may become zero.* The integral expressions for the superposition of the generalized eigenfunctions may be improper integrals. These integrals will be considered as existing if a finite neighborhood of the singularity exists for which the integral converges as the limit of the neighborhood converges to zero. In other words, we admit Cauchy principal-value type integrals over higher dimensions.

VII. **Orthogonality relations and class representations of eigenvectors and eigenfunctions.** To obtain some interesting integral relations, equation (2.5) is written in the form

$$\Lambda \cdot \Omega\Phi_\Lambda = \Phi_\Lambda - c\kappa/4\pi. \tag{7.1}$$

Here, Φ_Λ stands for any proper (Φ_Λ^0) or improper (Φ_{Λ,Ω_0}) eigenfunction. Operating with $\int \cdot \Phi_\Lambda \, d_2\Omega$ on both sides of (7.1), and remembering (2.6), we get

(7.2) $$\int \Lambda \cdot \Omega \Phi_\Lambda \Phi_{\Lambda'} \, d_2\Omega = \int \Phi_\Lambda \Phi_{\Lambda'} \, d_2\Omega - c\kappa^2/4\pi.$$

Interchanging Λ and Λ' in this equation, and taking the difference between these equations, yields a *general orthogonality relation*,

(7.3) $$(\Lambda - \Lambda') \cdot \int \Omega \Phi_\Lambda \Phi_{\Lambda'} \, d_2\Omega = 0.$$

Now, let $\Lambda' = \zeta\Lambda$, where $\zeta = \xi + i\eta$ is a nonvanishing complex scalar. Then (7.3) and (7.2), respectively, lead to the "*class orthogonality relations*,"

(7.4)

$$\int \Lambda \cdot \Omega \Phi_\Lambda \Phi_{\zeta\Lambda} \, d_2\Omega = N_\Lambda \delta(\zeta - 1) \quad \text{and} \quad \int \Phi_\Lambda \Phi_{\zeta\Lambda} \, d_2\Omega = \frac{c\kappa^2}{4\pi} + N_\Lambda \delta(\zeta - 1).$$

Equations (7.4) as written above are only symbolic. Their exact meaning is explained below in equations (7.23), (7.24), (7.26), and (7.31) and the remarks following. The class orthogonality relations (7.4) permit an elegant determination of the expansion coefficients in some important cases, as illustrated below and in §VIII. The representation of the general solution of §VI is further simplified in these cases if the integrations are performed according to "classes of eigenvectors."

DEFINITION 7.1. The eigenvectors Λ and Λ' belong to the same class if they are linearly dependent over the field of complex numbers.

Thus, if Λ and Λ' belong to the same class, there exists a complex number $\zeta = \xi + i\eta \neq 0$, such that $\Lambda' = \zeta\Lambda$. We select as base elements for the classes the quasi-proper vectors Λ^q, which are defined as follows.

DEFINITION 7.2. Quasi-proper vectors, Λ^q, are eigenvectors that satisfy the relations

(7.5) $$(\Lambda^q)^2 = \lambda_0^2,$$

where λ_0 is a solution of (5.2).

Thus, a quasi-proper vector, Λ^q, whose real part has magnitude less than unity ($\lambda_1^q < 1$) is a proper eigenvector, Λ^0, and vice versa. We concentrate on the case $c \neq 1$. Then, the following lemma is true.

LEMMA 7.1. *Let $c \neq 1$ in equation* (2.5). *Then* (1) *each quasi-proper vector and its negative define one and only one class;* (2) *the set of all quasi-proper vectors defines all possible classes uniquely;* (3) *if one element of a class is real* (*or pure imaginary*), *then all other elements are either real or pure imaginary.*

This lemma, therefore, asserts that each improper eigenvector has a unique representation $\Lambda = \zeta\Lambda^q$.

To prove the lemma, we project the real and complex parts of the vectors into E_3 as was previously done in §III. Then from

(7.6) $$\Lambda' = \Lambda_1' + i\Lambda_2'; \qquad \Lambda = \Lambda_1 + i\Lambda_2; \qquad \zeta = \xi + i\eta$$

follows the transformation of $\Lambda' = \zeta\Lambda$ in the projection space E_3:

$$\Lambda_1' = \xi\Lambda_1 - \eta\Lambda_2; \qquad \Lambda_2' = \eta\Lambda_1 + \xi\Lambda_2. \tag{7.7}$$

Thus, if $\Lambda_1 \neq 0$ and $\Lambda_2 \neq 0$, both components Λ_1' and Λ_2' lie in the plane spanned by Λ_1 and Λ_2 in E_3. If either $\Lambda_1 = 0$ or $\Lambda_2 = 0$, then Λ_1', Λ_2', and Λ are parallel to each other in E_3.

After these general remarks, we turn to the proof of part (3) of the lemma. Assume to the contrary that $\Lambda_1' \neq 0$, $\Lambda_2' \neq 0$. Since they are parallel to each other, $1 - \Lambda' \cdot \Omega \neq 0$ for all Ω. This implies that Λ' is a proper eigenvector, and from (5.5) follows $\Lambda_1' \cdot \Lambda_2' = 0$, which contradicts the fact that Λ_1' and Λ_2' are parallel. Thus either $\Lambda_1' = 0$ or $\Lambda_2' = 0$, Q.E.D.

To prove part (2) of the lemma, we demonstrate that for any vector Λ there exist unique ζ_q, Λ^q such that, except for sign, $\Lambda = \zeta_q\Lambda^q$. Then $\Lambda^q = \zeta\Lambda$, where $\zeta = \zeta_q^{-1}$. Let Λ' in (7.6) be a quasi-proper vector $\Lambda' = \Lambda^q$. Then by (7.5), $(\Lambda_1^q)^2 - (\Lambda_2^q)^2 = \lambda_0^2$ and $\Lambda_1^q \cdot \Lambda_2^q = 0$. Substituting in these equations the right sides of (7.7) yields

$$(\xi^2 - \eta^2)(\Lambda_1^2 - \Lambda_2^2) - 4\xi\eta\Lambda_1 \cdot \Lambda_2 = \lambda_0^2, \tag{7.8}$$

and

$$(\xi^2 - \eta^2)\Lambda_1 \cdot \Lambda_2 + \xi\eta(\Lambda_1^2 - \Lambda_2^2) = 0, \tag{7.9}$$

respectively. From these two equations follows

$$(\xi^2 - \eta^2) = \lambda_0^2(\Lambda_1^2 - \Lambda_2^2)/[(\Lambda_1^2 - \Lambda_2^2)^2 + 4(\Lambda_1 \cdot \Lambda_2)^2] \equiv 2a, \tag{7.10}$$

$$\xi\eta = -\lambda_0^2(\Lambda_1 \cdot \Lambda_2)/[(\Lambda_1^2 - \Lambda_2^2)^2 + 4(\Lambda_1 \cdot \Lambda_2)^2] \equiv b, \tag{7.11}$$

or, observing that $\xi^2 > 0$, $\eta^2 > 0$, and letting $\varepsilon = \pm 1$,

$$\xi = \varepsilon[a + (a^2 + b^2)^{1/2}]^{1/2}; \qquad \eta = \varepsilon[-a + (a^2 + b^2)^{1/2}]^{1/2}. \tag{7.12}$$

Thus $\Lambda^q = \zeta\Lambda$ with $\zeta = \xi + i\eta$ as given by (7.12) is uniquely determined, except for the sign, Q.E.D.

To prove part (1) of the lemma, we assume that in $\Lambda' = \xi\Lambda$ of (7.7), both Λ' and Λ are quasi-proper vectors. We must show that necessarily $\zeta = \pm 1$. We proceed as in the proof for part (2), but since Λ is also a proper eigenvector, we have the additional conditions $\Lambda_1^2 - \Lambda_2^2 = \lambda_0^2$ and $\Lambda_1 \cdot \Lambda_2 = 0$ to impose on (7.10) and (7.11). It follows that $2a = 1$ and $b = 0$; hence $\xi = \pm 1$, $\eta = 0$. This concludes the proof of lemma 7.1.

We have just proved that each improper eigenvector has a unique representation, $\Lambda = \zeta\Lambda^q$.[8] We ask now for the range of ζ such that for a given Λ^q, $\zeta\Lambda^q = \Lambda$ is an improper eigenvector. The transformation (7.7) takes the form

$$\Lambda_1 = \xi\Lambda_1^q - \eta\Lambda_2^q, \qquad \Lambda_2 = \eta\Lambda_1^q + \xi\Lambda_2^q, \tag{7.13}$$

[8] The sign of ζ is fixed as soon as the base of the class, say Λ^q, is chosen.

subject to the conditions of equations (7.5) for Λ^q,

$$\lambda_1^2 - \lambda_2^2 = \lambda_0^2, \qquad (\lambda_1 = |\Lambda_1^q|, \lambda_2 = |\Lambda_2^q|); \qquad \Lambda_1^q \cdot \Lambda_2^q = 0, \tag{7.14}$$

and of Corollary 3.1 for the improper eigenvector, Λ,

$$\Lambda_1^2 \geq 1 \quad \text{for} \quad \Lambda_2 = 0 \tag{7.15a}$$

or

$$\Lambda_1^2 \sin^2 \gamma \geq 1 \quad \text{for} \quad \Lambda_2 \neq 0. \tag{7.16a}$$

First assume (7.15a) to hold. Then, by Lemma 7.1, part (3), either $\Lambda_2^q = 0$ and $\Lambda = \Lambda_1 = \xi\Lambda_1^q$, or $\Lambda_1^q = 0$ and $\Lambda = \Lambda_1 = -\eta\Lambda_2^q$.

Thus $\zeta = \xi$ or $\zeta = i\eta$, where

$$\xi^2\lambda_1^2 \geq 1 \quad \text{or} \quad \eta^2\lambda_2^2 \geq 1 \tag{7.15b}$$

in order that $|\Lambda_1| > 1$. Now, assume (7.16a) to hold. Multiplying (7.16a) by Λ_2^2 yields

$$\Lambda_1^2\Lambda_2^2 - (\Lambda_1 \cdot \Lambda_2)^2 \geq \Lambda_2^2.$$

Substituting (7.13) in this equation, simplifying, and rearranging terms, leads to the condition

$$(\xi^2 + \eta^2)^2 \geq \xi^2/\lambda_1^2 + \eta^2/\lambda_2^2. \tag{7.16b}$$

This is a quadratic form in ξ^2 and η^2 and contains the condition of equation (7.15b) as a special case when either $\xi = 0$ or $\eta = 0$. By the usual methods of analytic geometry it can be shown that (7.16b) represent the exterior of a parabola, which under the additional conditions $\xi^2 > 0$ and $\eta^2 > 0$ defines the range for $\zeta = \xi + i\eta$. For computational purposes in applications, this range is explicitly given by

$$\begin{aligned} \xi^2 &\geqq -\eta^2 + (1 + (1 + 4\eta^2\lambda_0^2\lambda_1^2/\lambda_2^2)^{1/2})/2\lambda_1^2 \quad \text{for} \quad \eta^2 \leq 1/\lambda_2^2, \\ &\geqq 0 \quad \text{for} \quad \eta^2 \geq 1/\lambda_2^2, \end{aligned} \tag{7.16c}$$

or equivalently

$$\begin{aligned} \eta^2 &\geqq -\xi^2 + (1 + (1 - 4\xi^2\lambda_0^2\lambda_2^2/\lambda_1^2)^{1/2})/2\lambda_2^2 \quad \text{for} \quad \xi^2 \leq 1/\lambda_1^2, \\ &\geqq 0 \quad \text{for} \quad \xi^2 \geq 1/\lambda_1^2, \end{aligned} \tag{7.16d}$$

where λ_0, λ_1, and λ_2 are defined in (7.14). These equations contain (7.15b) as a special case. Therefore, the integration in (2.15) can be performed over the set of all quasi-proper vectors defined by (7.5) and the range of the complex plane as given by any one of equations (7.16b–d). Since to each complex improper Λ belong two singular directions, Ω_0^+ and Ω_0^- as shown in Figure 3.2 [and calculated by (3.9)], it is necessary to specify for each ζ which Ω_0 is used in $A(\Lambda, \Omega_0) = A(\zeta\Lambda^q, \Omega_0^\pm)$. Furthermore, when $\Lambda = \zeta\Lambda^q$ is real, there are infinitely many singular directions Ω_0 associated with Λ. The entire problem is simplified if, instead of calculating Ω_0 from ζ, we calculate ζ from Ω_0. Since $\Lambda = \zeta\Lambda^q$ is an improper eigenvector, equation (2.12) requires

$$\Lambda \cdot \Omega_0 = \zeta(\Lambda^q \cdot \Omega_0) = 1, \tag{7.17}$$

and thus

$$\zeta = 1/\Lambda^q \cdot \Omega_0 = (\Lambda_1^q - i\Lambda_2^q) \cdot \Omega_0/[(\Lambda_1^q \cdot \Omega_0)^2 + (\Lambda_2^q \cdot \Omega_0)^2]. \tag{7.18}$$

Hence, equation (2.15) takes the form

$$\begin{aligned} \Psi(r, \Omega) = &\int_{\{(\Lambda^0)^2=\lambda_0^2\}} A^0(\Lambda^0)\Psi_{\Lambda^0}\, d_4\Lambda^0 \\ &+ \int_{\{(\Lambda^q)^2=\lambda^2\}} d_4\Lambda^q \int_{\{\Omega_0\}} d_2\Omega_0 A\left(\frac{\Lambda^q}{\Lambda^q \cdot \Omega_0}, \Omega_0\right)\Psi_{\Lambda^q/\Lambda^q\cdot\Omega_0} \end{aligned} \tag{7.19}$$

where $\Psi_{\Lambda^q/\Lambda^q\cdot\Omega_0}$ is short for $\Psi_{\Lambda^q/\Lambda^q\cdot\Omega_0,\Omega_0}(r, \Omega)$. This second term corresponds to (6.8) and (6.9) when the order of integration over Λ and Ω_0 is interchanged.

Consider now the case when $\Psi(r, \Omega)$ can be represented by eigenfunctions of only one class, such that

$$\Psi(r, \Omega) = \Psi(r, \Omega; \Lambda^0) = a_{\Lambda^0}\Psi_{\Lambda^0} + a_{-\Lambda^0}\Psi_{-\Lambda^0} + \int_{\{\Omega_0\}} d_2\Omega_0 A_{\Lambda^0}(\Omega_0)\Psi_{\Lambda^0/\Lambda^0\cdot\Omega_0} \tag{7.20}$$

if the base vector is the proper eigenvector Λ^0, or such that

$$\Psi(r, \Omega) = \Psi(r, \Omega; \Lambda^q) = \int_{\{\Omega_0\}} d_2\Omega_0 A_{\Lambda^q}(\Omega_0)\Psi_{\Lambda^q/\Lambda^q\cdot\Omega_0}(r, \Omega) \tag{7.21}$$

if the base vector is a quasi-proper vector Λ^q with $\lambda_1^q \geq 1$. [By Lemma 7.1, no proper eigenfunctions exist for (7.21).] Now, let $\Psi(0, \Omega)$ in (7.20) be given. Then because of the assumption that (7.20) is valid,

$$\Psi(0, \Omega) = a_{\Lambda^0}\Phi_{\Lambda^0}(\Omega) + a_{-\Lambda^0}\Phi_{-\Lambda^0}(\Omega) + \int_{\{\Omega_0\}} d_2\Omega_0 A_{\Lambda^0}(\Omega_0)\Phi_{\Lambda^0/\Lambda^0\cdot\Omega_0}(\Omega). \tag{7.22}$$

It follows that the coefficients in equation (7.20) can be calculated, making use of the class orthogonality relations (7.4). Substituting (7.18) in (7.4), yields explicitly

$$\int_{\{\Omega\}} \Lambda^0 \cdot \Omega[\Phi_{\pm\Lambda^0}(\Omega)]^2\, d_2\Omega = N_{\lambda_0}, \tag{7.23a}$$

$$\begin{aligned} &\int_{\{\Omega\}} \Lambda^0 \cdot \Omega\Phi_{\Lambda^0}(\Omega)\Phi_{-\Lambda^0}(\Omega)\, d_2\Omega = 0, \\ &\int_{\{\Omega\}} \Lambda^0 \cdot \Omega\Phi_{\pm\Lambda^0}(\Omega)\Phi_{\Lambda^0/\Lambda^0\cdot\Omega_0}(\Omega)\, d_2\Omega = 0, \end{aligned} \tag{7.23b}$$

and

$$\int \Lambda^q \cdot \Omega\Phi_{\Lambda^q/\Lambda^q\cdot\Omega_0}(\Omega)\Phi_{\Lambda^q/\Lambda^q\cdot\Omega_1}(\Omega)\, d_2\Omega = N_{\lambda_1^q}(\Lambda^q \cdot \Omega_1)\delta(\Omega_0 - \Omega_1). \tag{7.24}$$

Therefore, by (7.22) and (7.23),

$$a_{\pm\Lambda^0} = \frac{1}{N_{\lambda_0}} \int_{\{\Omega\}} \Lambda^0 \cdot \Omega\Phi_{\pm\Lambda^0}(\Omega)\Psi(0, \Omega). \tag{7.25}$$

Carrying out (7.23a) with methods analogous to those used in §§III and IV yields

(7.26) $$N_{\lambda_0} = (c^2\kappa^2/4\pi)\lambda_0^2/(1-\lambda_0^2).$$

To obtain $A_{\Lambda^0}(\Omega)$, we consider first the evaluation of the integral

(7.27a) $$U = \int d_2\Omega \int d_2\Omega_0 \Lambda^q\cdot\Omega \Phi_{\Lambda^q/\Lambda^q\cdot\Omega_1}(\Omega)\Phi_{\Lambda^q/\Lambda^q\cdot\Omega_0}(\Omega)A_{\Lambda^q}(\Omega_0).$$

We make use of the explicit forms (2.11) for Φ_{Λ,Ω_0}, (2.13) for $G(\Lambda)$, and (3.1) for I_Λ, respectively,

(7.28a) $$\Phi_{\Lambda^q/\Lambda^q\cdot\Omega_0} = \frac{\kappa c}{4\pi}\frac{1}{1-\Lambda^q\cdot\Omega/\Lambda^q\cdot\Omega_0} + \kappa G\left(\frac{\Lambda^q}{\Lambda^q\cdot\Omega}\right)\delta(\Omega-\Omega_0),$$

(7.28b) $$G\left(\frac{\Lambda^q}{\Lambda^q\cdot\Omega}\right) = 1 - \frac{c}{4\pi}\int\frac{d_2\Omega_1}{1-\Lambda^q\cdot\Omega_1/\Lambda^q\cdot\Omega} = G_{\lambda_1^q}(\Lambda^q\cdot\Omega),$$

and

(7.28c) $$I_{\Lambda^q/\Lambda^q\cdot\Omega} = \int\frac{d_2\Omega_1}{1-\Lambda^q\cdot\Omega_1/\Lambda^q\cdot\Omega} = I_{\lambda_1^q}(\Lambda^q\cdot\Omega).$$

The last expression, (7.28c), could be evaluated using (4.33). There I_Λ depends only on λ_1, λ_2, and γ. From (7.18) and (7.7) it follows that λ_1, λ_2, and γ of the vector $\Lambda = \Lambda^q/\Lambda^q\cdot\Omega_1$ depend on λ_1^q (or λ_2^q), $\Lambda_1^q\cdot\Omega_1$ and $\Lambda_2^q\cdot\Omega_1$ [λ_2^q follows from the relation $(\lambda_1^q)^2-(\lambda_2^q)^2=\lambda_0^2$]. Thus, for any fixed class characterized by Λ^q, $I_{\Lambda^q/\Lambda^q\cdot\Omega_1}$ and $G_{\lambda_1^q}(\Lambda^q\cdot\Omega_1)$ are functions of $\Lambda^q\cdot\Omega_1$ only. Appropriate substitutions of (7.28) in (7.27a) and some algebraic manipulations yield

(7.27b) $$U = N_{\lambda_1^q}(\Lambda^q\cdot\Omega_1)\cdot A_{\Lambda^q}(\Omega_1),$$

where

(7.29a) $$N_{\lambda_1^q}(\Lambda^q\cdot\Omega_1) = \Lambda^q\cdot\Omega_0\kappa^2\{[G_{\lambda_1^q}(\Lambda^q\cdot\Omega_1)]^2 + H_{\lambda_1^q}(\Lambda^q\cdot\Omega_1)\},$$

(7.29b)

$$H_{\lambda_1^q}(\Lambda^q\cdot\Omega_1) = \left(\frac{c}{4\pi}\right)^2\int d_2\Omega\int d_2\Omega_0\left\{\frac{1}{(1-\Lambda^q\cdot\Omega/\Lambda^q\cdot\Omega_1)(1-\Lambda^q\cdot\Omega_0/\Lambda^q\cdot\Omega)} - \frac{1}{(1-\Lambda^q\cdot\Omega_0/\Lambda^q\cdot\Omega_1)(1-\Lambda^q\cdot\Omega/\Lambda^q\cdot\Omega_0)}\right\}.$$

An explicit form for $N_{\lambda_1^q}(\Lambda^q\cdot\Omega_1)$ is

(7.29c)

$$2N_{\lambda_1^q}(\Lambda^q\cdot\Omega_1) = \Lambda^q\cdot\Omega_1\kappa^2\left\{2 - \frac{c}{\pi}I_{\Lambda^q/\Lambda^q\cdot\Omega_1} + \left(\frac{c}{2\pi}\right)^2\int\frac{I_{\Lambda^q/\Lambda^q\cdot\Omega_1} - I_{\Lambda^q/\Lambda^q\cdot\Omega}}{1-\Lambda^q\cdot\Omega/\Lambda^q\cdot\Omega_1}d_2\Omega\right\}.$$

From this expression it follows that $N_{\lambda_1^q}(\Lambda^q\cdot\Omega_1)$ is bounded and analytic in the complex variable $\nu_1 = \Lambda^q\cdot\Omega_1$. To derive a shorter expression for $N_{\lambda_1^q}$ and for later reference, let us consider briefly the two limit cases for $\Lambda^q = \Lambda^0$: $\Lambda_1^0 = 0$ or

$\Lambda_2^0 = 0$; i.e. Λ^0 is either real or pure imaginary. Then $\Lambda^0 \cdot \Omega = \nu$, where $|\nu| \leq |1/\lambda_0|$. Equations (7.29a, b) take the form

$$N_{\lambda_1^0}(\nu_1) = \nu_1\kappa^2\biggl\{[G_{\lambda_1^0}(\nu_1)]^2 + \frac{c^2}{4}\biggl(\int d\nu_0 \int d\nu \frac{\nu_1\nu_0}{(\nu_1-\nu_0)(\nu_0-\nu)} - \int d\nu \int d\nu_0 \frac{\nu_1\nu_0}{(\nu_1-\nu_0)(\nu_0-\nu)}\biggr)\biggr\}.$$

Because of the Poincaré-Bertrand transformation formula **[40]**, this is

$$N_{\lambda_1^0}(\nu_1) = \nu_1\kappa^2\biggl\{[G_{\lambda_1^0}(\nu_1)]^2 + \biggl[\frac{c\pi}{2}\nu_1\biggr]^2\biggr\}, \qquad \nu_1 = \Lambda^0 \cdot \Omega_1. \tag{7.30}$$

Since $N_{\lambda_1^q}(\Lambda^q \cdot \Omega_1)$ depends only on λ_1^q and $\Lambda^q \cdot \Omega_1$, and since it is analytic in $\Lambda^q \cdot \Omega_1$, we conclude by reason of analytic continuation from (7.30) and (7.29) that

$$N_{\lambda_1^q}(\Lambda^q \cdot \Omega_1) = \Lambda^q \cdot \Omega_1\kappa^2\biggl\{[G_{\lambda_1^q}(\Lambda^q \cdot \Omega_1)]^2 + \biggl[\frac{c\pi}{2}\Lambda^q \cdot \Omega_1\biggr]^2\biggr\}. \tag{7.31}$$

We return now to the determination of $A_{\Lambda^0}(\Omega)$, and treat $A_{\Lambda^q}(\Omega)$ of (7.21) concurrently. Because of (7.20), (7.21), and the orthogonality relations (7.23),

$$\int \Lambda^q \cdot \Omega \Phi_{\Lambda^q/\Lambda^q\cdot\Omega_1}(\Omega)\Psi'(0, \Omega)\, d_2\Omega = -(0 + 0) + U,$$

where $\Psi'(0, \Omega)$ stands for $\Psi(0, \Omega; \Lambda^0) - (a_{\Lambda^0}\Phi_{\Lambda^0} + a_{-\Lambda^0}\Phi_{-\Lambda^0})$ when $\Lambda^q = \Lambda^0$, and for $\Psi(0, \Omega; \Lambda^q)$ when $\Lambda^q \notin \{\Lambda^0\}$. Thus by (7.27b),

$$A_{\Lambda^q}(\Omega_1) = \frac{1}{N_{\lambda_1^q}(\Lambda^q \cdot \Omega_1)} \int \Lambda^q \cdot \Omega \Phi_{\Lambda^q/\Lambda^q\cdot\Omega_1}(\Omega)\Psi'(0, \Omega)\, d_2\Omega. \tag{7.32}$$

If we interchange the order of integration in (7.27a) and consider (7.24), then

$$U = \int d_2\Omega_0 N_{\lambda_1^q}(\Lambda^0 \cdot \Omega_0)\delta(\Omega_0 - \Omega_1)A_{\Lambda^q}(\Omega_0) = N_{\lambda_1^q}(\Lambda^q \cdot \Omega_1)A_{\Lambda^q}(\Omega_1).$$

Therefore, the "orthogonality" condition (7.24) "legalizes" the interchange of variables in (7.27a). Thus, when we operate on (7.22) (or its equivalent for $\Lambda^q \notin \{\Lambda^0\}$) by $\int d_2\Omega\Lambda^q \cdot \Omega\Phi_{\Lambda^q/\Lambda^q\cdot\Omega_1}(\Omega)\,\cdot$, and treat the integrals as if they were proper integrals, and apply then the orthogonality relations (7.23b) and (7.24), we obtain the correct answer for the coefficient function $A_{\Lambda^q}(\Omega)$.

Inserting (7.28a) into (7.20) and (7.21) for $r = 0$ yields

$$G_{\lambda_1^q}(\Lambda^q \cdot \Omega)A_{\Lambda^q}(\Omega) + \frac{c}{4\pi}\int \frac{d_2\Omega_0 A_{\Lambda^q}(\Omega_0)}{1 - \Lambda^q \cdot \Omega/\Lambda^q \cdot \Omega_0} = \Psi'(0, \Omega). \tag{7.33}$$

This is a singular integral equation of the second kind for $A_{\Lambda^q}(\Omega)$. Therefore the function space whose elements $\Psi(r, \Omega)$ have a representation (7.20) or (7.21) and are simultaneously elements of the solution space of the transport equation (1.4), consists of all $\Psi(r, \Omega)$ such that a solution $A_{\Lambda^q}(\Omega)$ exists for (7.33). However, a solution exists if the right side of (7.32) is meaningful, or explicitly, if the expression

(7.34)
$$A_{\Lambda^q}(\Omega_1) = \kappa \frac{\Lambda^q \cdot \Omega_1 \left[G_{\lambda_1^q}(\Lambda^q \cdot \Omega_1)\Psi'(0, \Omega_1) + \frac{c}{4\pi} \int \frac{\Lambda^q \cdot \Omega \Psi'(0, \Omega)}{\Lambda^q \cdot \Omega_1 - \Lambda^q \cdot \Omega} d_2\Omega \right]}{N_{\lambda_1^q}(\Lambda^q \cdot \Omega_1)}$$

has a meaningful interpretation. In the one-dimensional case, it is sufficient for $\Psi(0, \Omega)$ to satisfy the H condition.

Let arbitrary functions $\Psi(0, \Omega; \Lambda^q)$ be given such that (7.34) exists for all quasi-proper vectors $\{\Lambda^q\}$. Then for each Λ^0 and $\Lambda^q \notin \{\Lambda^0\}$ exist, respectively, representations (7.20) and (7.21), namely $\Psi(r, \Omega; \Lambda^q)$. A superposition of these solutions $\Psi(r, \Omega; \Lambda^q)$ for all Λ^q yields the general solution of the transport equation (1.4) given in the form of equation (7.19). Now, take the special case when $\Psi(0, \Omega)$ is a given function and $\Psi(0, \Omega; \Lambda^q) = \Psi(0, \Omega)$ is true for given sets $\{\Lambda^0\}_g$ and $\{\Lambda^q \neq \Lambda^0\}_g$. An arbitrary weighted superposition of these $\Psi(r, \Omega; \Lambda^q)$ belonging to $\Psi(0, \Omega)$ yields again a solution of the transport equation with the given $\Psi(0, \Omega)$ as the angular distribution at the origin. Such a solution can be expressed as

(7.35)
$$\begin{aligned} \Psi(r, \Omega) = &\int_{\{\Lambda^0\}_g} w(\Lambda)\Psi(r, \Omega; \Lambda)\, d_4\Lambda \Big/ \int_{\{\Lambda^0\}_g} w(\Lambda)\, d_4\Lambda \\ &+ \int_{\{\Lambda^q \neq \Lambda^0\}_g} w(\Lambda)\Psi(r, \Omega; \Lambda)\, d_4\Lambda \Big/ \int_{\{\Lambda^q \neq \Lambda^0\}_g} w(\Lambda)\, d_4\Lambda. \end{aligned}$$

Here $w(\Lambda^q)$ is an arbitrary weight function depending on Λ^q (including Λ^0) only.

We turn now to the case when $c = 1$ in (2.5). By (5.3), the corresponding eigenvalue is $\lambda_0 = 0$. Thus by (7.11) and (7.12), the components ξ and η of the factor $\zeta = \xi + i\eta$, which transforms an arbitrary improper eigenvector Λ into the quasi-proper vector Λ^q of its class, are both zero, unless $\Lambda_1^2 - \Lambda_2^2 = 0$ and $\Lambda_1 \cdot \Lambda_2 = 0$. In other words, if $|\Lambda_1| \neq |\Lambda_2|$ and/or $\Lambda_1 \cdot \Lambda_2 \neq 0$, $\zeta\Lambda = \Lambda^0 \Rightarrow \xi = \eta = 0$, $\Lambda^0 = 0$. But if $|\Lambda_1| = |\Lambda_2|$ and $\Lambda_1 \cdot \Lambda_2 = 0$, Λ is itself a quasi-proper vector. It will be shown that all quasi-proper vectors whose components Λ_1^q and Λ_2^q lie in a given plane of the projection space E_3 belong to one of two classes, which are complex conjugates of each other. Let

(7.36)
$$\Lambda^q = \Lambda_1^q + i\Lambda_2^q, \qquad |\Lambda_1^q| = |\Lambda_2^q|, \qquad \Lambda_1^q \cdot \Lambda_2^q = 0$$

be the base vector. Then $\zeta\Lambda^q$, where ζ is any complex number, is also quasi-proper. Using (7.36) in (7.7) yields $(\Lambda_1')^2 = (\Lambda_2')^2 = (\xi^2 + \eta^2)(\Lambda_1^q)^2$ and $\Lambda_1' \cdot \Lambda_2' = 0$. The same result is obtained when we replace (7.36) by its complex conjugate, $\overline{\Lambda}^q = \Lambda_1^q - i\Lambda_2^q$. However, $\overline{\Lambda}^q = \zeta\Lambda^q$ has no solution as is easily verified. Thus we have

LEMMA 7.2. *Let $c = 1$ in equation* (2.5) *and consider any plane through the origin in E_3. Then*

(1) *All quasi-proper vectors of the given plane belong to exactly one of two mutually conjugate complex classes, and*

(2) *The improper eigenvectors in the given plane that are not quasi-proper cannot be generated from the quasi-proper vectors by multiplication with a complex number.*

Since the classification of the eigenfunctions according to quasi-proper vectors breaks down in this case, we have several choices for a substitute.

One such choice is to replace the set of base vectors $\{\Lambda^q\}$ defined by equation (7.5) by a new set of base vectors $\{\Lambda^c\}$ such that $\Lambda^c = \Lambda_1^c + i\Lambda_2^c$, $\Lambda_1^c \cdot \Lambda_2^c = 0$ and either ($|\Lambda_1^c| = 1$, $|\Lambda_2^c| = 0$) or ($|\Lambda_1^c| \geq 1$, $|\Lambda_2^c| = 1$); i.e.

(7.37) $\{\Lambda^c\} \equiv \{\Lambda : \Lambda_1 \cdot \Lambda_2 = 0 \text{ and } \Lambda_1^2 = 1, \Lambda_2 = 0 \text{ or } \Lambda_1^2 \geq 1, \Lambda_2^2 = 1\}$.

The class of improper eigenvectors is represented by $|\Lambda_1^c| = |\Lambda_2^c| = 1$. Finally, consider the case $c = 0$ in (2.5). Here, no proper eigenfunction exists, and the improper eigenvectors are given by

$$\Phi_{\Lambda,\Omega_0}^{c=0} = \kappa\delta(\Omega - \Omega_0) \tag{7.38a}$$

with

$$\Psi_{\Lambda,\Omega}^{c=0} = \kappa e^{-\Lambda\cdot r}\delta(\Omega - \Omega_0). \tag{7.38b}$$

In §VIII we shall see that the sets $\{\Lambda^0\}$ and $\{\Lambda : \Lambda \cdot \Omega_0 = 1\}$ and consequently the general solution space $\{\Psi\}$ of (1.4) are often considerably simplified when the solutions Ψ are subject to geometrical invariance conditions.

VIII. **Reduction of the general solution to special cases.**

A. *Slab geometry with and without axial symmetry.* The requirement that $\Psi(r, \Omega)$ be a plane solution of equation (1.4) means, by definition, that $\Psi(r, \Omega)$ be constant for all r whose end points lie on parallel planes. We rotate our coordinate system so that these planes are orthogonal to the z-axis. It follows, then, from equations (2.10) and (2.14), that

$$\partial\Psi_\Lambda^0/\partial x = \partial\Psi_\Lambda^0/\partial y = 0$$

and

$$\partial\Psi_{\Lambda,\Omega_0}/\partial x = \partial\Psi_{\Lambda,\Omega_0}/\partial y = 0,$$

if and only if

$$\Lambda_x = \Lambda_y = 0. \tag{8.1}$$

Hence the elementary solutions have the form

$$\begin{aligned} \Psi_\Lambda^0 &= e^{-\Lambda_z z}\Phi_\Lambda^0(\Omega) \qquad \text{(regular expansion modes);} \\ \Psi_{\Lambda,\Omega_0} &= e^{-\Lambda_z z}\Phi_{\Lambda,\Omega_0}(\Omega) \quad \text{(generalized expansion modes).} \end{aligned} \tag{8.2}$$

We must now submit the regular eigenfunctions to the conditions (5.5) that resulted from the characteristic equation. Because of equations (3.2), (3.3), and (8.1),

$$|\mathrm{Re}\,\Lambda_z| = \lambda_1 \quad \text{and} \quad |\mathrm{Im}\,\Lambda_z| = \lambda_2.$$

We show that either λ_1 or λ_2 must be zero. From equation (8.1) it follows that $\sin\gamma = 0$, since Λ_{1z} is necessarily parallel to Λ_{2z} when all other components of Λ are zero. This means $\cos\gamma = \pm 1$. But then the second equation of (5.5) requires

$$\lambda_1\lambda_2 = 0, \qquad \text{Q.E.D.}$$

From equations (5.3) and the first equation of (5.5), we deduce that

$$\begin{array}{lll}
\text{(a) } \lambda_2 = 0 & \text{for } \; 0 < c < 1 & (\lambda_0 \text{ real});\\
\text{(b) } \lambda_1 = \lambda_2 = 0 & \text{for } \; c = 1 & (\lambda_0 = 0);\\
\text{(c) } \lambda_1 = 0 & \text{for } \; 1 < c < \infty & (\lambda_0 \text{ imaginary}).
\end{array}$$

Hence the regular expansion modes are

$$\Psi'^{0}_{\Lambda} = e^{\pm\lambda_0 z}\Phi^0_\Lambda(\Omega),$$

where

$$\Phi^0_\Lambda = \frac{\kappa c}{4\pi}\frac{1}{1 \pm \lambda_0\Omega_z} = \frac{\kappa c}{4\pi}\frac{1}{1 \pm \lambda_0\mu} \tag{8.3}$$

if Ω is defined as in (3.4) and $\mu = \cos\theta$.

Next we must reduce the generalized eigenfunctions to the case of plane geometry.

From equation (8.1) it follows, as before, that $\sin\gamma = 0$ since Λ_{1z} and Λ_{2z} are necessarily parallel. But then equations (6.4) and (6.5) cannot be satisfied simultaneously in E_3, unless $\lambda_2 = 0$. Thus, by Lemma 7.1, equations (7.18), (7.28a), (3.4), (4.1), (7.28b), and (6.10),

$$\Phi_{\Lambda^0/\Lambda^0\cdot\Omega_0} = \frac{\kappa c}{4\pi}\frac{1}{1 - \mu/\mu_0} + \kappa G(\mu_0)\delta(\mu - \mu_0)\delta(\phi - \phi_0). \tag{8.4}$$

Again by Lemma (7.1) and the remarks above, no other class of quasi-proper vectors exists. Therefore the general solution of (1.4) in slab geometry *without axial symmetry* has the form (7.20) and is in the assumed coordinate system,

$$\begin{aligned}\Psi(z, \mu, \phi) &= a_+ e^{-\lambda_0 z}\Phi_{\Lambda^0}(\mu) + a_- e^{+\lambda_0 z}\Phi_{-\Lambda^0}(\mu)\\
&\quad + \int_0^{2\pi} d\phi_0 \int_{-1}^{1} d\mu_0 A(\mu_0, \phi_0) e^{-z/\mu_0}\Phi_{\Lambda^0/\Lambda^0\cdot\Omega_0}(\mu, \phi).\end{aligned} \tag{8.5}$$

The corresponding orthogonality relations are given by (7.23), (7.26), (7.24), and (7.30), respectively. In the case of *axial symmetry* $\Psi(z, \mu, \phi) = \Psi(z, \mu)$ and $A(\mu, \phi) = A(\mu)$ are independent of ϕ. If we further define

$$\Phi_{\pm\lambda_0} = 2\pi\Phi_{\pm\Lambda^0}, \qquad \Phi_{1/\mu_0}(\mu) = \int_0^{2\pi} \Phi_{\Lambda^0/\Lambda^0\cdot\Omega_0}(\mu, \phi)\, d\phi_0, \tag{8.6a}$$

then (8.5) reduces to

(8.6b) $$\Psi(z,\mu) = a_+\Phi_{\lambda_0}(\mu)e^{-\lambda_0 z} + a_-\Phi_{-\lambda_0}(\mu)e^{\lambda_0 z} + \int_{-1}^{1} d\mu_0 A(\mu_0)\Phi_{1/\mu_0}(\mu)e^{-z/\mu_0}$$

which is identical to the results obtained in **[4]** and **[12]**.

B. *Spherical symmetry*. For spherical symmetry, the angular distribution ψ is invariant under rotation of the (r, Ω)-system and therefore is only a function of the distance from the origin, $\rho = (r \cdot r)^{1/2}$, and the cosine of the angle between r and Ω, $\mu = \Omega \cdot r/\rho$. The transport equation (1.4) reduces under these symmetry conditions to

(8.7) $$\mu \frac{\partial}{\partial \rho}\psi(\rho,\mu) + \frac{1-\mu^2}{\rho}\frac{\partial}{\partial \mu}\psi(\rho,\mu) + \psi(\rho,\mu) = \frac{c}{2}\int_{-1}^{+1}\psi(\rho,\mu')\,d\mu'.$$

Necessary and sufficient for (2.15) to be the general solution of (8.7) is the condition that the expansion coefficients $A^0(\Lambda)$ and $A(\Lambda, \Omega)$ are independent of the (complex) directions of Λ, which implies

(8.8a) $$A^0(\Lambda) = A^0(\lambda_1),$$

where $\Lambda \in \{\Lambda^0\}$; i.e. $\lambda_1 = |\Lambda_1| < 1$, $|\lambda_2| = \lambda_2$, $\lambda_1^2 = \lambda_2^2 + \lambda_0^2$, $\cos\gamma = 0$ (from (5.5)), and

(8.8b) $$A(\Lambda, \Omega_0) = A(\lambda_1, \lambda_2, \gamma; \Omega_0^\pm),$$

where Λ is an improper eigenvector, satisfying either $\lambda_1 \sin\gamma \geq 1$ or $\lambda_1 \geq 1$, $\lambda_2 = 0$ (2.12 and Corollary 3.1). Preceding Corollary 3.1 and in Figure 3.2 we have seen that for each given Λ, ($\lambda_2 \neq 0$) two singular directions, Ω_0^+ and Ω_0^- exist, such that, $\Lambda_1, \Lambda_2, \Omega_0^+$ and $\Lambda_1, \Lambda_2, \Omega_0^-$ form a right-handed and left-handed coordinate system. respectively. Conversely, given any Ω_0^+, λ_1, λ_2, γ or Ω_0^-, λ_1, λ_2, γ specifies the orientation of all associated $\Lambda_1, \Lambda_2, \Omega_0$ systems uniquely as right- or left-handed. Thus, equation (2.15) takes the form

(8.9a) $$\begin{aligned}\psi(\rho,\mu) = &\int_{\lambda_1 = \left\{\substack{\lambda_0 \text{ for } c\leq 1 \\ 0 \text{ for } c \geq 1}\right.} d\lambda_1 A^0(\lambda_1)\psi_{\lambda_1}(\rho,\mu) \\ &+ \int_1^\infty d\lambda_1 \int_0^\infty d\lambda_2 \int_{\sin\gamma \geq 1/\lambda_1} d\gamma \{A(\lambda_1,\lambda_2,\gamma;\Omega_0^+)\psi_{\lambda_1,\lambda_2,\gamma;\Omega_0}{}^+(\rho,\mu) \\ &\qquad + A(\lambda_1,\lambda_2,\gamma;\Omega_0^-)\psi_{\lambda_1,\lambda_2,\gamma;\Omega_0}{}^-(\rho,\mu)\}\end{aligned}$$

where

(8.9b) $$\psi_{\lambda_1}(\rho,\mu) = \int_{\mathscr{R}} d_3R\,\psi_{R\Lambda^0}(r,\Omega),$$

and

(8.9c) $$\psi_{\lambda_1,\lambda_2,\gamma;\Omega_0}{}^\pm(\rho,\mu) = \int_{\{\Omega_0\}} d_2\Omega_0^\pm \int_{\mathscr{R}\Omega_0} dR^{\Omega_0^\pm}\psi_{R\Lambda',\Omega_0}{}^\pm(r,\Omega).$$

The first term on the right side of (8.9a), together with (8.9b), corresponds to (5.12) and the remarks preceding (5.12) apply. The integration over λ_1, instead of λ_2 as in (5.12), is just a matter of changing variables of integration, and is an example of an alternate approach to (5.12). That $\psi_{\lambda_1}(\rho, \mu)$ is a function of λ_1, ρ, and μ only shall be explained briefly for the sake of completeness, although it may appear obvious. Let us select our Cartesian reference system such that

$$r = \rho(0, 0, 1) \tag{8.10a}$$

and

$$\Omega = (\sin\theta, 0, \cos\theta), \tag{8.10b}$$

where

$$\cos\theta = \mu. \tag{8.10c}$$

Furthermore, let a rotated eigenvector $R\Lambda^0$ in this system be given by $R\Lambda^0 = \Lambda'$,

$$\begin{aligned} \Lambda_1' &= \lambda_1(\cos\phi_1 \sin\theta_1, \sin\phi_1 \sin\theta_1, \cos\theta_1) \\ \Lambda_2' &= \lambda_2(\cos\phi_2 \sin\theta_2, \sin\phi_2 \sin\theta_2, \cos\theta_2), \end{aligned} \tag{8.10d}$$

under the orthogonality constraint $\Lambda_1' \cdot \Lambda_2' = 0$,

$$\cos(\phi_1 - \phi_2) = \cot\theta_1 \cot\theta_2. \tag{8.10e}$$

Then

$$\psi_{R\Lambda^0}(r, \Omega) = \frac{c\kappa}{4\pi} \frac{e^{-r\cdot\Lambda'}}{1 - \Lambda'\cdot\Omega}, \tag{8.11a}$$

where

$$\left\{\begin{aligned} r\cdot\Lambda' &= \rho(\lambda_1 \cos\theta_1 + i\lambda_2 \cos\theta_2), \\ &\text{and} \\ \Lambda'\cdot\Omega &= \lambda_1(\sin\theta\cos\phi_1 \sin\theta_1 + \cos\theta\cos\theta_1) \\ &\quad + i\lambda_2(\sin\theta\cos\phi_2\sin\theta_2 + \cos\theta\cos\theta_2). \end{aligned}\right. \tag{8.11b}$$

The integration of (8.9b) in this coordinate system is now explicitly

$$\psi_{\lambda_1}(\rho, \mu) = \frac{c\kappa}{4\pi} \int_0^{2\pi} d\phi_1 \int_0^{\pi} d\theta_1 \int_0^{2\pi} d\phi_2 \frac{e^{-r\cdot\Lambda'}}{1 - \Lambda'\cdot\Omega}. \tag{8.12a}$$

Therefore, $\psi_{\lambda_1}(\rho, \mu)$ is by (8.12a) and (8.11b) a function of ρ, λ_1, λ_2, $\sin\theta$, and $\cos\theta$, but because of (8.10c) and (5.5), the independent variables reduce to λ_1, ρ, and μ only. If we let $\lambda_1 \to \lambda_0$ for $c \leq 1$ and $\lambda_1 \to 0$ for $c \geq 1$, then (8.9c) becomes **[38]**

$$\psi_{\lambda_1}(\rho, \mu) = \text{const.} \int_{-1}^{+1} \frac{\exp(\pm\lambda_0 \rho \nu\mu)}{1 \pm \lambda_0 \nu} I_0(\rho\lambda_0 \sqrt{(1 - \mu^2)(1 - \nu^2)})\, d\nu, \tag{8.12b}$$

where $\eta = \cos\theta$ ($\theta = \theta_1$ for $c \leq 1$, $\theta = \theta_2$ for $c \geq 1$), and $I_0(z)$ is the modified Bessel function of the first kind.

The second term on the right of (8.9a), together with (8.9c), corresponds to (6.8) and the remarks following it. The integration over λ_1 instead of λ_1^Ω, as in (6.8), is just a matter of changing variables of integration [cf. (6.11)] and is an example of an alternate approach to (6.8). To show that $\psi_{\lambda_1,\lambda_2,\gamma;\Omega_0^\pm}(\rho, \mu)$ is a function of the indicated parameters and variables only, we use the same reference system as before [equations (8.10)]. The integrand is given by

$$\text{(8.13)} \quad \psi_{R\Lambda^I,\Omega_0^\pm}(r, \Omega) = e^{-r\cdot\Lambda}\left[\frac{c\kappa}{4\pi}\frac{1}{1-\Lambda'\cdot\Omega} + \kappa G(\lambda_1, \lambda_2, \gamma)\delta(\mu-\mu_0^\pm)\delta(\phi-\phi_0^\pm)\right].$$

Here Λ^I is a convenient reference vector as in §VI (e.g. Λ_1^I is taken co-planar to r and Ω). The δ-function corresponds to (6.10), and the integral [referring to the second term in (8.13)]

$$\int d_2\Omega_0^\pm \int dR^{\Omega_0\pm} e^{-r\cdot\Lambda}\kappa G\delta(\mu-\mu_0^\pm)\delta(\phi-\phi_0^\pm)$$

corresponds to the first integral in (6.9) and to (6.13). We note that $\mathscr{R}^{\Omega_0}$ represents the plane rotation group with Ω_0^+ as axis of rotation.

Referring to the first term in (8.13), the integral

$$\int d_2\Omega_0 \int dR^{\Omega_0}\frac{c\kappa}{4\pi}\frac{e^{-r\cdot\Lambda'}}{1-\Lambda'\cdot\Omega}$$

corresponds to the second part of (6.9). They have the same structure as (8.9b), since in our notation $\int d_2\Omega_0 \int dR^{\Omega_0} = \int d_3R$. The same reasoning as was applied to (8.9b) can be applied to (8.9c) to prove that $\psi_{\lambda_1,\lambda_2,\gamma,\Omega_0^\pm}(\rho, \mu)$ is indeed a function of λ_1, λ_2, γ, σ, and μ only. This proves that the equations (8.8) are sufficient for spherical geometry. The proof that they are also necessary to yield the general solution of (8.7) is omitted.

C. *Cylindrical symmetries.* The space vector r in cylindrical geometry will be specified by the coordinates (ρ, ϕ_ρ, z), and the direction of motion Ω by the two angles (θ, ϕ_Ω). If z is taken as the axis of the cylinder, then cylindrical symmetry requires that ψ remains invariant under *rotation* about this axis. Thus ψ can depend only on the difference $\phi = \phi_\Omega - \phi_\rho$ of the azimuthal angles. Under these conditions, the transport equation (1.4) becomes

$$\text{(8.14)} \quad \begin{aligned} &\sin\theta\left[\cos\phi\frac{\partial\psi(\rho,\theta,\phi,z)}{\partial\rho} - \frac{\sin\phi}{\rho}\frac{\partial\psi(\rho,\theta,\phi,z)}{\partial\phi}\right] + \cos\theta\frac{\partial\psi(\rho,\theta,\phi,z)}{\partial z} \\ &\quad + \psi(r,\theta,\phi,z) = \frac{c}{4\pi}\int_0^\pi \sin\theta'\,d\theta'\int_0^{2\pi}\psi(r,\theta',\phi',z)\,d\phi'. \end{aligned}$$

Necessary and sufficient for (2.15) to be the general solution of (8.14) is to have the expansion coefficients, $A^0(\Lambda)$ and $A(\Lambda, \Omega)$, satisfy

$$\text{(8.15a)} \quad A^0(\Lambda) = A^0(\lambda_1, \Lambda_z^\pm),$$

where $\Lambda \in \{\Lambda^0\}$ as in (8.8a) and

$$\text{(8.15b)} \quad A(\Lambda, \Omega) = A(\lambda_1, \lambda_2, \gamma, \Lambda_z^\pm, \Omega_0^\pm),$$

where Λ is an improper eigenvector as in (8.8b). The Λ_z^+ and Λ_z^- identify the two possible cases where the vector triples, consisting of the z-axis, Λ_1, and Λ_2, form right-handed or left-handed systems, respectively. With these coefficients, (2.15) takes the form

$$\psi(\rho, \theta, \phi, z) = \int_{\lambda_1=\left\{ {\lambda_1 \text{ for } c\le 1 \atop 0 \text{ for } c \ge 1} \right.}^{1} d\lambda_1 \int_{-\lambda_1}^{+\lambda_1} d\Lambda_{1z}$$

$$\times \int_{-[(\lambda_1^2-\lambda_0^2)(\lambda_1^2-\Lambda_{1z}^2)]^{1/2}/\lambda_1}^{+[(\lambda_1^2-\lambda_0^2)(\lambda_1^2-\Lambda_{1z}^2)]^{1/2}/\lambda_1} d\Lambda_{2z}[A^0(\lambda_1, \Lambda_z^+)\psi_{\lambda_1,\Lambda_z^+} + A^0(\lambda_1, \Lambda_z^-)\psi_{\lambda_1,\Lambda_z^-}]$$

$$\text{(8.16a)} \qquad + \int_1^\infty d\lambda_1 \int_{-\lambda_1}^{+\lambda_1} d\Lambda_{1z} \int_0^\infty d\lambda_2 \int_{\left\{ {\lambda_1 \sin \gamma \ge 1 \atop 0 \le \gamma \le 2\pi} \right\}} d\gamma$$

$$\times \int_{(\Lambda_{1z}\cos\gamma-(\lambda_1^2-\Lambda_{1z}^2)^{1/2}\sin\gamma)\lambda_2/\lambda_1}^{(\Lambda_{1z}\cos\gamma+(\lambda_1^2-\Lambda_{1z}^2)^{1/2}\sin\gamma)\lambda_2/\lambda_1} d\Lambda_{2z}[A(\lambda_1, \lambda_2, \gamma, \Lambda_z^+, \Omega_0^+)\psi_{\lambda_1,\lambda_2,\gamma,\Lambda_z^+,\Omega_0^+}$$

$$+ A(\lambda_1, \lambda_2, \gamma, \Lambda_z^-, \Omega_0^+)\psi_{\lambda_1,\lambda_2,\gamma,\Lambda_z^-,\Omega_0^+} + A(\lambda_1, \lambda_2, \gamma, \Lambda_z^+, \Omega_0^-)\psi_{\lambda_1,\lambda_2,\gamma,\Lambda_z^+,\Omega_0^-}$$
$$+ A(\lambda_1, \lambda_2, \gamma, \Lambda_z^-, \Omega_0^-)\psi_{\lambda_1,\lambda_2,\gamma,\Lambda_z^-,\Omega_0^-}],$$

where

$$\text{(8.16b)} \qquad \psi_{\lambda_1,\Lambda_z^\pm} \equiv \psi_{\lambda_1,\Lambda_z^\pm}(\rho, \theta, \phi, z) = \int_0^{2\pi} d\phi_1 \psi_{\Lambda^\pm}; \qquad \Lambda^\pm \in \{\Lambda^0\};$$

$$\text{(8.16c)} \qquad \psi_{\lambda_1,\lambda_2,\gamma,\Lambda_z^\pm,\Omega_0^\pm} \equiv \psi_{\lambda_1,\lambda_2,\gamma,\Lambda_z^\pm,\Omega_0^\pm}(\rho, \theta, \phi, z) = \int_0^{2\pi} d\phi_1 \psi_{\Lambda^\pm,\Omega_0^\pm};$$

and $\Lambda^\pm$ refers to those vectors for which the triple (Λ_1, Λ_2, z-axis) form right- and left-handed systems, respectively. In these formulas, we have the Cartesian coordinate system chosen so that r lies in the (z, x) plane; thus,

$$\text{(8.17a)} \qquad r = (\rho, 0, z), \qquad \Omega = (\cos\phi \sin\theta, \sin\phi \sin\theta, \cos\theta)$$

and

$$\text{(8.17b)} \quad \Lambda_j = (\cos\phi_j(\lambda_j^2 - \Lambda_{jz}^2)^{1/2}, \sin\phi_j(\lambda_j^2 - \Lambda_{jz}^2)^{1/2}, \Lambda_{jz}), \qquad j = 1, 2$$

under the condition $\Lambda_1 \cdot \Lambda_2 = \lambda_1\lambda_2 \cos\gamma$,

$$\text{(8.17c)} \quad \cos(\phi_1 - \phi_2) = (\lambda_1\lambda_2\cos\gamma - \Lambda_{1z}\Lambda_{2z})/[(\lambda_1^2 - \Lambda_{1z}^2)(\lambda_2^2 - \Lambda_{2z}^2)]^{1/2}.$$

Thus, in the course of integration for each choice λ_1, Λ_{1z}, λ_2, γ, Λ_{2z}, and ϕ_1, two possibilities exist for Λ_2, namely Λ_2^+ on the positive side of the (z, Λ_1)-plane and Λ_2^- on the negative side of the (z, Λ_1)-plane. Once the relative position of Λ_2 is fixed, we know two singular directions Ω_0^+ and Ω_0^- exist. All possible combinations account for the different terms under the integral signs in (8.16a).

To show that the elementary solutions for cylindrical symmetry, given by equations (8.16b,c), are functions of the indicated arguments only, we write ψ_{Λ^0}

and ψ_{Λ,Ω_0} explicitly and insert the explicit expressions for $r \cdot \Lambda$ and $\Lambda \cdot \Omega$. It is then seen in a similar manner as for spherical symmetry that here ψ depends on ρ, $\sin\theta$, $\cos\theta$, $\sin\phi$, $\cos\phi$, and z only, which proves the assertion.

Often, additional invariance properties are imposed on the solutions of (8.14) such as (a) invariance under reflection on any plane through the z-axis, (b) invariance under translation along the z-axis, and (c) symmetry with respect to a plane orthogonal to the z-axis.

(a) Let us impose on the solution ψ of (8.16a) the additional requirement of symmetry with respect to any plane through the axis of symmetry. This means ψ is required to be an even function in ϕ in our adopted coordinate system. It is easily verified that when $\psi(\rho, \theta, \phi, z)$ is a solution, $\psi(\rho, \theta, -\phi, z)$ is also a solution, and the sum of these two solutions is an even function in ϕ. The implication of this remark on the coefficient functions, A^0 and A, is seen by considering the explicit expressions of $r \cdot \Lambda$ and $\Omega \cdot \Lambda$ as above. It is observed that all terms in $r \cdot \Lambda$ and $\Omega \cdot \Lambda$ are even in ϕ, ϕ_1, and ϕ_2, except

$$\sin\theta \sin\phi[\sin\phi_1(\lambda_1^2 - \Lambda_{1z}^2)^{1/2} + i \sin\phi_2(\lambda_2^2 - \Lambda_{2z}^2)^{1/2}]. \tag{8.18}$$

Replacing ϕ by $-\phi$ in (8.18) is equivalent to replacing ϕ_1 by $-\phi_1$ and ϕ_2 by $-\phi_2$ while all other terms in $r \cdot \Lambda$ and $\Omega \cdot \Lambda$ remain unchanged. This transformation is achieved by a reflection with respect to the (z, r)-plane of the vectors Λ. If the original vector was Λ^+, the reflected one is Λ^-. If the vector is an improper eigenvector, then Ω_0^+ reflected becomes Ω_0^-. Similar remarks hold for Λ^- and Ω_0^-. It follows now that in (8.16) the elementary solutions are

$$\psi_{\lambda_1,\Lambda_z^+}(\rho, \theta, \phi, z) = \psi_{\lambda_1,\Lambda_z^-}(\rho, \theta, -\phi, z), \tag{8.19a}$$

and

$$\psi_{\lambda_1,\lambda_2,\gamma,\Lambda_z^+,\Omega_0^\pm}(\rho, \theta, \phi, z) = \psi_{\lambda_1,\lambda_2,\gamma,\Lambda_z^-,\Omega_0^\mp}(\rho, \theta, -\phi, z). \tag{8.19b}$$

Thus the conditions for (8.16a) to be an even function in ϕ are

$$A^0(\lambda_1, \Lambda_z^+) = A^0(\lambda_1, \Lambda_z^-), \tag{8.20a}$$

and

$$A(\lambda_1, \lambda_2, \gamma, \Lambda_z^+, \Omega_0^\pm) = A(\lambda_1, \lambda_2, \gamma, \Lambda_z^-, \Omega_0^\mp), \tag{8.20b}$$

where the upper and lower symbols + and − correspond, respectively.

It is of interest to know the behavior of ψ in (8.16a) when $\Lambda_{1z} = \lambda_1$. The elementary solutions (8.16b,c) will satisfy (8.19) and (8.20). In this case, the solution is simply obtained by

(8.21a)

$$\psi(\rho, \theta, \phi, z) = \int_{\lambda_1=\lambda_0 \text{ for } c\le 1;\, =0 \text{ for } c\ge 1}^{1} d\lambda_1 A^0(\lambda_1)\psi_{\lambda_1} + \int_1^\infty d\lambda_1 \int_0^\infty d\lambda_2 \int_{\{\lambda_1 \sin\gamma \ge 1\}} d\gamma$$

$$\cdot [A(\lambda_1, \lambda_2, \gamma, \Omega_0^+)\psi_{\lambda_1,\lambda_2,\gamma,\Omega_0^+} + A(\lambda_1, \lambda_2, \gamma, \Omega_0^-)\psi_{\lambda_1,\lambda_2,\gamma,\Omega_0^-}]$$

$$(\Lambda_{1z} = \lambda_1, \Lambda_{1x} = \Lambda_{1y} = 0),$$

where

$$\psi_{\lambda_1} = \int_0^{2\pi} d\phi_2 \psi_\Lambda, \qquad \psi_{\lambda_1,\lambda_2,\gamma,\Omega_0^\pm} = \int_0^{2\pi} d\phi_2 \psi_{\Lambda,\Omega_0^\pm}. \tag{8.21b,c}$$

(b) In the case of cylindrical symmetry with translational invariance with respect to the z-axis, Λ_z must be set equal to zero. Equation (8.16a) simplifies to

$$\begin{aligned}\psi(\rho, \theta, \phi) = &\int_{\lambda_1=\lambda_0 \text{ for } c\le 1;\, =0 \text{ for } c\ge 1}^{1} d\lambda_1 \{A_+^0(\lambda_1)\psi_{\lambda_1}^+ + A_-^0(\lambda_1)\psi_{\lambda}^- \}\\ &+ \int_1^\infty d\lambda_1 \int_0^\infty d\lambda_2 \int_{\left\{\substack{\lambda_1 \sin\gamma\ge 1\\ 0\le\gamma\le\pi}\right\}} d\gamma [A_+(\lambda_1, \lambda_2, \gamma, \Omega_0^+)\psi_{\lambda_1,\lambda_2,\gamma,\Omega_0^+}^+\\ &+ A_+(\lambda_1, \lambda_2, \gamma, \Omega_0^-)\psi_{\lambda_1,\lambda_2,\gamma,\Omega_0^-}^+\\ &+ A_-(\lambda_1, \lambda_2, \gamma, \Omega_0^+)\psi_{\lambda_1,\lambda_2,\gamma,\Omega_0^+}^- + A_-(\lambda_1, \lambda_2, \gamma, \Omega_0^-)\psi_{\lambda_1,\lambda_2,\gamma,\Omega_0^-}^-],\end{aligned} \tag{8.22}$$

where the $A_\pm^0$, $A_\pm$, and $\psi_{\lambda_1,\lambda_2,\gamma,\Omega_0^\pm}^\pm$ have the meaning of (8.15a,b) and (8.16b,c), respectively, with $\lim \Lambda_z^\pm = 0$.

However, if $\Lambda_z = 0$, all Λ_1 and Λ_2 are co-planar, and the solutions of (8.14) are suitable to be represented in Λ^q-classes as given by (7.19). To give an explicit representation, we choose the coordinate system as in (8.17). As base vectors for the Λ^q-classes, we take the reference vector $\Lambda^{q,0}$ such that their real components $\Lambda_1^{q,0}$ are along the y-axis and their imaginary components along the x-axis. The other base vectors are then obtained from these by rotation around the z-axis. If we denote an element of the set $\{\Lambda^{q,0}\}$ by $\Lambda^{\lambda,0}$, then

$$\Lambda^{\lambda,0} = \lambda(0, 1, 0) + i[+(\lambda^2 - \lambda_0^2)^{1/2}](1, 0, 0) \tag{8.23}$$

where the parameter λ is a point in the interval defined by

$$\begin{aligned}|\lambda| &\geqq \lambda_0 \quad \text{for} \quad c \leqq 1,\\ &\geqq 0 \quad \text{for} \quad c \geqq 1.\end{aligned} \tag{8.24a}$$

It follows from this that $|\lambda| = |\Lambda_1^{\lambda,0}| = \lambda_1$ and $+(\lambda^2 - \lambda_0^2)^{1/2} = |\Lambda_2^{\lambda,0}| = \lambda_2$ in the usual notation. Letting λ take positive and negative values enables us to obtain a left-handed and right-handed reference system in one representation only.[9] All other base vectors can be obtained from (8.23) by rotation of angle ϕ_1 such that $R_{\phi_1}\Lambda^{\lambda,0} = \Lambda^{\lambda,\phi_1}$; i.e.

$$\Lambda^{\lambda,\phi_1} = \lambda(-\sin\phi_1, \cos\phi_1, 0) + i\lambda_2(\cos\phi_1, \sin\phi_1, 0), \tag{8.24b}$$

where

$$\lambda_2 = +(\lambda^2 - \lambda_0^2)^{1/2}, \qquad 0 \le \phi_1 \le 2\pi, \tag{8.24c}$$

[9] In a similar way, we could have introduced "negative lengths" in (8.16) and (8.9) to avoid the distinction between Ω_0^+ and Ω_0^-.

and the range of λ is given by (8.24a). From (8.17a) and (8.24b) follows

(8.25a) $$r \cdot \Lambda^{\lambda,\phi_1} = \rho(-\lambda \sin \phi_1 + i\lambda_2 \cos \phi_1),$$

(8.25b) $$\Omega \cdot \Lambda^{\lambda,\phi_1} = \sin \theta[\lambda \sin (\phi - \phi_1) + i\lambda_2 \cos (\phi - \phi_1)],$$

(8.25c) $$\zeta_{\Omega_0,\phi_1} = 1/\Omega_0 \cdot \Lambda^{\lambda,\phi_1},$$

(8.26a) $$\psi_{\Lambda^{\lambda,\phi_1}} = \frac{c\kappa}{4\pi} \frac{\exp(-r \cdot \Lambda^{\lambda,\phi_1})}{1 - \Omega \cdot \Lambda^{\lambda,\phi_1}}, \qquad |\lambda| < 1,$$

and

(8.26b) $$\psi_{\Lambda^{\lambda,\phi_1},\Omega_0} = \frac{c\kappa}{4\pi} \frac{\exp(-r \cdot \Lambda^{\lambda,\phi_1}/\Omega_0 \cdot \Lambda^{\lambda,\phi_1})}{1 - (\Omega \cdot \Lambda^{\lambda,\phi_1}/\Omega_0 \cdot \Lambda^{\lambda,\phi_1})} + \kappa G(\lambda, \mu_0, \phi_0 - \phi_1)\delta(\mu - \mu_0)\delta(\phi - \phi_0),$$

where

(8.26c) $$\mu = \cos\theta \quad \text{and} \quad \mu_0 = \cos\theta_0$$

as before. Considerations similar to those in the previous paragraphs show that the coefficient functions in (8.15) now become

(8.27a) $$A^0(\Lambda) = A^0(\lambda)$$

and

(8.27b) $$A(\Lambda, \Omega_0) = A(\Lambda^{\lambda,\phi_1}/\Omega_0 \cdot \Lambda^{\lambda,\phi_1}, \Omega_0) = A(\lambda, \mu_0, \phi_0),$$

and (8.22) is replaced by

(8.28a) $$\psi(\rho, \theta, \phi) = \int_{\substack{1>|\lambda| \geq \lambda_0 \text{ for } c \leq 1 \\ \geq 0 \text{ for } c \geq 1}} d\lambda A^0(\lambda)\psi_\lambda + \int_{|\lambda| \geq 1} d\lambda \int_0^{2\pi} d\phi_0 \int_{-1}^{+1} d\mu_0 A(\lambda, \mu_0, \phi_0)\psi_{\lambda,\phi_0,\mu_0},$$

where

(8.28b) $$\psi_\lambda = \psi_\lambda(\rho,\theta,\phi) = \int_0^{2\pi} d\phi_1 \psi_{\Lambda^{\lambda,\phi_1}},$$

and

(8.28c) $$\psi_{\lambda,\phi_0,\mu_0} = \psi_{\lambda,\phi_0,\mu_0}(\rho, \theta, \phi) = \int_0^{2\pi} d\phi_1 \psi_{\Lambda^{\lambda,\phi_1},\Omega_0}.$$

If we impose the additional condition on ψ to be even in ϕ,

(8.29a) $$\psi(\rho, \theta, \phi) = \psi(\rho, \theta, -\phi),$$

then

(8.29b,c) $$A^0(\lambda) = A^0(-\lambda); \qquad A(\lambda, \mu_0, \phi_0) = A(-\lambda, \mu_0, -\phi_0).$$

That this is true follows from the fact that equations (8.25) and (8.26) remain invariant under the simultaneous substitution of λ, ϕ, and ϕ_1 by $-\lambda$, $-\phi$, and $-\phi_1$, respectively.

(c) The case where we require symmetry with respect to a plane orthogonal to the z-axis, in addition to cylindrical symmetry, is obtained from (8.15) by letting

(8.30)

$$A^0(\lambda_1, \Lambda_z^{\pm}) = A^0(\lambda_1, -\Lambda_z^{\pm}) \quad \text{and} \quad A(\lambda_1, \lambda_2, \gamma, \Lambda_z^{\pm}, \Omega_0^{\pm}) = A(\lambda_1, \lambda_2, \gamma, -\Lambda_z^{\pm}, \Omega_0^{\mp}).$$

These equations can then be applied to (8.16), (8.19), and (8.20) to obtain symmetry with respect to planes through the z-axis in addition to cylindrical symmetry and symmetry with respect to a plane orthogonal to the z-axis.

ACKNOWLEDGMENTS. The authors acknowledge the fine research atmosphere at Argonne and appreciate the cheerfulness and efficiency of Mrs. Doris Haight in transcribing the manuscript.

REFERENCES

1. S. Albertoni and B. Montagnini, *On the spectrum of neutron transport equation in finite bodies*, J. Math. Anal. Appl. **13** (1966), 19–48.

2. E. H. Bareiss, *Flexible transport theory routines for nuclear reactor design*, David Taylor Model Basin Report 1030, 1956.

3. ———, *A survey and classification of transport theory calculation techniques*, Proc. 2nd United Nations Inter. Conf. on Peaceful Uses of Atomic Energy, **16**, United Nations, Geneva (1958), 503–516.

4. ———, *A spectral theory for the stationary transport operator in slab geometry*, J. Math. Anal. Appl. **13** (1966), 53–91. See also ANL-6940 (Dec. 1964).

5. ———, *Decomposition of the stationary isotropic transport operator in three independent space variables*, ANL-6914 (Aug. 1964).

6. E. H. Bareiss and C. P. Neuman, *Singular integrals and singular integral equations with a Cauchy kernel and the method of symmetric pairing*, ANL-6988 (Jan. 1965).

7. R. J. Bednorz and J. R. Mika, *Energy-dependent Boltzmann equation in plane geometry*, J. Math. Phys. **4, 9** (1963), 1285.

8. R. E. Bellman, R. E. Kalaba, and M. C. Prestrud, *Invariant imbedding and radioactive transfer in slabs of finite thickness*, American Elsevier, New York, 1963.

9. J. H. Bennett, *Integral equation methods for the transport problems: Some numerical results*, Numer. Math. **6** (1964), 49–54.

10. G. Birkhoff, *Reactor criticality in neutron transport theory*, Rendiconti di Mathematica **22** (1963), 102–126.

11. B. Carlson, *Numerical theory for solving transient and steady state neutron transport problems*, Los Alamos Sci. Lab. Report LA-2260, 1959.

12. K. M. Case, *Elementary solutions of the transport equation and their applications*, Ann. Phys. **9** (1960), 1–23.

13. K. M. Case, F. de Hoffmann and G. Placzek, *Introduction to the theory of neutron diffusion*, Los Almos Scientific Lab., June 1953.

14. K. M. Case and P. F. Zweifel, *Existence and uniqueness theorems for the neutron transport equation*, J. Mathematical Phys. **4** (1963), 1376.

15. S. Chandrasekhar, *Radiative transfer*, Clarendon Press, Oxford, 1950.

16. N. R. Corngold, P. Michael and W. Wollman, *The time decay constants in neutron thermalization*, Nucl. Sci. Eng. **15** (1963), 13.

17. B. Davison, *Angular distribution due to an isotropic point source and spherically symmetrical eigensolutions of the transport equation*, Canadian Report MT-112, 1945.

18. A. Douglis, "The solution of multidimensional generalized transport equations and their calculation by difference methods" in *Numerical solutions of partial differential equations*, Academic Press, New York, 1966, 197–256.

19. J. H. Fertziger and A. Leonard, *Energy-dependent neutron transport theory*, Ann. Phys. **22** (1963), 192–209.

20. G. Girand, *Equations à intégrales principales étude suivie d'une application*, Ann. Sci. École Norm. Sup. (3) **51** (1943), 251.

21. E. W. Hobson, *The theory of spherical and ellipsoidal harmonics*, Cambridge Univ. Press, New York, 1931.

22. E. Hopf, *Mathematical problems of radiative equilibrium*, Cambridge Tracts, No. 31, 1934.

23. F. D. Judge, *Variational method in the calculation of reactor neutron flux density*, Knolls Atomic Power Lab., KAPL-2151, 1961.

24. N. G. Van Kampen, *On the theory of stationary waves in plasmas*, Physica **21** (1955), 949.

25. W. Kofink, *New solutions of the Boltzmann equation for mono-energetic neutron transport in spherical geometry*, Oak Ridge Nat. Lab., ORNL-3216, 1961.

26. ———, *Studies of the spherical harmonic method in neutron transport theory*. Part II, Oak Ridge Nat. Lab., ORNL-2358, 1957.

27. V. Kourganoff, *Basic methods in transfer problems*, Dover, New York, 1963.

28. A. Kuszell, *The critical problems for multi-layer slab systems*, Acta Phys. Polon. **27** (1961), 567.

29. K. D. Lathrop and B. G. Carlson, *Discrete ordinates angular quadrature of the neutron transport equation*, Los Almos Sci. Lab. Report LA-3186, 1965.

30. A. Leonard and J. H. Ferziger, *Energy-dependent neutron transport theory in plane geometry II and III*, Nucl. Sci. Eng. **26** (1966), 170–191.

31. A. Leonard and T. W. Mullikin, *Solutions to the criticality problems for spheres and slabs*, The Rand Corp. Memo RM-2356-PR, July 1962.

32. ———, *Spectral analysis of the anisotropic neutron transport kernel in slab geometry with applications*, J. Mathematical Phys. **5** (1964), 399–411.

33. C. Mark, *The spectral harmonic method, 1*, CRT-340, Chalk River, Ontario, 1957.

34. J. T. Marti, *Mathematical foundations of kinetics in neutron transport theory*, Nukleonik **8** (1966), 159–163.

35. M. B. Maslennikov, *The characteristic equation of the theory of radiative transfer*, Dokl. Akad. Nauk. SSSR **168** (1966), 747. Soviet Math. Dokl. **7** (1966), 723.

36. ———, *Milne problem with an arbitrary indicatrix and with consideration of azimuthal inhomogeneity of the solution*, Dokl. Akad. Nauk. SSSR **168** (1961), 1001 = Soviet Math. Dokl. **7** (1966), 768.

37. J. R. Mika, *Neutron transport with anisotropic scattering*, Nucl. Sci. Eng. **11** (1961), 415–427.

38. G. J. Mitsis, *Transport solutions to the monoenergetic critical problems*, ANL-6787, Nov. 1963.

39. T. W. Mullikin, *Estimates of the critical dimensions of spherical and slab reactors*, J. Math. Anal. Appl. **5** (1962), 184.

40. N. I. Muskhelishvili, *Singular integral equations*, Noordhoff, Groningen, 1953.

41. S. Pahor, *A new approach to half-space transport problems*, Nucl. Sci. Eng. **26** (1966), 192.

42. I. Pollack and E. H. Bareiss, *Spectral representation and criticality problem of a two-region cell transport operator*, ANL-6590, Mar. 1965.

43. D. H. Sattinger, *The eigenvalues of an integral equation in anisotropic neutron transport theory*, J. Math. and Phys. **45** (1966), 188–196.

44. ———, *A singular eigenfunction expansion in anisotropic transport theory*, J. Math. Anal. Appl. **15** (1966), 497–511.

45. F. Tricomi, *Equazioni integrali contenenti il valor principale di un integrale doppio*, Math. Z. **27** (1928), 87–133.

46. V. S. Vladimirov, *Mathematical problems in the one-velocity theory of particle transport*, AECL-1661, Chalk River, Ontario, 1963.

47. E. P. Wigner, *Mathematical problems of nuclear reactor theory*, Proc. Symp. Appl. Math. **11** (1961), 89.

48. A. H. Wilson, *The general properties of the transport equation and its uses in determining the multiplication in bodies having spherical symmetry*, A.E.R.E. Harwell Report, M.S. 105A, Dec. 1950.

49. G. M. Wing, *An Introduction to transport theory*, Wiley, New York, 1962.

50. R. Zelazny, R. Kuszell and J. Mika, *Solution of the one-velocity Boltzmann equation with the first order anisotropic scattering of neutrons in plane geometry*, Polish Academy of Sciences, Institute of Nuclear Research Report No. 216/IX, Warsaw, 1961.

Argonne National Laboratory

Some recent results in the theory of the transport of thermal neutrons

Noel Corngold

Introduction. The transport equation for thermal neutrons is

$$(1) \qquad \frac{\partial \Psi}{\partial t} + \boldsymbol{v} \cdot \nabla \Psi = \int d\boldsymbol{v}' v' \Sigma_s(\boldsymbol{v}', \boldsymbol{v}) \Psi - v\Sigma(v)\Psi + Q.$$

$\Psi \equiv \Psi(\boldsymbol{r}, \boldsymbol{v}, t)$ is the distribution function for neutrons, and Q is the source function. $\Sigma_s(\boldsymbol{v}', \boldsymbol{v})$, a positive function, is called the "scattering kernel," and $v\Sigma(v)$ is the total reaction rate. If scattering is the only process which occurs, $\Sigma(v) \equiv \int d\boldsymbol{v}' \Sigma_s(\boldsymbol{v}, \boldsymbol{v}')$. When solutions to (1) are sought in a bounded spatial domain, they are required to satisfy the condition $\Psi(\boldsymbol{r}_s, \boldsymbol{v}, t) = 0$ when $(\boldsymbol{v} \cdot \hat{\boldsymbol{n}}) < 0$. In the condition, $\boldsymbol{r}_s$ denotes a point on the surface, and $\hat{\boldsymbol{n}}$ the unit, outward normal.

In its form, equation (1) closely resembles the linearized Boltzmann equation of the kinetic theory of gases (in the case of a cut-off potential). However, the physical phenomena described by the two equations are rather different. The particles which comprise the neutron gas never collide with each other; their diffusion is due to collisions with atoms of a background or moderating material. The moderator may be solid, liquid, or gas. The scattering law is described by the kernel $\Sigma_s(\boldsymbol{v}', \boldsymbol{v})$, which is usually quite complex. Indeed, for liquids, and anharmonic solids, it is unknown. We shall see in the following that the greater variety of mathematical forms possible for the kernel in the neutron case leads to a greater variety of results in the spectral theory of the transport operator. On the other hand, the gas-kinetic equation, which describes collisions between atoms of the same substance, is capable of describing "collective" motions, e.g. hydrodynamics, the propagation of sound, which have no counterparts in neutron diffusion. In other terms, the dispersion relation for neutron "modes" has fewer branches than the gas-kinetic equation, and these modes are all of a dissipative kind.

The scattering kernel. Since the scattering kernel obeys the condition of "detailed balance"

$$v' \exp(-v'^2)\Sigma_s(\boldsymbol{v}', \boldsymbol{v}) = v \exp(-v^2)\Sigma_s(\boldsymbol{v}, \boldsymbol{v}'), \tag{2}$$

it may be symmetrized.[1] We can then rewrite equation (1) so that it contains a real, symmetric kernel, $K_s(\boldsymbol{v}', \boldsymbol{v})$,

$$\frac{\partial \psi}{\partial t} + \boldsymbol{v} \cdot \nabla \psi = \int d\boldsymbol{v}' K_s(\boldsymbol{v}', \boldsymbol{v})\psi - w(v)\psi + Q_1, \tag{3}$$

where $K_s(\boldsymbol{v}', \boldsymbol{v}) = \exp\{\frac{1}{2}(v^2 - v'^2)\}v'\Sigma_s(\boldsymbol{v}', \boldsymbol{v})$, $\psi = \exp(\frac{1}{2}v^2)\Psi$, and $w = v\Sigma(v)$. An important question then arises: Is K_s a compact, or completely continuous operator in the Hilbert space of complex square integrable functions, $N(\boldsymbol{v})$? Its answer appears to depend upon the physical state of the moderator **[1]**.

The simplest moderator is an ideal gas of atoms; its scattering kernel is closely related to the gas-kinetic kernel for hard spheres. Indeed, the analysis of the latter case may be carried over with little effort to show that K_s is compact **[2]**. ($\|K_s\|^2$ diverges, but $\|K_s^3\|^2$ is finite.) A more complex moderator is the harmonic solid, which is a highly "ordered" system. The long-range order of the crystal lattice makes "coherent" scattering possible. The coherent scattering, which manifests itself in the laboratory as Laue spots or Debye-Scherrer rings, appears in the mathematics as a series of delta-function singularities in K_s. In a real, poly-crystalline moderator, most of the coherent scattering is elastic and is concentrated into a series of rings. Further, the reaction rate which stems from the kernel is an irregular, and discontinuous function of v. The singular terms may be extracted from the kernel, leaving the so-called "incoherent kernel" behind. However, the transport equation is then in the form of an integro-differential-difference equation, and no progress has yet been made in its solution. In the following, we shall indicate the effects of coherent scattering in an approximate manner. We take the coherent scattering to be elastic and *isotropic*. Thus,

$$K_s(\boldsymbol{v}', \boldsymbol{v}) = w_c(v)\delta(v'^2 - v^2)/2\pi v + K_i(\boldsymbol{v}', \boldsymbol{v}), \tag{4}$$

where $w_c(v)$ is the reaction-rate for coherent scattering and K_i is the kernel for incoherent scattering. $w_c(v)$ will be chosen to have suitable discontinuities. $w_i(v)$ will denote $w - w_c$.

The kernel for incoherent scattering has no singularities in $0 < v', v < \infty$ which are not integrable. Its behavior at large velocities is similar to the ideal gas, while at small velocities it vanishes more rapidly than the gas. Thus, it too is a compact

[1] We use a unit of velocity such that $v^2 = 1$ corresponds to $v^2 = kT$ in the usual notation.

kernel.[2] The final type of kernel, that corresponding to a monatomic liquid, will be treated briefly. Its precise mathematical form is not known, but it lies somewhere between the two cases already considered, for the system is said to possess "short-range" order. Delta-function singularities do not occur, and the behavior of the kernel at large velocities is similar to that for the gas. However, $\|K_s\|^2$ diverges at the origin of the $(\boldsymbol{v}, \boldsymbol{v}')$ space. A proof of its compactness does not yet exist, but one conjectures—from the rather mild divergence—that an iterate of K_s is square integrable.

Special solutions. In the Eleventh Symposium of this series, considerable attention was devoted to the First Problem of neutron thermalization—the steady-state distribution in energy in a medium of infinite extent, excited by a uniform, isotropic source, at high energy. This situation has no counterpart in gas-kinetics, for establishment of a steady-state distribution depends upon the "absorption" of particles. The general features of the distribution are easy enough to understand, and precise agreement with experiment is achieved through great refinement of the form of the scattering kernel. We shall not discuss that aspect of the problem here, since it is a problem in physics rather than applied mathematics [3].

The transient behavior of neutron distributions presents a more difficult mathematical problem. Experiments which probe these distributions fall into two types, the initial value, or "pulsed" experiment, whose analysis suggests special solutions: $\Psi(\boldsymbol{r}, \boldsymbol{v}, t) = \exp(i\boldsymbol{B} \cdot \boldsymbol{r} + \lambda t)n(\boldsymbol{v})$, and the modulated source, or "neutron-wave" experiment, which suggests $\Psi(\boldsymbol{r}, \boldsymbol{v}, t) = \exp(\boldsymbol{\varkappa} \cdot \boldsymbol{r} + i\omega t)h(\boldsymbol{v})$. Equation (3) becomes, in the two cases,

$$\int d\boldsymbol{v}' K_s(\boldsymbol{v}', \boldsymbol{v})n(\boldsymbol{v}') - (w(v) + i\boldsymbol{B} \cdot \boldsymbol{v})n(\boldsymbol{v}) = \lambda n(\boldsymbol{v}), \tag{5}$$

$$\int d\boldsymbol{v}' K_s(\boldsymbol{v}', \boldsymbol{v})h(\boldsymbol{v}') - (w(v) + i\omega)h(\boldsymbol{v}) = (\boldsymbol{\varkappa} \cdot \boldsymbol{v})h(\boldsymbol{v}). \tag{6}$$

We are to think of equation (5) as an eigenvalue equation for λ, given the wave-vector, $\boldsymbol{B}$, and of equation (6) as an equation for the (complex) propagation vector $\boldsymbol{\varkappa}$, given the frequency, ω. Equation (6) is the formal counterpart of the equation for the propagation of sound waves in gases. Also, when $\omega = 0$, it becomes the equation for the relaxation lengths ("diffusion lengths") for steady-state neutron distributions. Of course, the dispersion laws, $\lambda(B^2)$ and $\kappa^2(\omega)$, are not unrelated. The complete analysis of *either* (5) or (6) for complex frequencies or wave-numbers would suffice. We prefer, however, to deal with two separate equations.

Pulsed experiment—general remarks. We begin with equation (5). In order to work with a compact kernel, we introduce the form given in equation (4). We

[2] Elastic, incoherent scattering also exists. Its angular distribution is smooth, and it may be approximated by a term similar to the first term of (4). However, its amplitude is quite small in all practical cases, other than moderation by a hydrogeneous solid. The reader who wishes to include the effect should simply replace the subscripts c (coherent) and i (incoherent) by e (elastic) and i (inelastic).

may then specialize to the gas, or liquid, by setting $w_c(v) \equiv 0$. We shall also assume that the scattering described by K_i is isotropic, and use the same notation for the isotropic kernel. That is,

$$(7) \qquad \int dv' K_i(\boldsymbol{v}', \boldsymbol{v}) \cdots \equiv \int_0^\infty dv' K_i(v', v) \int \frac{d\Omega}{4\pi} \cdots .$$

The approximation simplifies the mathematics considerably, without disturbing the physical content significantly. Equation (5) becomes

$$(8) \qquad \int_0^\infty dv' K_i(v', v) \int \frac{d\hat{\Omega}}{4\pi} n(\boldsymbol{v}') - \left(i\boldsymbol{B} \cdot \boldsymbol{v} + w(v) - w_c(v) \int \frac{d\hat{\Omega}}{4\pi} \right) n(\boldsymbol{v}) = \lambda n(\boldsymbol{v})$$

or

$$(9) \qquad (K - W - \lambda) n(\boldsymbol{v}) = 0.$$

The spectrum of $K - W$ may be defined through the singularities of the resolvent operator $(K - W - \lambda)^{-1}$. Indeed, these points play a crucial role in the solution of the initial-value problem via Laplace transform. If K is absent, the singularities may be located easily. Then, if one can associate—in a Hilbert space of complex square integrable functions—a weakly convergent sequence $\{f_n\}$ with each λ such that

$$(10) \qquad \|(W + \lambda) f_n\| \to 0, \qquad \|f_n\| > 0,$$

it follows easily that these singularities will remain when a compact K is added to W. (A compact operator converts a weakly convergent sequence into one that is strongly convergent.) Thus, we consider first the spectrum of W, i.e. the equation

$$(11) \qquad \left[iB \cdot \boldsymbol{v} + w(v) - w_c(v) \int \frac{d\hat{\Omega}}{4\pi} \right] n(\boldsymbol{v}) = -\lambda n(\boldsymbol{v})$$

or

$$(11a) \qquad (A - C) n = -\lambda n.$$

The simple form of the operator C, the angular integration, enables us to write the resolvent associated with (11a) as

(12)

$$(A - C + \lambda)^{-1} = \frac{1}{w + i\boldsymbol{B} \cdot \boldsymbol{v} + \lambda} \left\{ 1 + \frac{w_c(v)}{1 - w_c(v) F(v, \lambda, B^2)} \int \frac{d\Omega}{4\pi} \frac{1}{w + i\boldsymbol{B} \cdot \boldsymbol{v} + \lambda} \right\}$$

where

$$(13) \qquad F = \int_{-1}^{1} \frac{d\mu}{2} \frac{1}{\lambda + iBv\mu + w(v)}, \qquad B = \sqrt{\boldsymbol{B}^2},$$

$$(14) \qquad = \frac{1}{2iBv} \ln \left\{ \frac{\lambda + w(v) + iBv}{\lambda + w(v) - iBv} \right\}.$$

Note that if $(A - C + \lambda)^{-1}$ operates upon functions of v^2 only, we have

(12a) $$(A - C + \lambda)^{-1} = \frac{1}{w + i\boldsymbol{B}\cdot\boldsymbol{v} + \lambda} \cdot \frac{1}{1 - w_c(v)F}.$$

In any case, it is clear that the singularities of $(A - C + \lambda)^{-1}$ may be obtained from those of $(A + \lambda)^{-1}$ (denoted by G) and from the zeros of $1 - w_c F$ (denoted Γ). We shall first discuss the case Γ.

For B^2 fixed, $1 - w_c(v)F(v, \lambda, B^2) = 0$ describes one, or several, curves $\lambda(v)$. $\lambda(v)$ will be real, and the conditions $\lambda + w(v) > 0$, $Bv/w_c(v) \equiv B/\Sigma_c < \pi/2$ must be satisfied. Then,

(15) $$-\lambda = \left[w(v) - \left\{ \frac{B/\Sigma_c}{\tan B/\Sigma_c} \right\} w_c(v) \right] > 0 \qquad (0 \leq v < \infty)$$

and it is clear that $-\lambda > w_i(v)$ when $B^2 > 0$.

A precise description of the segment(s) of the negative real axis indicated by (15) depends upon the $w(v)$ and $w_c(v)$ which we choose, and upon B^2. In any case, $w_c(v)$ vanishes for $v < v_0 < 1$. Above v_0, it varies in an irregular manner, possessing points of discontinuity in slope, and finally decreases to zero. In most cases, however, it is approximated by a smoothly decreasing function of v. Even this is not critical; on occasion, it is permitted to increase, in proportion to v. The total reaction rate, $w(v)$, plays a more important role. It is almost constant below v_0 (min $w(v) = w(0) \equiv \lambda^*$), increases discontinuously at v_0, and is proportional to v for large v. Its minimum, for $v > v_0$, is denoted λ^{**}. It is often convenient to take $w(v)$ to be a smoothly increasing function of v, so that $\lambda^{**} = w(v_0^+)$. Finally, it should be borne in mind that in "practical" cases $\lambda^{**} \approx 10\lambda^*$.

Using these functions, one finds that when B^2 is "small," Γ is a single segment, extending from a point, γ_1, slightly more negative than $-\lambda^*$, to a point, γ_2, defined through the equations $-\gamma_2 = w(v_2)$, $\Sigma_c(v_2) = (2/\pi)B$. $-\gamma_2$ is considerably larger than λ^{**}. As B^2 is increased, $-\gamma_1$ increases and $-\gamma_2$ decreases. For large B^2, when v_2 lies in the range of rapid variation of w_c, Γ may become several disconnected segments. Finally, when $(2/\pi)B > \max \Sigma_c(v)$, Γ disappears altogether.[3]

The remaining collection of singularities (G) is, clearly, the set of all values assumed by the function $i\boldsymbol{B}\cdot\boldsymbol{v} + w(v)$. This set is broken into two separate parts by the discontinuity of $w(v)$ at the Bragg cut-off, v_0. Since it is a good approximation to reality to consider $w(v)$ as constant and $= \lambda^*$ for $v < v_0$, one part of G will be a line (denoted G_1) extending from $-\lambda^* - iBv_0$ to $-\lambda^* + iBv_0$, while the remainder of the set, G_2, describes an area (see Figure 1).

That the singularities in G are, indeed, points of the continuous spectrum may be seen from the following construction. For $\lambda \in G$ we can write

$$w(v_\alpha) + iBv_\alpha\mu_\beta = -\lambda, \qquad 0 \leq v_\alpha < \infty, \quad -1 \leq \mu_\beta \leq 1.$$

[3] Γ shrinks to a point, and disappears, at $-\lambda^{**}$.

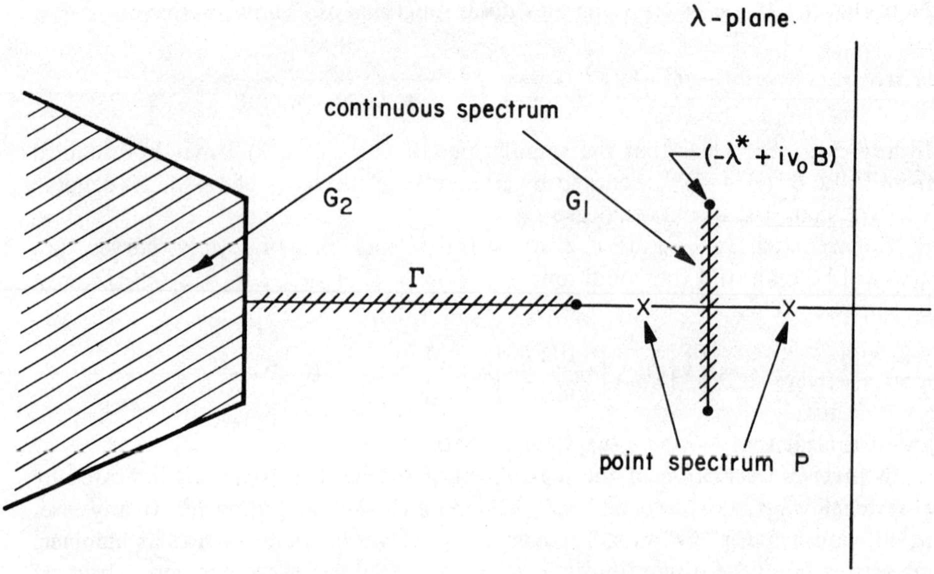

FIGURE 1. λ-plane, pulsed experiment.

Then, if ψ_n is one of a sequence we shall describe presently,

$$\|(A + \lambda)\psi_n\| = 2\pi \int_0^\infty dv v^2 \int_{-1}^{1} d\mu \, |\psi_n|^2 \, \{[w(v) - w(v_\alpha)]^2 + B^2[v\mu - v_\alpha \mu_\beta]^2\}. \tag{16}$$

We choose $\psi_n(v, \hat{\Omega})$ as

$$\begin{aligned} \psi_n &= \left(\frac{n}{2\pi}\right)^{1/2} \quad \text{when} \quad |v - v_\alpha| < \frac{1}{2n^{1/2}}, \quad |\mu - \mu_\beta| < \frac{1}{2n^{1/2}}. \\ &= 0 \quad \text{elsewhere.} \end{aligned} \tag{17}$$

Clearly, $\|\psi_n\| \geq C > 0$, independent of n. If $f = g + ih$ is an arbitrary element in the L^2 space, it is easy to see that $(f, \psi_n) \to 0$. Then, returning to (16), and limiting ourselves to those v_α for which the derivative $w'(v)$ exists, we introduce $\rho \cos \theta = v - v_\alpha$, $\rho \sin \theta = \mu - \mu_\beta$. Thus

$$\begin{aligned} \|(A + \lambda)\psi_n\| &= C_1 n \int_0^{\rho_n} d\rho \rho^3 \int_0^{2\pi} d\theta \\ &\quad \times \{(w')^2 \cos^2 \theta + (v_\alpha \sin \theta + \mu_\beta \cos \theta + \rho \sin \theta \cos \theta)^2\} \to 0, \end{aligned} \tag{18}$$

since ρ_n is proportional to $1/n^{1/2}$. Thus, the points $w(v_\alpha) + iBv_\alpha \mu_\beta$ $(w'(v_\alpha) < \infty)$ belong to the continuous spectrum. Since the action of the operator C upon the sequence ψ_n is such that $\|C\psi_n\| \to 0$, $\lambda \in G$ lies in the continuous spectrum of $A - C$. A similar argument may be used to identify the points of Γ.

To summarize the discussion of this section, we have elucidated the spectrum $(G_1 + G_2)$, of the multiplicative operator A, and the effect upon that spectrum of C, the coherent term. The spectrum of $(A - C \equiv W)$ is thus $(G_1 + G_2 + \Gamma)$. Since the kernel, K, is compact, we are finally able to assert that the essential spectrum of the full operator, $K - W$, differs from that of W only by a point spectrum. We turn, now, to its consideration.

Pulsed experiment—point spectrum. It is easy enough to show that the discrete eigenvalues must lie on the negative, real axis **[4]**. (We shall not discuss the case of a discrete λ sitting in G or Γ; in practice, the case is dealt with as it arises.) The presence of the gap between Γ and G_1 is a feature unique to the coherent moderator. Eigenvalues located in the gap are surely responsible for the "quasi-asymptotic" modes seen in experiments.

The distribution of point eigenvalues has received considerable discussion in the literature, in the "incoherent" case, when they occur in a single segment of the real axis **[1]**. Then, one finds a finite number of eigenvalues when the moderator is in the form of gas or solid, and $B^2 > 0$. The liquid, on the other hand, can exhibit a countable infinity of eigenvalues for $B^2 > 0$. This unusual behavior is due, in part, to the "quasi-elastic" scattering, which is a reflection of the short-range order present in liquids. As the "quasi-elastic" scattering becomes truly elastic, the infinity of eigenvalues doubtless becomes the elastic continuum, Γ. For all moderators, a critical value of B^2, denoted B^2_*, exists, such that the point spectrum having $\operatorname{Re}\lambda > -\lambda^*$ is empty, when $B^2 > B^2_*$ **[4]**.

The behavior of the eigenvalue curves $\lambda_k(B^2)$ in the neighborhood of $-\lambda^*$ is of particular interest. Consider first the fundamental mode, $n_0(\boldsymbol{v})$, with its associated λ_0. In an obvious notation, we may rewrite equation (8) as

$$n_0(\boldsymbol{v}) = \frac{1}{\lambda + i\boldsymbol{B}\cdot\boldsymbol{v} + w(v)}\,\frac{1}{1 - w_c(v)F}\,K_i \cdot N_0(v^2) \tag{19}$$

or

$$n_0(\boldsymbol{v}) = \frac{1}{\lambda + i\boldsymbol{B}\cdot\boldsymbol{v} + w(v)}\,\Phi_0(v^2), \tag{20}$$

$$N_0(v^2) = F(\lambda, B^2, v^2)\Phi_0(v^2). \tag{21}$$

The discrete eigenfunctions associated with (20) or (21) are integrable. The fundamental mode, $N_0(v^2)$, is real and positive. Thus, we are able to normalize the integral of $v^2 N_0(v^2)$ to unity. Upon integration, (20)–(21) yield a dispersion relation:

$$1 = \frac{1}{4\pi}\int d\boldsymbol{v}\,\frac{1}{\lambda + i\boldsymbol{B}\cdot\boldsymbol{v} + w(v)}\,\Phi_0(v^2). \tag{22}$$

Our assumption of $w(v) = \text{const}$ for $v < v_0$ causes (22) to have a particularly transparent form. We divide the velocity space into two parts by means of a spherical surface of radius v_0. For points inside the sphere, $w(v) = \lambda^*$, and we find:

$$(23)\qquad 1 = \frac{1}{4\pi}\int_{s_0} d\boldsymbol{v}\,\frac{1}{\lambda + \lambda^* + i\boldsymbol{B}\cdot\boldsymbol{v}}\,\Phi_0(v^2) + \frac{1}{4\pi}\int' d\boldsymbol{v}\ldots,$$

$$(24)\qquad 1 = \frac{1}{4}\int_{-v_0}^{v_0} dv_3\,\frac{1}{\lambda + \lambda^* + iBv_3}\,\Phi_1(v_3^2) + \frac{1}{4\pi}\int' d\boldsymbol{v}\ldots,$$

where

$$(25)\qquad \Phi_1(v_3^2) = \int_0^{v_0^2 - v_3^2} d\eta\,\Phi_0(\eta + v_3^2).$$

We assume that the function Φ_1, which depends upon λ as well as v_3^2, satisfies a suitable Hölder condition [**5**] on v_3 for every λ under consideration. Then, the first integral, a Cauchy integral, defines a function which may be analytically continued across the line G_1, in the λ-plane. The second integral in equation (24) defines a function analytic in the half-plane Re $\lambda > -\lambda^{**}$. Thus, the fundamental dispersion curve, $\lambda_0(B^2)$, may be continued onto an adjacent sheet of the λ-plane. We do not yet know just how smooth the continued dispersion relation is, or how useful the continuation will be, in practice.[4]

A calculation has shown that at a value of B^2 slightly less than B_*^2, an eigenvalue, $\lambda_0^+ \geq -\lambda^*$, "emerges" from G_1, and—so to speak—replaces λ_0 for $B^2 > B_*^2$, when the latter has disappeared. That this phenomenon is not peculiar to the calculational model is borne out by experiment [**6**]; a fundamental eigenvalue, λ_0, is observed for a range of B^2 which includes and exceeds B_*^2. In the experiments, no discontinuities are seen in the neighborhood of $-\lambda^*$. We can estimate the separation of the two segments $\lambda_0^-(B^2)$ and $\lambda_0^+(B^2)$, by the following argument. For the present, let us call the limiting B^2 approached from below B_+^2, rather than B_*^2. The value of B^2 at which the new segment (λ_0^+) appears is denoted B_-^2. Then, if we denote the lower, fundamental, segment by λ_0^-, we have $\lambda_0^-(B_+^2) = \lambda_0^+(B_-^2) = \lambda^*$. To estimate $B_+^2 - B_-^2$, consider equation (21), rewritten as

$$(26)\qquad [F^{-1}(v, \lambda, B^2) - w_c(v)]N_0(v^2) = K_i \cdot N_0.$$

Let u_1 and u_2 be two solutions to (26), corresponding to different pairs of (λ, B^2). Then, the symmetry of K_i leads to

$$(27)\qquad \int_0^\infty dv\{F^{-1}(v, \lambda_1, B_1^2) - F^{-1}(v, \lambda_2, B_2^2)\}u_1u_2 = 0.$$

If $\lambda_1 = \lambda_2 = -\lambda^*$ and $B_1^2 = B_+^2$, $B_2^2 = B_-^2$, (27) becomes

$$(28)\qquad \frac{2}{\pi}(B_+ + B_-)\int_0^{v_0} dv v N_0^+ N_0^- + \int_{v_0}^\infty dv\{F_+^{-1} - F_-^{-1}\}N_0^+N_0^- = 0.$$

[4] This question is discussed in the 1968 Caltech doctoral thesis of Robert Conn.

For $\lambda = -\lambda^*$, and v fixed outside the Debye sphere, F^{-1} is a smooth function of B^2. Thus, $F^{-1}(\ldots B_+^2) \approx F^{-1}(\ldots B_-^2) + (F^{-1})'(B_+^2 - B_-^2)$, with the result

$$(29) \qquad B_+ - B_- = -\frac{2}{\pi}\int_0^{v_0} dv v N_0^+ N_0^- \Big/ \int_{v_0}^{\infty} dv \frac{\partial}{\partial B^2}\left(\frac{1}{F}\right)_- N_0^+ N_0^-.$$

Closer examination shows that $\partial(1/F)/\partial B^2 > 0$, that the numerator is negative, and $B_+ - B_- > 0$. Thus, the density N_0^- is negative inside the Bragg sphere. This peculiar behavior, for $B^2 > B_*^2$, is compensated by the contribution from G_1, when one inverts the Laplace transform to obtain the solution to the initial value problem. Thus, the jump in buckling is proportional to an averaged square-density of particles in the Debye sphere and is likely to be small. (When K_i is a degenerate kernel, $N_0^\pm$ are constant near $v = 0$. Thus $B_+ - B_-$ is $O(v_0^2)$.)

Another relationship connecting $B_\pm$ with the particle density may be obtained from (23) upon taking the limit. If $f_\pm$ are the fractions of the (normalized) population contained in the Bragg sphere, we find

$$(30) \qquad \begin{aligned} f_\pm &= \pm(\pi/4B_\pm)\Phi_1(0; \lambda^*, B_\pm^2) \\ &= \pm(\pi/4B_\pm)\int_0^{v_0{}^2} d\eta \Phi_0(\eta; \lambda^*, B_\pm^2). \end{aligned}$$

It is not difficult to predict the behavior of $\lambda^+(B^2)$ when B^2 is large. Let the eigenvalue equation be written

$$(31) \qquad N_0(v^2) = (F/(1 - w_c F))K_i \cdot N_0 \equiv \tilde{K}_i \cdot N_0.$$

Then,

$$(32) \qquad (N_0, N_0) \le (N_0, N_0)^{1/2}(\tilde{K}N_0, \tilde{K}N_0)^{1/2},$$

and, if $\tilde{K}$ is bounded so that

$$(33) \qquad (\tilde{K}N_0, \tilde{K}N_0) < C(\lambda, B^2)(N_0, N_0),$$

a necessary condition for the existence of a solution is $1 < C(\lambda, B^2)$. Since K_i is compact, it is also bounded. To bound $\tilde{K}_i$ we need only bound $F(1 - w_c F)^{-1}$, as a function of v. A little curve sketching shows that this is possible; the bound may be decreased (for fixed λ) by increasing B^2, and increased (for fixed B^2) by causing λ to approach γ_1, the edge of the line Γ. Thus, if a solution of (31) is to exist for large B^2, we must have (in most models)

$$(34) \qquad -\lambda_0^+ \to w(v_0^+) - (B/\Sigma_c(v_0^+)/\tan B/\Sigma_c(v_0^+))w_c(v_0^+).$$

Before closing our discussion of the dispersion relation, we should point out an aspect of it that is particularly interesting from the point of view of applied mathematics. The first term of (23) became a Cauchy-integral, because of our assumption, $w(v) = \lambda^*$. The second term remains a double integral. To appreciate its structure, we consider a slightly different problem. Say $w_c(v) \equiv 0$ and $w(v)$ is positive, increasing, for all v. Then, the dispersion relation has the form

$$1 = \int_0^\infty dv \int_{-1}^{1} \frac{d\mu}{2} \frac{1}{\lambda + iBv\mu + w(v)} \Phi_2(v). \tag{35}$$

If we write $x = -w(v)$, $y = -Bv\mu$, $z = x + iy$,

$$1 = -\frac{1}{\pi} \iint_G dx\, dy \frac{1}{z - \lambda} \Phi_3(x, y) = T_G \cdot \Phi_3. \tag{36}$$

The operator T_G plays a central role in the theory of generalized analytic functions [7], [8]. For our purposes we need only remark that if Φ_3 satisfies some "reasonable" conditions inside G, $T_G\Phi_3$ is analytic in the half-plane Re $\lambda > -\lambda^*$. Further, the function $T_G\Phi_3$ is defined, but not analytic, inside G; indeed it is continuous throughout the entire λ-plane. These properties make the following lines of research possible:

(a) One can "continue" the dispersion relation "into" the area-continuum G. However, since one cannot easily deform an inversion contour to pass through G, it will be difficult to distinguish between the contribution from the "pole" and from the continuous singularities.

(b) One can relax the requirement that the sub-Bragg reaction rates be constant. If $w(v)$ varies slightly from λ^*, the line, G, will be changed into a narrow strip. Then, there will be a narrow range of B^2, above B_*^2, wherein a discrete λ_0 cannot be precisely resolved. When B^2 has transcended the range, λ_0 will emerge from the shadow.

(c) One will have to deal with the denominator, $1 - w_c F$, introduced by the elastic scattering. This term destroys the "canonical form" (36) and can introduce new singularities into G.

The neutron-wave problem. The analysis of the neutron-wave experiment proceeds in a similar manner [9], [10]. We do not have as much information about quantitative details of the spectrum, because in the wave problem we seek complex eigenvalues, $\boldsymbol{\kappa}$, rather than real, negative eigenvalues λ. The key operators in the elucidation of the discrete spectrum are no longer Hermitian.

To begin, we note that for given positive frequency ω, and our assumption of an isotropic scattering kernel, eigenvalues occur in pairs, $\pm\boldsymbol{\kappa}$. Also, if $\boldsymbol{\kappa}(\omega)$ and $h(\boldsymbol{v}, \omega)$ are solutions, $\boldsymbol{\kappa}^*$ and h^* are solutions for $-\omega$. We shall, therefore, limit our considerations to positive frequencies, and the first quadrant of the complex κ_3 plane.

As before, our kernel consists of a compact, inelastic part, and an elastic part. Both are taken to be isotropic. We consider the operator

$$U = w(v) - w_c \int \frac{d\Omega}{4\pi} + i\omega + \boldsymbol{\kappa} \cdot \boldsymbol{v}, \tag{37}$$

whose inverse, U^{-1}, may be shown to be

$$U^{-1} = \frac{1}{w(v) + \boldsymbol{\varkappa} \cdot \boldsymbol{v} + i\omega} \left\{ 1 + \frac{w_c(v)}{1 - w_c \mathscr{F}} \int \frac{d\hat{\Omega}}{4\pi} \frac{1}{w + \boldsymbol{\varkappa} \cdot \boldsymbol{v} + i\omega} \right\}. \tag{38}$$

In equation (38)

$$\mathscr{F}(v, \omega, \kappa^2) = \int \frac{d\hat{\Omega}}{4\pi} \frac{1}{w + \boldsymbol{\varkappa} \cdot \boldsymbol{v} + i\omega}. \tag{39}$$

When U^{-1} acts upon functions of v^2 only, it assumes the simple form

$$U^{-1} = \frac{1}{w(v) + \boldsymbol{\varkappa} \cdot \boldsymbol{v} + i\omega} \frac{1}{1 - w_c \mathscr{F}}. \tag{40}$$

The singularities of U^{-1} in the κ_3-plane ($\boldsymbol{\varkappa} \cdot \boldsymbol{v} = \kappa_3 \mu v$) are the κ_3 causing the denominator of the first factor or the quantity $1 - w_c \mathscr{F}$ to vanish. The first factor generates an area, A, defined by

$$\kappa_3 = -(w(v) + i\omega)/v\mu, \qquad 0 \leq v < \infty, \quad -1 \leq \mu \leq +1.$$

When $w(v)$ is chosen to be constant, and equal to λ^* in $0 \leq v < v_0$, A consists of two disjoint parts, a line, or ray A_1, corresponding to velocities inside the Bragg sphere, and an area, A_2, describing the remainder. The point in A_2 lying on the real axis, closest to the origin, is $\lim_{v \to \infty} \Sigma(v)$.

The zeros of $1 - w_c \mathscr{F}$ are more difficult to describe than those of $1 - w_c F$. They generate a curve, $\mathscr{A}$, one end of which is joined to the point $\lim_{v \to \infty} \Sigma(v)$. The shape of the "whisker", $\mathscr{A}$, resembles the letter 'C'. Computations based upon specific models indicate that it does not intersect the ray, A_1. The curve, $\mathscr{A}$, is generated by the elastic scattering. It does not appear to play as important a role in the analysis of the wave experiment as its counterpart, Γ, does in the pulsed experiment. One must emphasize, in closing this short discussion, that the location of the singularities changes as the frequency is altered. As ω is increased, the area A_2 expands, the direction and starting-point of the ray, A_1, is altered, and the shape of the "whisker," $\mathscr{A}$, is changed (Figure 2, p. 90).

One proves that the addition of a compact, inelastic scattering operator to U does not alter its spectrum by an argument similar to the one associated with equations (16) and (17). When we turn to consider the full problem, we are most interested in the discrete spectrum. The relevant equation here is

$$h_0(v^2) = \frac{\mathscr{F}(v, \omega, \kappa^2)}{1 - w_c(v)\mathscr{F}} K_i \cdot h_0, \tag{41}$$

and, unfortunately, both $\mathscr{F}$ and h_0 are complex—in a nontrivial manner. In the pulsed-problem, we know much about the discrete spectrum when $B^2 = 0$, and we can easily follow the eigenvalues into $B^2 \neq 0$. The analogous situation, $\omega = 0$, defines the problem of the static diffusion-length. Few general results exist there, but indications are that the point spectrum is scanty **[11]**. We do not know whether the number of eigenvalues changes as ω is increased.

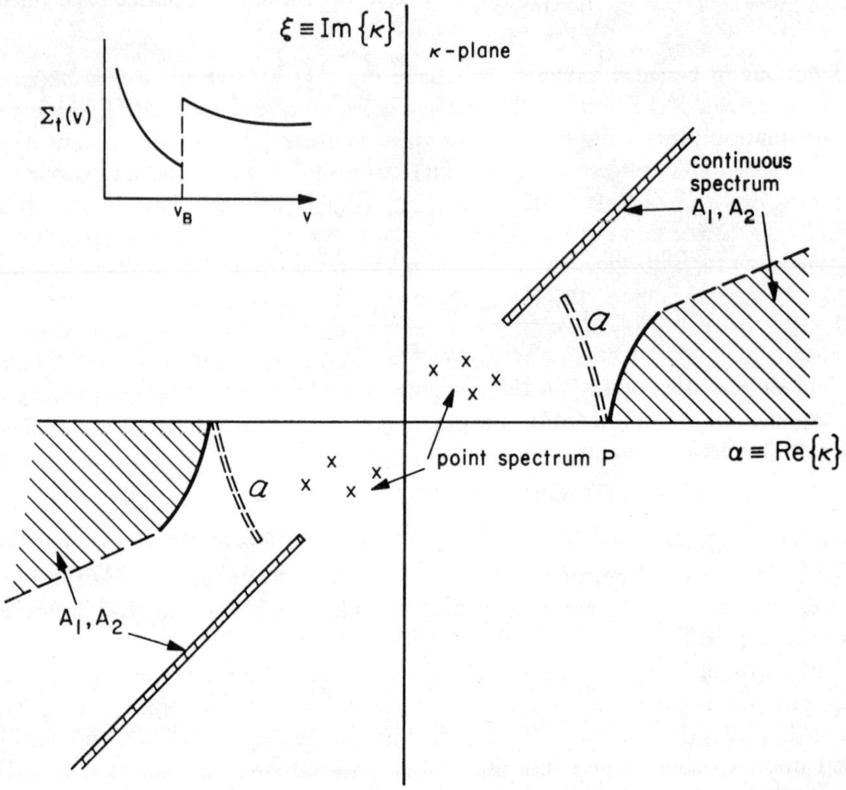

FIGURE 2. κ_3-plane, wave experiment.

Calculations performed with simple versions of the compact K_i suggest that the fundamental eigenvalue, κ_3, strikes the ray, A_1. Simultaneously, another eigenvalue emerges from the other side of A_1. Thus, the situation is quite similar to the one discussed with equations (22)–(30). One sees that analytic continuation of the dispersion relation is possible, and that one can define the integral over the area A_2 when $\kappa_3 \in A_2$. The jump in the dispersion curve when it "crosses" A_1 may be estimated in a manner similar to equations (27)–(28). Suppose that we describe points on A_1 through

$$\kappa_3 = \eta(\lambda^* + i\omega)/v_0 \qquad 1 \leq \eta < \infty.$$

Then, using the formula

$$\lim_{\epsilon \to 0} \mathscr{F}\left(v, \omega, \eta\left(\frac{\lambda^* + i\omega}{v_0}\right)(1 \pm i\epsilon)\right) = \frac{1}{2v\kappa_3}\left\{\ln \frac{\eta v + v_0}{\eta v - v_0} \pm \pi i\right\} \tag{42}$$

for $v \leq v_0$, one can construct formulas similar to equation (28). These complex expressions are not particularly useful; they serve to tell us that the jump is proportional to certain averages performed over the Debye sphere. Since the associated eigenfunctions are not as large, in this region, as are those in the pulsed experiment, the jump should be correspondingly smaller.

Solutions in bounded volume. We have devoted almost all of this paper to special solutions of the Boltzmann equation in an infinite moderator. Experiments are, of course, carried out in finite samples of material—sometimes sufficiently small that boundary effects must be taken seriously. Yet, the special exponential solutions, by virtue of their simplicity and remarkable accuracy, have been retained as the basis for the comparison of theory and experiment. Recently, papers have appeared in which the eigenvalues of the neutron transport operator (in bounded geometry) are considered [**12**]–[**15**]. These papers treat incoherent scattering only, and are concerned primarily with the discrete spectrum and its existence when the moderator samples are small. However, they deal with the Boltzmann equation supplanted by the correct boundary conditions, and the analysis is conducted with considerable rigor. The results of the exponential and the rigorous analyses coincide when the moderator dimensions are large; their relation at intermediate sizes is an interesting subject for future research.

ACKNOWLEDGMENT. Some of the material discussed herein is the subject of research carried out by the author in collaboration with Mr. Robert Conn and Mr. James Duderstadt of the California Institute. The author acknowledges gratefully the help and stimulus afforded by the collaboration.

REFERENCES

1. I. Kuščer and N. Corngold, *Discrete relaxation times in neutron thermalization*, Phys. Rev. **139** (1965), A981–A990.

2. R. Dorfman, *Note on the linearized Boltzmann integral equation for rigid sphere molecules*, Proc. Nat. Acad. Sci. U.S.A. **50** (1963), 804–806.

3. J. A. Young, "Atomic motion in moderators," in *Reactor physics in the resonance and thermal regions*, M.I.T. Press, Cambridge, Mass., 1966.

4. N. Corngold, *Some transient phenomena in thermalization.* I, Nucl. Sci. and Engrg. **19** (1964), 80–90.

5. N. I. Muskhelishvili, *Singular integral equations*, Noordhof, Groningen, 1953.

6. N. Corngold and P. Michael, *Some transient phenomena in thermalization.* II, Nucl. Sci. and Engrg. **19** (1964), 91–94.

7. I. N. Vekua, *Generalized analytic functions*, Pergamon, New York, 1962.

8. C. Cercignani, *Unsteady solutions of kinetic models*, Ann. Physics **40** (1966), 454–468.

9. J. Duderstadt, *On the continuous spectrum of the neutron wave experiment*, Nucl. Sci. and Engrg. **29** (1967), 156–57.

10. M. N. Moore, *The dispersion law of a moderator*, Nucl. Sci. and Engrg. **26** (1966), 354–62.

11. M. M. R. Williams, *Eigenvalues of the Boltzmann equation with a synthetic kernel*, Nucl. Sci. and Engrg. **26** (1966), 262–270.

12. S. Ukai, *Eigenvalues of the neutron transport operator for a homogeneous finite moderator*, J. Math. Anal. Appl. **18** (1967), 297–314.

13. S. Albertoni and B. Montagnini, *Some special properties of the transport operator*, Pulsed Neutron Research, IAEA, Vienna, 1966.

14. R. Bednarz, *Spectrum of the Boltzmann operator*, Pulsed Neutron Research, IAEA, Vienna, 1966.

15. A. Mockel, *Disappearance of discrete decay constants in slabs*, Nucl. Sci. and Engrg. **26** (1966), 279–280.

CALIFORNIA INSTITUTE OF TECHNOLOGY

II. Numerical Neutron Transport

Invariant imbedding and computational methods in radiative transfer

Richard Bellman[1]

1. **Introduction.** With the advent of the digital computer, the possibility of using the complex equations of mathematical physics to obtain numerical answers to numerical questions became a probability. The scientific importance of this operational procedure resides in the fact that numerical results thus derived can be compared with experimental data to provide a systematic means of accepting or rejecting plausible a priori hypotheses. That the process has not proceeded at the expected pace is due to the fact that effective computational solution of functional equations requires on the whole a high level of mathematical sophistication, higher than that which can reasonably be expected of the working scientist, or, indeed, of the average working mathematician.

It has thus been recognized that if the computer is to become the standard laboratory tool that it should, it will be necessary first to develop a number of powerful and versatile numerical algorithms that can be easily learned, and as easily used. The responsibility for providing these scientific tools falls upon the mathematician. Ad hoc ingenuity in mathematics is delightful to behold, but like a brilliant combination in chess, it is usually unnecessary if the proper preliminary moves have been made. In any case, it is tiresome to be forced repeatedly to be ingenious.

As we shall see below, the theory of invariant imbedding furnishes a number of methods which fulfill the requirements stated above. These methods can be applied to a number of the basic processes of mathematical physics, providing both

[1] This work was supported by the National Science Foundation under Grant No. GP-6154.

analytic and computational results. Here we shall restrict ourselves to radiative transfer [1], [2], [3]. The reader interested in applications to neutron transport theory, random walk, wave propagation, differential equations, or analytical mechanics may wish to refer to the following papers and books, where many additional references will be found: [4]–[22].

2. **Imbedding.** Although we do not wish to enter into any general discussion of the theory of invariant imbedding, we feel that a few preliminary remarks are in order to orient the reader. One of the basic concepts of mathematics is that of imbedding. An idea used with equal force in algebra, analysis, and geometry is that of imbedding a particular problem within a family of related problems. This enveloping can be accomplished in many different ways, as might be imagined. This means that some procedures will be more suitable for certain purposes than others. After all, there are many different objectives in a mathematical formulation of a physical process. It is to be expected that no single formulation will satisfy all of the conceptual, analytical, and computational requirements, and that it may be desirable from several points of view to have a variety of approaches available.

In the early days of science, mathematicians and scientists were so delighted to discover one meaningful formulation of a physical process that they ignored the possibility of alternate formulations. As time went by, there developed a natural, but lamentable, tendency to regard the existing analytic formulation as the only such. From this it is a simple step to identify the classical mathematical equation with the underlying physical process. Thus, in the literature, we see references to "the" heat equation, "the" potential equation, "the" wave equation, "the" transport equation, and so forth. It is important to emphasize the fact that any mathematical equation represents a projection of the physical phenomena on some conceptual axes. There is never a one-to-one correspondence. It is particularly important when we remember that the classical formulations go back to precomputer days, back to a time when a million multiplications a second would have been considered miraculous.

There is no uniform prescription for performing an imbedding. We can, however, distinguish various important types of procedures. Consider, to illustrate, a multiple scattering process of the type that is fundamental in radiative transfer. A particle traversing a medium travels in a straight line until an interaction with the medium occurs. This results in a change of direction, and so on, as indicated in Figure 1. Suppose that a stream of particles of this nature is

FIGURE 1

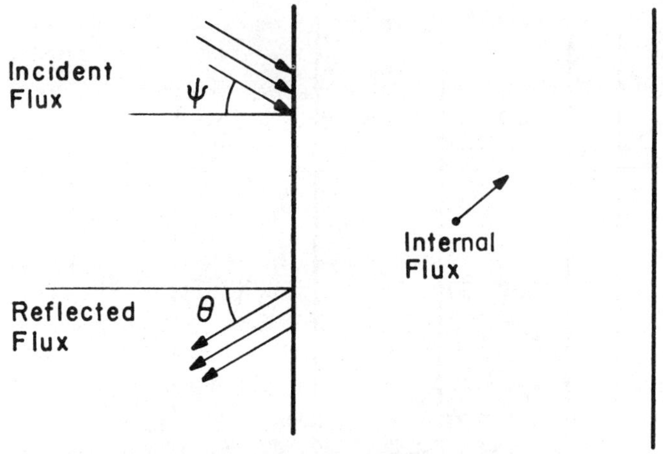

FIGURE 2

incident upon a plane-parallel slab, as indicated in Figure 2. We wish to determine the intensity of reflected flux at a specified angle. The classical approach to this question is to imbed this specific problem within the family of problems represented by the determination of the intensity of flux at an arbitrary direction at an arbitrary point within the medium. Clearly, if we can answer the general question, we can answer the specific question.

A systematic way to answer the general question is to obtain relations connecting contiguous members in this family of problems. Analysis of input-output relations for an infinitesimal neighborhood of a point provides the classical transport equation, which may be linear or nonlinear depending upon the types of interactions that are allowed.

An entirely different imbedding arises from the examination of the behavior of the intensity of reflected flux as the region of interactions is varied. In the case of a plane-parallel slab we can consider the family of regions obtained by allowing the thickness to vary from zero on up. This is a particular illustration of "invariant imbedding." The intensity of reflected flux is now regarded as a function of the angle of incidence, the angle of reflection, and the thickness of the slab. This point of view is closely related to the procedure that might be used by an experimentalist to study a process of this nature. It places an emphasis upon observables. See Figure 3.

The theory of invariant imbedding owes an important part of its development to the pioneering "principles of invariance" of Ambarzumian and Chandrasekhar **[23]**, **[24]**, and another important part to the "point of regeneration" technique of the theory of branching processes **[25]**, **[26]**. However, by far the most important part is a heritage of the theory of dynamic programming where the concepts of a multistage process and invariance play such a fundamental role **[27]**, **[28]**, **[29]**. The first paper devoted to invariant imbedding, an application to radiative

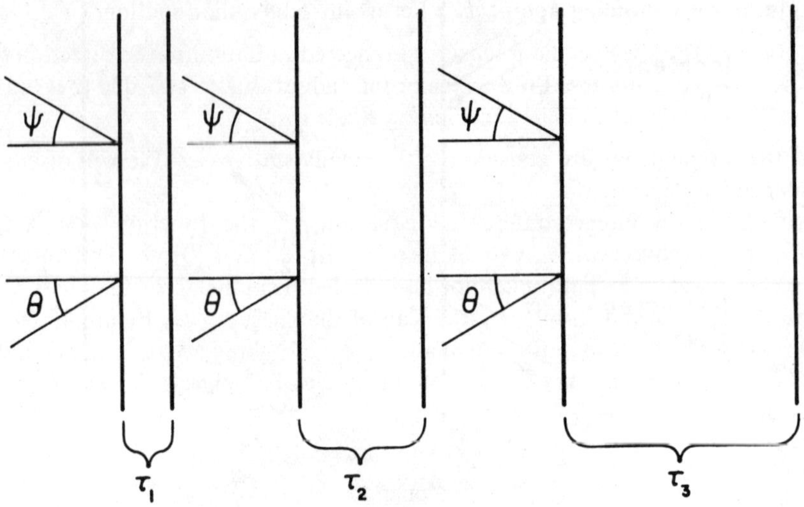

FIGURE 3

transfer, appeared in 1957 **[30]**, and the first extensive numerical results were given in **[1]**, **[2]**.

We shall describe how to obtain equations connecting contiguous members of this new family of problems below, after we have described the physical background of the radiative transfer process we will consider.

3. **A radiative transfer model.** Let us briefly describe the physical background of the radiative transfer process we will treat below. Referring to Figure 2, we suppose that parallel rays of light of uniform intensity are incident at an angle ψ upon an inhomogeneous slab of finite thickness x composed of material that both absorbs and scatters light. It is desired to determine the intensity of diffusely reflected light at angle θ.

As far as the properties of the medium and its effect upon the light are concerned, we suppose the following:

1. In traversing a distance d in any direction in the slab, the intensity of the beam is reduced by absorption to a fraction $e^{-\sigma d} + o(d)$ of the original intensity. We will use this only for infinitesimal d and usually write it in the form $1 - \sigma d + o(d)$.

2. A fraction c of the absorbed intensity is reradiated. This fraction is called the albedo for single scattering.

3. Radiation is scattered isotropically without changing the polarization. The light that is scattered and reradiated is taken to consist of photons treated as point particles.

Taking cognizance of the symmetry of the situation, azimuthal angles may be neglected. The quantities σ and c referred to above are taken to depend only upon the depth inside the slab. In other words, we are dealing with a plane-stratified medium.

4. Invariant imbedding approach. Let us introduce the function

(1) $\rho(\psi, \theta, x) =$ the specific intensity of reflected radiation in the direction θ per unit area on the face of the slab of thickness x due to a beam of unit intensity incident at angle ψ.

At the moment, we are considering the steady-state case. We will discuss the time-dependent case below.

We obtain an integro-differential equation for the function $\rho(\psi, \theta, x)$ by decomposing the interval $[0, x + \Delta]$ into $[x, x + \Delta]$ and $[0, x]$. The underlying idea is that incident radiation cannot be reflected from a slab of thickness $x + \Delta$ without being reflected from the inner slab of thickness x; see Figure 4.

On the basis of the assumptions made in §3 and taking account of the possible interactions in the slab $[x, x + \Delta]$ of infinitesimal thickness Δ, we obtain the following equation

$$
\begin{aligned}
\rho(\psi, \theta, x + \Delta) = {} & \left[1 - \frac{\sigma(x)}{\cos \psi} \Delta\right] \rho(\psi, \theta, x) \left[1 - \frac{\sigma(x)}{\cos \theta}\right] + \frac{\sigma(x) c(x) \Delta}{2\pi \cos \psi} \\
& + 2\pi \frac{\sigma(x) c(x) \Delta}{4\pi \cos \psi} \int_0^{\pi/2} \rho(\psi', \theta, x) \sin \psi' \, d\psi' \\
& + 2\pi \int_0^{\pi/2} \frac{\sigma(x) c(x) \Delta}{4\pi \cos \theta'} \rho(\psi, \theta', x) \sin \theta' \, d\theta' \qquad (2) \\
& + (2\pi)^2 \int_0^{\pi/2} \rho(\psi, \theta', x) \frac{\sigma(x) c(x) \Delta}{4\pi \cos \theta'} \sin \theta' \, d\theta' \\
& \times \int_0^{\pi/2} \rho(\psi', \theta, x) \sin \psi' \, d\psi' + 0(\Delta).
\end{aligned}
$$

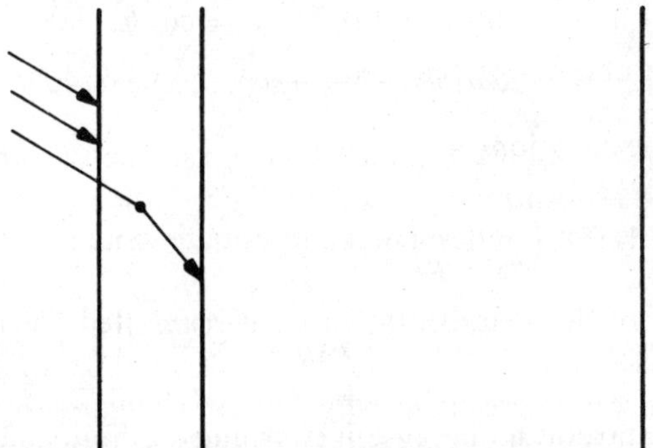

FIGURE 4

The different terms on the right of (2) arise in the following ways. The first is due to absorption losses in passing through the layer of thickness Δ on the way in and the way out. The second is the contribution due to direct scattering from the layer of thickness Δ, while the third is due to light that is scattered in the slab of thickness Δ and then reflected from the slab of thickness x. The fourth arises from light reflected from the slab of thickness x and then scattered in the slab of thickness Δ. The last, the nonlinear term, represents the contribution of light that is reflected from the slab of thickness x, scattered in the Δ-slab, and then rereflected from the slab of thickness x.

All other interactions lead to terms of order Δ^2, and thus are ignored. It is rather interesting to point out how we can treat a multiple-interaction process with a very small amount of bookkeeping through the introduction of the appropriate functions. Passing to the limit as $\Delta \to 0$, we obtain the equation

$$
\begin{aligned}
\rho_x = {} & \frac{\sigma(x)c(x)}{4\pi \cos \psi} - \sigma(x)\left(\frac{1}{\cos \theta} + \frac{1}{\cos \psi}\right)\rho + \frac{\sigma(x)c(x)}{2 \cos \psi} \int_0^{\pi/2} \rho(\psi', \theta, x) \sin \psi' \, d\psi' \\
& + \frac{\sigma(x)c(x)}{2} \int_0^{\pi/2} \rho(\psi, \theta', x) \frac{\sin \theta'}{\cos \theta'} \, d\theta' \qquad (3) \\
& + \pi\sigma(x)c(x) \int_0^{\pi/2} \frac{\sin \theta'}{\cos \theta'} \rho(\psi, \theta', x) \, d\theta' \int_0^{\pi/2} \rho(\psi', \theta, x) \sin \psi' \, d\psi'.
\end{aligned}
$$

The initial condition is

$$
(4) \qquad \rho(\psi, \theta, 0) = 0,
$$

which is to say, no reflection from a slab of zero thickness. Some simple changes of variable are helpful. Set

$$
(5) \qquad \begin{aligned} \mu_0 &= \cos \psi, \qquad \mu = \cos \theta, \\ R(\mu, \mu_0, x) &= \pi \rho(\psi, \theta, x)\mu_0/4. \end{aligned}
$$

Then R satisfies the equation

$$
(6) \qquad \begin{aligned} & \frac{1}{\sigma(x)c(x)} \left[R_x + \sigma(x)\left(\frac{1}{\mu_0} + \frac{1}{\mu}\right) R \right] \\ & \qquad = \left[1 + \frac{1}{2} \int_0^1 R(\mu, \mu', x) \frac{d\mu'}{\mu'}\right]\left[1 + \frac{1}{2} \int_0^1 R(\mu'', \mu_0, x) \frac{d\mu''}{\mu''}\right]. \end{aligned}
$$

The function R is introduced because it is symmetric in μ_0 and μ, a point of some analytic and computational import. The initial condition is again

$$
(7) \qquad R(\mu, \mu_0, 0) = 0, \qquad 0 \le \mu, \mu_0 \le 1.
$$

We now have the equation and initial condition that we will use for computational purposes.

5. **Discussion.** The equation for R is a nonlinear partial differential-integral equation, as opposed to the classical transport equation,

$$\mu \frac{\partial}{\partial x} \psi + \sigma\psi = \frac{c\sigma}{2} \int_{-1}^{1} \psi(x, \mu')\, d\mu', \tag{1}$$

which is linear. The important point, however, to note is that the equation obtained by invariant imbedding techniques is subject to an *initial value* condition, while the traditional equation is subject to a two-point boundary condition. This two-point condition arises in a natural way from the information concerning the fluxes incident on both boundaries of the slab. On the other hand, the invariant imbedding approach automatically leads to an initial-value problem because we start with the case of a slab of zero thickness for which the information concerning reflected and transmitted fluxes is immediate.

What is particularly interesting to note is the fact that an adroit combination of the two approaches (utilizing the X-functions and Y-functions of Chandrasekhar [**31**]) can be used to reduce significantly the dimension of the system of differential equations used to obtain a numerical solution; see [**32**], [**33**]. This is a striking illustration of the advantage accruing of having more than one formulation available.

The equation for R was obtained in the preceding section by means of a direct use of the mathematical model of the physical process, a technique often called "particle counting." There is a great advantage to doing this, since the derivation makes explicit the physical meaning of each term in the equation, which in turn means that various approximations are readily made. Furthermore, this technique makes apparent how to obtain difference approximations which preserve non-negativity and the conservation laws [**34**], [**35**], [**36**]. However, it is also essential to have available automatic techniques for deriving the invariant imbedding equations from the classical equations and the converse. It turns out that the equation for R may be regarded as a consequence of the linearity of perturbation equations; see [**37**], [**38**]. Alternatively, as pointed out by P. Lax, the invariant imbedding equations may be conceived as examples of the rule for differentiating an inverse operator.

$$\frac{d}{dt}(X^{-1}) = -X^{-1} \frac{dX}{dt} X^{-1}. \tag{2}$$

This, in turn, is connected with the fact that the equation for R is a generalized Riccati equation; see [**39**], [**40**], [**41**] for further consequences of this.

Finally, let us note that the equation for R is a limiting form of a set of semigroup relations connecting reflection and transmission functions. New types of semigroups arise from invariant imbedding; see [**42**].

6. **Quadrature techniques.** It remains to introduce some approximation methods which will replace the partial differential-integral equation for R by an equation which can be readily handled by a digital computer. This means that we want to end up with a system of ordinary differential equations subject to an initial

condition. To achieve this, we use a quadrature technique in the conventional manner. Write

$$\int_0^1 g(w)\,dw = \sum_{k=1}^{N} c_k g(w_k) \tag{1}$$

where the quadrature points w_k and the coefficients c_k have been chosen in some efficient fashion. For example, we may use Gaussian quadrature, which means that these quantities are determined by the condition that the relation in (1) is exact for any polynomial of degree at most equal to $2N - 1$. We followed this route originally in order to take advantage of the fact that the w_k and c_k are well tabulated for Gaussian quadrature.

As of today, with the speed and capacity of the current computer, it is better to combine this standard quadrature technique with an adaptive procedure in which the kind of quadrature formula used is made to depend upon preliminary knowledge of the function $g(w)$. Obviously, we can expect to obtain improved accuracy in this way. What is more important in the treatment of realistic physical processes is that we can keep the dimensionality of the approximating equations under control by using these more sophisticated procedures.

With the aid of Gram-Schmidt methods, we can readily form the orthogonal polynomials corresponding to any assigned weight function. Since we rarely wish to use quadrature techniques of order higher than ten or eleven, it is an equally simple matter to determine the roots of these polynomials which yield the quadrature points. What we wish to emphasize is that what was formerly a major undertaking in itself, the determination of quadrature formulas, is now a single simple step in a computational process requiring negligible time. The way is therefore clear for the systematic use of adaptive computational procedures, but little has been done in this area; see, however, **[43]**.

Returning to (4.5), the use of Gaussian quadrature yields the approximate system

$$\begin{aligned}\frac{1}{c(x)}\left[\frac{\partial R(\mu,\mu_0,x)}{\partial x} + \left(\frac{1}{\mu_0}+\frac{1}{\mu}\right)R(\mu,\mu_0,x)\right] &= 1 + \frac{1}{2}\left[\sum_{k=1}^{N}\frac{c_k}{w_k}R(w_k,\mu_0,x)\right] \\ &+ \frac{1}{2}\left[\sum_{k=1}^{N}\frac{c_k}{z_k}R(\mu,z_k,x)\right] + \frac{1}{4}\left[\sum_{k=1}^{N}\frac{c_k}{w_k}R(w_k,\mu_0,x)\right]\left[\sum_{k=1}^{N}\frac{c_k}{z_k}R(\mu,z_k,x)\right].\end{aligned} \tag{2}$$

Setting

$$f_{ij}(x) = R(w_i, z_j, x), \tag{3}$$

we obtain the N^2 ordinary differential equations

$$\begin{aligned}&\frac{1}{c(x)}\left[\frac{df_{ij}(x)}{dx} + \left(\frac{1}{z_i}+\frac{1}{w_j}\right)f_{ij}(x)\right] \\ &= 1 + \frac{1}{2}\left[\sum_{k=1}^{N}\frac{c_k}{z_k}f_{kj}(x)\right] + \frac{1}{2}\left[\sum_{k=1}^{N}\frac{c_k}{w_k}f_{ik}(x)\right] + \frac{1}{4}\left[\sum_{k=1}^{N}\frac{c_k}{z_k}f_{kj}(x)\right]\left[\sum_{k=1}^{N}\frac{c_k}{w_k}f_{ik}(x)\right],\end{aligned} \tag{4}$$

subject to the initial conditions

(5) $$f_{ij}(0) = 0, \qquad i, j = 1, 2, \ldots, N.$$

7. **Computational feasibility.** Since we can expect a significant improvement in accuracy as N, the order of the quadrature, is increased, there is a strong incentive to increase N. On the other hand, the number of differential equations is N^2, which means that there is considerable motivation for keeping N of moderate size. How large a system can we handle?

With computers of current vintage, 1967, we can handle $N = 10$ with ease and go up to $N = 30$ with no particular effort. With the aid of those soon to be available, $N = 100$ becomes feasible. Taking advantage of the symmetry of f_{ij}, we can reduce N^2 to $N(N+1)/2$, a considerable simplification; e.g. for $N = 100$ we are led to 5,500 instead of 10,000. With the computers envisaged ten years hence, we can easily think of $N = 1000$. If the special-purpose computers for solving large systems of differential equations are built, as they should be, systems of orders of several million will be handled in a commonplace fashion. All of this means that computer capacity is not a serious constraint.

In practice, we have found that for the foregoing simple case there is no need to use large values of N. Values such as $N = 9$ or 11 yield highly accurate results. Let us note that by carrying out the calculation for both values, $N = 9$ and $N = 11$, we obtain an important test of inner consistency. Furthermore, there is no cost in time, since the calculations are done simultaneously as if there were one large system of order 20. This is a point to note, since in the large complex calculations contemplated for the treatment of realistic models of physical processes it will be necessary to use check points at various stages in order to guarantee the accuracy of the final results. This is connected with the subject of parallel computation, into which we do not want to enter. It is important, however, to observe that not only do we frequently want to compare the results of variants of the same method, but also the results of entirely different methods. This is one of the advantages of possessing a number of alternate analytic formulations. The days of naive allegiance to one equation are over.

Let us further observe that even small values of N such as $N = 3$ or $N = 5$ yield useful numerical results which clearly exhibit the qualitative behavior of the solution. This is useful to keep in mind since very often we are more interested in the ability to reject a reasonable physical model on the basis of some quick calculations than in any highly accurate numerical investigation. This has always been the advantage of possessing an analytic approximation. Unfortunately, in many cases it is not at all easy to find these convenient analytic expressions, which forces us to use numerical probes. In general, little attention has been paid to deriving numerical results which are inaccurate by specified amounts, e.g. ten percent, one percent, and so forth. Mathematics is a very difficult tool to control in the sense that it usually forces us to learn more than we wanted to about a process. To obtain a result to two significant figures, we must often calculate it to ten significant figures.

An essential part of the discussion of feasibility is that of numerical stability.

Can we obtain highly accurate numerical results? Suppose, for example, that we wished to determine the reflection properties of a semi-infinite atmosphere, which is to say an atmosphere which has a thickness of ten or twenty mean free paths assuming realistic values for the albedo, and suppose that we wished to obtain this information using an invariant imbedding formulation. This means that we have to integrate a system of differential equations over a long range of the independent variable. This is always a risky affair if nothing is known about the analytic structure of the equations. Fortunately, the equations of invariant imbedding possess certain fundamental conservation properties which provide a built-in stability. This will be discussed below.

Let us mention in passing that taking advantage of the known asymptotic behavior of the reflection function as the thickness of the slab is increased, we can use extrapolation procedures to estimate the reflection from a semi-infinite slab with great accuracy from the calculations for less than a mean free path. Semigroup relations can also be used. See **[44]**, **[42]**.

On the basis of extensive calculations we have carried out over the last ten years, we can unreservedly assert that the method is a useful one. It is easy to impart, easy to apply, and accurate. For detailed discussions of specific applications, see **[45]**–**[55]**. Comparisons with the P_n-method and Monte Carlo techniques will be found in **[45]**, **[53]**, **[54]**. For numerical treatment of an atmosphere backed up by a partial absorber (Chandrasekhar's planetary problem), see **[55]**.

8. **Conservation relations.** Since we know that the reflection function exists in the physical process for arbitrary thickness of the slab, and since we have a great deal of faith, based upon experience, in the underlying mathematical model, we suspect that the solution of the system of nonlinear ordinary differential equations obtained by quadrature exists for all $x > 0$. Furthermore, we suspect that $f_{ij}(x)$ possesses a limit as $x \to \infty$. It remains to establish these results.

One way is to return to the linear transport equation and analyze the connection between this equation and the invariant imbedding equation. Perhaps a simpler approach, and in any case an illuminating approach, is based upon the use of conservation relations. In addition to the reflection function, we can introduce the equally basic transmission function, and a dissipation function which measures the total energy lost while traversing the medium. Both of these new functions satisfy equations similar to that obtained for R, and derived in the same fashion. A difference is that R enters into the equations for these new functions. On this basis, R may be considered to be *the* fundamental function.

On physical grounds, it is easy to see that the total of the reflected, transmitted, and dissipated fluxes must be equal to the intensity of incident flux. It is not difficult to establish this analytically. This result combined with the nonnegativity of the reflection, transmission, and dissipation functions, results readily established, leads to the conclusion that the reflection function is uniformly bounded for all $x \geq 0$. This, in turn, yields the desired result that the solution of the equation exists for all $x \geq 0$. Details are given in **[35]**.

It is possible to obtain difference relations which preserve these conservation relations [36], [57]. This is an interesting type of problem which has not been much discussed.

What is most important, of course, is the numerical stability found in practice. Once observed, it is a challenge to ascertain the reasons.

9. **More realistic models of radiative transfer.** The plane-parallel slab problem has produced some very elegant analysis under the simplifying assumptions of stationarity and isotropy. The computational aspects of this physical model may be considered to be completely under control, using any of a number of different approaches. More challenging problems to the mathematician and his giant computer arise when we allow more realistic situations: anisotropy, time-dependence, and cylindrical and spherical geometries. Furthermore, as we indicate below, even the question of determining the internal flux in a plane-parallel slab is not without its features of interest.

Let us briefly indicate the current situation in these areas as far as invariant imbedding is concerned.

10. **Internal fluxes.** We have previously mentioned the fact that the classical transport equation, while linear, suffers from the disadvantage of being subject to a two-point boundary condition.

One way to circumvent this is to use invariant imbedding to determine the reflected flux, which yields the missing initial condition. Presumably the internal flux can now be obtained by solving an initial-value problem. In practice, this does not work. The reason is the following. When quadrature techniques are used to replace the linear partial differential-integral equation by a set of ordinary differential equations, we are led to a set of equations with constant coefficients with the property that the characteristic roots of the coefficient matrix assume both large negative and positive values. This results in numerical instability.

It turns out that it is just as easy to apply invariant imbedding techniques to the determination of the internal flux as it is to the determination of the reflected or transmitted flux. Furthermore, there is no difficulty in taking care of internal sources. See [31], [58], [59], [60].

11. **Anisotropic scattering.** In the previous discussion, we assumed that a particle had equal probability of going in any direction after scattering or when reradiated. Far more important for applications are mathematical models based upon anisotropic scattering.

Let $p(\cos \alpha)$ be the phase function, where α is the scattering angle between an incident ray and a ray resulting from an elementary act of scattering. Let c be the albedo for single scattering ($0 < c \leq 1$). The quantity $cp(\cos \alpha)\, d\Omega\, 4\pi$ is the fraction of absorbed energy which is scattered into a solid angle of magnitude $d\Omega$ around α. The phase function is normalized so that

$$\text{(1)} \qquad \frac{1}{4\pi}\int_0^{\pi} p(\cos \alpha) 2\pi \sin \alpha \, d\alpha = 1.$$

Let monodirectional flux of radiation be uniformly incident on the top of a horizontal, plane parallel medium. The incident flux is π per unit area normal to the direction of incidence, and its polar angle is arc cosine $(-\mu_0)$ measured from the upward vertical $(0 < \mu_0 \leq 1)$. Let all polar angles be measured from the upward vertical, and let the azimuth angle be measured such that it is zero for the incident flux. The optical thickness of the slab is x, as before.

The intensity of the diffusely reflected (multiply scattered) flux emerging from the top of the slab in the direction whose polar angle is arc cosine μ $(0 < \mu \leq 1)$ and whose azimuth is φ $(0 \leq \varphi \leq 2\pi)$ is defined to be $r(\mu, \varphi; \mu_0, \varphi_0, x)$, where φ_0 is written for the incident azimuth, $\varphi_0 = 0$. The scattering function S is introduced,

$$r(\mu, \varphi; \mu_0, \varphi_0, x) = S(\mu, \varphi; \mu_0, \varphi_0, x)/4\mu. \tag{2}$$

The type of reasoning used above leads to the equation

$$\begin{aligned}
&\partial S(\mu, \varphi; \mu_0, \varphi_0, x)/\partial x + (1/\mu + 1/\mu_0)S \\
&= c\Big\{p(\mu, \varphi; -\mu_0, \varphi_0) + \frac{1}{4\pi}\int_0^1\int_0^{2\pi} S(\mu', \varphi'; \mu_0, \varphi_0, x)p(\mu, \varphi; \mu', \varphi')\frac{d\mu'}{\mu'}\,d\varphi' \\
&+ \frac{1}{4\pi}\int_0^1\int_0^{2\pi} S(\mu, \varphi, \mu'', \varphi'', x)p(-\mu'', \varphi''; -\mu_0, \varphi_0)\frac{d\mu''}{\mu''}\,d\varphi'' \\
&+ \frac{1}{(4\pi)^2}\int_0^1\int_0^{2\pi}\int_0^1\int_0^{2\pi} S(\mu', \varphi'; \mu_0, \varphi_0, x) \\
&\times p(-\mu'', \varphi''; \mu', \varphi')S(\mu, \varphi; \mu'', \varphi'', x)\frac{d\mu'}{\mu'}\,d\varphi'\,\frac{d\mu''}{\mu''}\,d\varphi''\Big\}.
\end{aligned} \tag{3}$$

In the limit as the thickness approaches zero, we have the condition

$$S(\mu, \varphi; \mu_0, \varphi_0, 0) = 0 \tag{4}$$

when the lower boundary is a perfect absorber. This condition depends upon the assumption of the characteristics of the lower boundary. For example, in the case of a reflecting Lambert surface, the right-hand side of (4) will be nonzero.

Let the phase function be expanded in a series of Legendre polynomials consisting of $M + 1$ terms, where M is finite. In general, M will not be large. For strongly directional scattering, where many terms would be required, a different technique is employed. Let

$$p(\cos\alpha) = \sum_{m=0}^{M} c_m P_m(\cos\alpha) \tag{5}$$

and let the S function be expanded in the form

$$S(\mu, \varphi; \mu_0, \varphi_0, x) = \sum_{m=0}^{M} S^{(m)}(\mu, \mu_0, x)\cos m(\varphi - \varphi_0). \tag{6}$$

Each of the $S^{(m)}$ components then satisfies the equation

$$\frac{\partial S^{(m)}}{\partial x} + \left(\frac{1}{\mu} + \frac{1}{\mu_0}\right)S^{(m)} = c(2 - \delta_{0m})\sum_{i=m}^{M}(-1)^{i+m}c_i\,\frac{(i-m)!}{(i+m)!}\,\psi_i^m(\mu)\psi_i^m(\mu_0) \tag{7}$$

for $m = 0, 1, \ldots, M$, where

$$(8) \qquad \psi_i^m(\mu) = P_i^m(\mu) + \frac{(-1)^{i+m}}{2(2-\delta_{0m})}\int_0^1 S^{(m)}(\mu, \mu', x) P_i^m(\mu') \frac{d\mu'}{\mu'}$$

for $m = 0, 1, 2, \ldots, M$, and $i = m, m+1, \ldots, M$; δ_{0m} is the Kronecker delta function ($\delta_{0m} = 1$ if $m = 0$; otherwise $\delta_{0m} = 0$); and $P_i^m(\mu)$ is an associated Legendre polynomial. The initial conditions are

$$(9) \qquad S^{(m)}(\mu, \mu_0, 0) = 0, \qquad m = 0, 1, \ldots, M.$$

In order to produce a computational solution, we proceed as before. Introduce the functions of one variable,

$$(10) \qquad S_{ij}^{(m)}(x) = S^{(m)}(z_i, z_j, x)$$

where $z_1, z_2, \ldots, z_N$ are the roots in ascending order of the shifted Legendre polynomial, $P_N^*(z) = P_N(1-2z)$, and replace the definite integral in (8) by a sum according to the Gaussian quadrature method, with weights $w_1, w_2, \ldots, w_N$. Then we have the system of ordinary differential equations for $S_{ij}^{(m)}(x)$,

$$(11) \qquad \frac{dS_{ij}^{(m)}(x)}{dx} = -\left(\frac{1}{z_i} + \frac{1}{z_j}\right) S_{ij}^{(m)} + c(2-\delta_{0m}) \sum_{k=m}^{M} (-1)^{k+m} \frac{(k-m)!}{(k+m)!} c_k \psi_{ki}^m \psi_{kj}^m,$$

$$m = 0, 1, \ldots, M; \quad i = 1, 2, \ldots, N; \quad j = 1, 2, \ldots, N,$$

where

$$(12) \qquad \psi_{kl}^m = P_k^m(z_l) + \frac{(-1)^{k+m}}{2(2-\delta_{0m})} \sum_{j=1}^{N} S_{lj}^{(m)} P_k^m(z_j) \frac{w_j}{z_j}.$$

The initial conditions are

$$(13) \qquad S_{ij}^{(m)}(0) = 0.$$

It should be borne in mind that the independent variable is the thickness x.

Observe that there are $M + 1$ sets of independent simultaneous systems of order N^2. This means that there is no strain on the capacity of the computer. The reason for this miraculous decoupling is, of course, the nature of the linear transport equation.

If M is not large, the $(M+1)N^2$ equations are solved simultaneously rather than in sequence. For further details, see **[61]**.

12. **Time-dependent radiative transfer.** Let us now turn to the question of time-dependent processes. There is no difficulty in deriving a partial differential-integral equation for the intensity of flux per unit time. It is

$$(1) \quad g_x + (\mu_0^{-1} + \mu^{-1})\left(\frac{\partial}{\partial t} + 1\right) g$$
$$= c\left[\frac{H(t)}{4\mu} + \frac{1}{2}\int_0^1 g(\mu, \mu', x, t)\frac{d\mu'}{\mu'} + \frac{1}{2\mu}\int_0^1 g(\mu'', \mu_0, x, t)\, d\mu'' \right.$$
$$\left. + \int_0^t dt' \int_0^1 g(\mu'', \mu_0, x, t')\, d\mu'' \int_0^1 g_t(\mu, \mu', x, t-t')\frac{d\mu'}{\mu'}\right].$$

The question is: What to do with it? The combination of two partial derivatives and the convolution term makes for a formidable problem.

Fortunately, there is a single device which eliminates one of the partial derivatives and the convolution terms at one blow, the Laplace transform. Introducing the function P defined by

$$P(\mu, \mu_0, x, s) = 4\mu L(g) \tag{2}$$

we find the familiar equation

$$\begin{aligned} P_x &+ (s+1)(\mu_0^{-1} + \mu^{-1})P \\ &= c\left\{\frac{1}{s} + \frac{1}{2}\int_0^1 P(\mu, \mu', x, s)\frac{d\mu'}{\mu'} + \frac{1}{2}\int_0^1 P(\mu'', \mu_0, x, s)\frac{d\mu''}{\mu''}\right. \\ &\quad \left. + \frac{s}{4}\int_0^1\int_0^1 P(\mu'', \mu_0, x, s)P(\mu, \mu', x, s)\frac{d\mu''}{\mu''}\frac{d\mu'}{\mu'}\right\}. \end{aligned} \tag{3}$$

There is thus no difficulty in obtaining the numerical values of the Laplace transform of the desired solution. It remains to use a numerical inversion technique to find the solution itself. Once again, quadrature techniques of the type described above can be used. The details will be found in **[2]**. Indeed, the method was so simple to use and provided such accuracy that we applied it to a number of other types of functional equations that arise in engineering and physics with great frequency. For the details, see **[43]**.

13. **Spherical shell.** Let us now turn to spherical atmospheres, an additional step toward a realistic description of radiative transfer processes. When conical flux is incident upon a spherical hollow shell of inner radius σ and outer radius z, the equation for the reflection function takes the form

$$\begin{aligned} &\frac{\partial S(z, \mu, \mu_0)}{\partial z} + \frac{1-\mu^2}{\mu z}\frac{\partial S}{\partial \mu} + \frac{1-\mu_0^2}{\mu_0 z}\frac{\partial S}{\partial \mu_0} + \left(\frac{1}{\mu} + \frac{1}{\mu_0}\right)S - \frac{\mu^2+\mu_0^2}{\mu^2\mu_0^2}\frac{S}{z} \\ &\qquad = c\left[1 + \frac{1}{2}\int_0^1 S(z, \mu, \mu')\frac{d\mu'}{\mu'}\right]\left[1 + \frac{1}{2}\int_0^1 S(z, \mu'', \mu_0)\frac{d\mu''}{\mu''}\right]. \end{aligned} \tag{1}$$

One approach to a numerical solution of this equation is to use a "quadrature" on the partial derivatives as well as on the integral terms.

Let us briefly sketch the idea. We wish to approximate to the first derivative of a function f at the usual quadrature points x by a linear combination of its values at the entire set,

$$f'(x_i) = \sum_{j=1}^{N} \sigma_{ij} f(x_j), \qquad i = 1, 2, \ldots, N. \tag{2}$$

To determine the coefficients σ_{ij} we ask that the formula be exact for all polynomials of degree $N - 1$ or less. Using the appropriate test functions, we readily determine the coefficients.

Using this procedure we once again can reduce the numerical solution of a complex functional equation to that of the numerical solution of a system of ordinary differential equations subject to initial conditions. Details will be found in [**62**], [**63**]. A different approach based upon perturbation techniques is described in [**64**].

14. **Conclusion.** In a number of papers listed in the bibliography we have presented the formulas for an invariant imbedding approach to a large number of other types of processes associated with radiative transfer and other physical processes ruled by similar equations. Although we have not systematically carried out the computational solution of all of the equations that appear in these papers, we are reasonably certain that these direct methods can be successfully carried through. In a forthcoming book [**3**] we will present a large number of numerical results which have not previously appeared either in published papers or in RAND reports.

References

1. R. Bellman, R. Kalaba, and M. Prestrud, *Invariant imbedding and radiative transfer in slabs of finite thickness*, American Elsevier, New York, 1963.

2. R. Bellman, H. Kagiwada, R. Kalaba, and M. Prestrud, *Invariant imbedding and time-dependent processes*, American Elsevier, New York, 1964.

3. R. Bellman, H. Kagiwada, R. Kalaba, and S. Ueno, *Computational methods in radiative transfer*, American Elsevier, New York (to appear).

4. G. M. Wing, *An introduction to transport theory*, John Wiley, New York, 1962.

5. R. N. Adams and E. D. Denman, *Wave propagation and turbulent media*, American Elsevier, New York, 1966.

6. R. Bellman, R. Kalaba, and R. Vasudevan, *Invariant imbedding theory of neutron transport: Correlation functions*, J. Math. Anal. Appl. **8** (1964), 225–231.

7. ———, *Invariant imbedding and the Townsend avalanche*, J. Math. Anal. Appl. **7** (1963), 264–270.

8. R. Bellman, R. Kalaba, and G. M. Wing, *Invariant imbedding and variational principles in transport theory*, Bull. Amer. Math. Soc. **67** (1961), 396–399.

9. ———, *Invariant imbedding and neutron transport theory*. V: *Diffusion as a limiting case*, J. Math. Mech. **9** (1960), 933–944.

10. ———, *Invariant imbedding and neutron transport in a rod of changing length*, Proc. Nat. Acad. Sci. U.S.A. **46** (1960), 128–130.

11. ———, *Invariant imbedding and neutron transport theory*. III: *Neutron-neutron collision processes*, J. Math. Mech. **8** (1959), 249–262.

12. ———, *Invariant imbedding and mathematical physics*. I: *Particle processes*, J. Math. Phys. **1** (1960), 280–308.

13. R. Bellman, *Invariant imbedding and random walk*, Proc. Amer. Math. Soc. **13** (1962), 251–254.

14. ———, "Invariant imbedding and wave propagation in stochastic media" in *Electromagnetic wave propagation*, Academic Press, New York, 1960.

15. R. Bellman and R. Kalaba, *Wave branching processes and invariant imbedding*. I, Proc. Nat. Acad. Sci. U.S.A. **47** (1961), 1507–1509.

16. R. Bellman, *Perturbation techniques in mathematics, physics, and engineering*, Holt, New York, 1964.

17. G. M. Wing, *Invariant imbedding and the asymptotic behavior of solutions to initial value problems*, J. Math. Anal. Appl. **9** (1964), 85–98.

18. W. A. Beyer, *Asymptotic phase and amplitude for a modified Coulomb potential in scattering theory: An application of invariant imbedding*, J. Math. Anal. Appl. **13** (1966), 348–360.

19. R. Bellman and R. Kalaba, *A note on Hamilton's equations and invariant imbedding*, Quart. Appl. Math. **21** (1963), 166–168.

20. ———, *Invariant imbedding and the integration of Hamilton's equations*, Rend. Circ. Mat. Palermo **12** (1963), 1–11.

21. R. Bellman, H. Kagiwada, and R. Kalaba, *Invariant imbedding and nonvariational principles in analytical dynamics*, Internat. J. Non-linear Mech. **1** (1965), 51–55.

22. R. Bellman, H. Kagiwada, R. Kalaba, and R. Sridhar, *Invariant imbedding and nonlinear filtering theory*, J. Astronom. Sci. **13** (1966), 110–115.

23. V. Ambarzumian, *Diffuse reflection of light by a foggy medium*, Dokl. Akad. Nauk SSSR **38** (1943), 229–232.

24. S. Chandrasekhar, *Radiative transfer*, Oxford Univ. Press, England, 1950.

25. R. Bellman and T. E. Harris, *On age-dependent binary branching processes*, Ann. of Math. (2) **55** (1952), 280–295.

26. T. E. Harris, *The theory of branching processes*, Springer-Verlag, Berlin, 1963.

27. R. Bellman, *Dynamic programming*, Princeton Univ. Press, Princeton, N.J., 1957.

28. ———, *Adaptive control processes: A guided tour*, Princeton Univ. Press, Princeton, N.J., 1961.

29. R. Bellman and S. Dreyfus, *Applied dynamic programming*, Princeton Univ. Press, Princeton, N.J., 1962.

30. R. Bellman and R. Kalaba, *On the principle of invariant imbedding and propagation through inhomogeneous media*, Proc. Nat. Acad. Sci. U.S.A. **42** (1956), 629–632.

31. H. Kagiwada and R. Kalaba, *A new initial value method for internal intensities in radiative transfer*, RM-4906-PR, The RAND Corporation, March 1966.

32. R. Bellman, H. Kagiwada, R. Kalaba, and S. Ueno, *Numerical results for Chandrasekhar's X and Y functions of radiative transfer*, J. Quant. Spectroscopy and Radiative Transfer **6** (1966), 479–500.

33. R. Bellman, H. Kagiwada, and R. Kalaba, *Numerical results for the auxiliary equation of radiative transfer*, J. Quant. Spectroscopy and Radiative Transfer **6** (1966), 291–310.

34. R. Bellman, R. Kalaba, and G. M. Wing, *Invariant imbedding, conservation relations, and nonlinear equations with two-point boundary values*, Proc. Nat. Acad. Sci. U.S.A. **46** (1960), 1258–1260.

35. R. Bellman, K. Cooke, R. Kalaba, and G. M. Wing, *Existence and uniqueness theorems in invariant imbedding*. I: *Conservation principles*, J. Math. Anal. Appl. **10** (1965), 234–244.

36. R. Bellman and S. Lehman, *Invariant imbedding, particle interaction and conservation relations*, J. Math. Anal. Appl. **10** (1965), 112–122.

37. R. Bellman, R. Kalaba, and G. M. Wing, *Invariant imbedding and the reduction of two-point boundary-value problems to initial value problems*, Proc. Nat. Acad. Sci. U.S.A. **46** (1960), 1646–1649.

38. R. Bellman and R. Kalaba, *On the fundamental equations of invariant imbedding*. I, Proc. Nat. Acad. Sci. U.S.A. **47** (1961), 336–338.

39. R. Redheffer, *Novel uses of functional equations*, J. Rational Mech. Anal. **3** (1954), 271–279.

40. ———, *The Mycielski-Pazkowski diffusion problem*, J. Rational Mech. Anal. **9** (1960), 607–621.

41. ———, "Difference and functional equations and transmission line theory" in *Modern mathematics for the engineer*. II, edited by E. F. Beckenbach, McGraw-Hill, New York, 1961, pp. 282–337.

42. R. Bellman and T. Brown, *A note on invariant imbedding and generalized semi-groups*, J. Math. Anal. Appl. **9** (1964), 394–396.

43. R. Bellman, R. Kalaba, and J. Lockett, *Numerical inversion of the Laplace transform*, American Elsevier, New York, 1966.

44. R. Bellman and R. Kalaba, *A note on nonlinear summability techniques in invariant imbedding*, J. Math. Anal. Appl. **6** (1963), 465–472.

45. D. A. Mathews, K. F. Hansen, and E. A. Mason, *Deep penetration of radiation by the method of invariant imbedding*, Nuclear Sci. & Engr. **27** (1967), 263–270.

46. R. E. Beissner, *The application of invariant imbedding to shielding problems*, General Dynamics, Fort Worth, Texas, MR-N-287 (NARF-61-41T), 1962.

47. ———, *An analysis of fast-neutron energy-angle distributions*, General Dynamics, Fort Worth, Texas, FZK-9-186 (NARF-63-5T), 1963.

48. J. O. Mingle, *Applications of the invariant imbedding method to monoenergetic neutron transport theory in slab geometry*, Nuclear Sci. & Engr. **28** (1967), 177–189.

49. D. S. Dodson and J. O. Mingle, *Escape probabilities by the method of invariant imbedding*, Trans. Amer. Nuclear Soc. **7** (1964), 26.

50. J. O. Mingle, *Escape probabilities by a generalized invariant imbedding kernel*, Trans. Amer. Nuclear Soc. **8** (1965), 230.

51. ———, *Reflection and transmission intensities from gray slabs utilizing invariant imbedding methods*, ASME-AIChE Joint Heat Transfer Conference at Los Angeles, CEP Symposium Series 62, 64, 250, AIChE, New York, 1966.

52. M. Aoki and A. Shimizu, *Reflection and transmission of gamma rays through slabs*, Academic Press, New York (to appear).

53. A. Shimizu and H. Mizuta, *Application of invariant imbedding to the reflection and transmission problem of gamma rays* (I), J. Nuclear Sci. Tech. **3** (1966), 57.

54. ———, *Application of invariant imbedding to the reflection and transmission problem of gamma rays* (II), J. Nuclear Sci. Tech. **3** (1966), 441.

55. P. B. Bailey and G. M. Wing, *A correction to some invariant imbedding equations of transport theory obtained by 'particle counting*,' J. Math. Anal. Appl. **8** (1964), 170–174.

56. R. Bellman, H. Kagiwada, R. Kalaba, and S. Ueno, *A computational approach to Chandrasekhar's planetary problem*, J. Franklin Inst. **282** (1966), 330–334.

57. R. Bellman and K. Cooke, *Existence and uniqueness theorems in invariant imbedding*. II: *Convergence of a new difference algorithm*, J. Math. Anal. Appl. **12** (1965), 247–253.

58. R. Bellman, H. Kagiwada, R. Kalaba, and S. Ueno, *Invariant imbedding and the computation of internal fields for transport processes*, J. Math. Anal. Appl. **12** (1965), 538–541.

59. R. Bellman, H. Kagiwada, and R. Kalaba, *Invariant imbedding and a reformulation of the internal intensity problem in transport theory*, Monthly Notices Roy. Astronom. Soc. **132** (1966), 183–191.

60. H. Kagiwada and R. Kalaba, *Numerical results for internal intensities in atmospheres illuminated by isotropic sources*, The RAND Corporation, RM-4958-PR, April 1966.

61. R. Bellman, H. Kagiwada, R. Kalaba, and S. Ueno, *Some mathematical aspects of multiple scattering in a finite inhomogeneous slab with anisotropic scattering*, The RAND Corporation, RM-4595-PR, May 1965.

62. ———, *Diffuse transmission of light from a central source through an inhomogeneous spherical shell with isotropic scattering*, J. Math. and Phys. (to appear).

63. R. Bellman, H. Kagiwada, and R. Kalaba, *Invariant imbedding and radiative transfer in spherical shells*, J. Comp. Phys. (to appear).

64. ———, *Invariant imbedding and perturbation techniques applied to diffuse reflection from spherical shells*, The RAND Corporation, RM-4730-NASA, August 1965.

University of Southern California,
Los Angeles, California

Direct and inverse problems for integral equations via initial-value methods

H. H. Kagiwada and R. E. Kalaba

I. **Introduction.** The mathematical theory of multiple scattering processes frequently leads to Fredholm integral equations. In radiative transfer, for example, one of the basic equations is the auxiliary equation which has the form **[1]**, **[2]**

$$\varphi(x) = g(x) + c(x) \int_0^b E_1(|x - y|)\varphi(y)\, dy, \qquad 0 \le x \le b, \tag{1}$$

where E_1 is an exponential-integral function. In this equation the function φ characterizes the radiation field, g the sources of the field, and c the medium. Direct problems involve finding φ, being given g and c. Inverse, or system identification, problems involve estimating properties of g and c, based on measured or assumed properties of φ, the radiation field. Recent progress in the analytical and computational treatment of these problems is to be described in subsequent sections, the methodology having grown out of earlier studies in invariant imbedding **[3]**.

Modern computing machines can integrate systems of several hundred or several thousand ordinary differential equations given a complete set of initial conditions. Our aim has been to develop analytical and numerical methods which harness this vast and rapidly growing capability **[4]**. In §II we consider the integral equation

$$\varphi(x) = g(x) + c(x) \int_0^b k(|x - y|)\varphi(y)\, dy, \qquad 0 \le x \le b, \tag{2}$$

and show how to convert it into an initial-value problem [5]. Then, a numerical procedure based on our new mathematical formalism is given. Results of some numerical experiments are described.

In §III we turn to a consideration of inverse problems. Of great importance in the theory of adaptive control processes are system identification problems of the form

$$\dot{x} = f(x, a), \qquad 0 \le t \le t_R, \tag{3}$$

$$\min \sum_{i=1}^{R} [x_1(t_i) - b_i]^2, \tag{4}$$

where the minimization is over the system parameter vector a and the initial vector $x(0)$. The quantities $b_1, b_2, \ldots, b_R$ represent noisy observations on the first component of x at times $t_1, t_2, \ldots, t_R$. Much previous experience with the numerical solution of such problems [6], [7] led us to the belief that once system identification problems in transport theory are converted from the integral equation form into initial-value problems, then they too could be successfully attacked. In this section a general class of inverse problems in radiative transfer is converted into initial-value form, making use of the results of §II. Then the quasilinearization technique for solution is described, and the results of numerical experiments are given. Special attention is paid to the degradation in estimates of the medium caused by errors in observations of the radiation field.

The initial-value treatments of integral equations sketched in §§II and III generalize along many different lines. In those sections and in the discussion of §IV, references are given to treatments of eigenvalue problems, of equations with more general kernels, and of the calculation of resolvents. Further numerical experiments on inverse problems have shown that measurements on reflected radiation fields can be used to estimate local anisotropic scattering properties, local source strengths, and the position of interfaces in stratified media.

Much work remains to be done to determine the full range of applicability of these new methods to direct and inverse problems of multiple scattering.

During the course of these investigations we were fortunate in having conversations and correspondence with Professors R. Bellman, A. Schumitzky, S. Ueno, and G. M. Wing.

II. **Conversion of Fredholm integral equations into initial-value problems [5], [8].** One of the basic functional equations of radiative transfer [1] is the Fredholm integral equation

$$J(\tau, \mu_0) = \frac{c}{4} e^{-(\tau_1-\tau)/\mu_0} + \frac{c}{2}\int_0^{\tau_1} E_1(|\tau - \tau'|)J(\tau', \mu_0)\, d\tau', \tag{5}$$

where E_1 is an exponential integral function

$$E_1(r) = \int_0^1 e^{-r/z} \frac{dz}{z}, \qquad r > 0. \tag{6}$$

The source function $J(\tau, \mu_0)$ represents the rate of production of scattered radiation per unit volume per unit solid angle at the optical altitude τ when a slab of optical thickness τ_1 is illuminated by uniform parallel rays of energy of net flux π having a direction cosine with respect to an inward normal of μ_0. The scattering is isotropic, and c is the albedo for single scattering. The intensity of the diffusely reflected radiation having a direction cosine of μ with respect to an outward normal is $r(\mu, \mu_0)$ and

$$(7) \qquad r(\mu, \mu_0) = \int_0^{\tau_1} J(\tau, \mu_0) \exp\left[-(\tau_1 - \tau)/\mu\right] \frac{d\tau}{\mu}.$$

Earlier work in radiative transfer showed that J and r are solutions of initial-value problems in which the independent variable is the slab thickness τ_1.

Let us now proceed to the general theory.

Consider the integral equation

$$(8) \qquad \varphi(\tau, x) = g(\tau) + c(\tau) \int_0^x k(\tau, y)\varphi(y, x)\, dy, \qquad 0 \le \tau \le x,$$

where the unknown function φ is explicitly viewed as a function of τ and x, the interval length. The kernel $k(\tau, y)$ is assumed to be a function of the absolute value of the difference of its arguments

$$(9) \qquad k(\tau, y) = k(|\tau - y|),$$

where $k(r)$ is representable in exponential form

$$(10) \qquad k(r) = \int_a^b e^{-rz}\, dW(z), \qquad r > 0,$$

and

$$(11) \qquad k(r) = k(-r), \qquad r > 0.$$

Differentiation of equation (8) with respect to x yields

$$(12) \qquad \varphi_x(\tau, x) = c(\tau)k(\tau - x)\varphi(x, x) + c(\tau)\int_0^x k(\tau - y)\varphi_x(y, x)\, dy.$$

This is viewed as an integral equation for the function ϕ_x having the solution

$$(13) \qquad \varphi_x(\tau, x) = \Phi(\tau, x)\varphi(x, x),$$

where the function Φ is the solution of the integral equation

$$(14) \qquad \Phi(\tau, x) = c(\tau)k(x - \tau) + c(\tau)\int_0^x k(\tau - y)\Phi(y, x)\, dy.$$

The two factors on the right-hand side of equation (13) will now be considered in more detail.

In view of equation (10) the above equation for the function Φ may be written in the form

$$(15) \qquad \Phi(\tau, x) = c(\tau)\int_a^b e^{-(x-\tau)z}\, dW(z) + c(\tau)\int_0^x k(\tau - y)\Phi(y, x)\, dy.$$

This suggests introducing the new function $J = J(\tau, x, z)$ as the solution of the integral equation

$$J(\tau, x, z) = c(\tau)e^{-(x-\tau)z} + c(\tau)\int_0^x k(\tau - y)J(y, x, z)\,dy. \tag{16}$$

Making use of the superposition principle, it is seen that the function Φ may be expressed in the form

$$\Phi(\tau, x) = \int_a^b J(\tau, x, z)\,dW(z). \tag{17}$$

A differential equation for J is now obtained. Differentiate equation (16) with respect to x to obtain the relation

$$\begin{aligned} J_x(\tau, x, z) = -zc(\tau)e^{-(x-\tau)z} &+ c(\tau)k(\tau - x)J(x, x, z) \\ &+ c(\tau)\int_0^x k(\tau - y)J_x(y, x, z)\,dy. \end{aligned} \tag{18}$$

This may be regarded as an integral equation for the function J_x having as its solution

$$J_x(\tau, x, z) = -zJ(\tau, x, z) + \Phi(\tau, x)J(x, x, z). \tag{19}$$

The above equation is a differential equation for $J(\tau, x, z)$, but it involves $J(x, x, z)$, a quantity with which we must now deal.

Return to equation (16) to see that

$$J(x, x, z) = c(x) + c(x)\int_0^x k(x - y)J(y, x, z)\,dy, \tag{20}$$

or

$$J(x, x, z) = c(x)\left[1 + \int_0^x\int_a^b e^{-(x-y)v}\,dW(v)J(y, x, z)\,dy\right]. \tag{21}$$

Through an interchange in the order of integration this becomes

$$J(x, x, z) = c(x)\left[1 + \int_a^b R(v, z, x)\,dW(v)\right], \tag{22}$$

where the new function $R(v, z, x)$ is defined to be

$$R(v, z, x) = \int_0^x e^{-(x-y)v}J(y, x, z)\,dy. \tag{23}$$

It will now be shown that the "reflection" function R satisfies a nonlinear differential-integral equation.

Differentiate equation (23) with respect to x to obtain

$$R_x(v, z, x) = J(x, x, z) - v\int_0^x e^{-(x-y)v}J(y, x, z)\,dy + \int_0^x e^{-(x-y)v}J_x(y, x, z)\,dy. \tag{24}$$

Next it is observed that the first integral in the right-hand side of the above equation is $R(v, z, x)$, and the second may be evaluated by referring back to equation (19).

The result is

$$R_x(v, z, x) = J(x, x, z) - vR(v, z, x) - z\int_0^x e^{-(x-y)v} J(y, x, z)\, dy \tag{25}$$
$$+ J(x, x, z)\int_0^x \Phi(y, x) e^{-(x-y)v}\, dy.$$

Making use of equations (17) and (23), this equation becomes

$$R_x(v, z, x) + (v + z)R(v, z, x)$$
$$= J(x, x, z)\left[1 + \int_0^x \int_a^b J(y, x, v_1)\, dW(v_1) e^{-(x-y)v}\, dy\right] \tag{26}$$
$$= J(x, x, z)\left[1 + \int_a^b R(v, v_1, x)\, dW(v_1)\right].$$

The final result, in view of equation (22), is

$$R_x(v, z, x) = -(v + z)R(v, z, x) + c(x)\left[1 + \int_a^b R(v_1, z, x)\, dW(v_1)\right] \tag{27}$$
$$\times \left[1 + \int_a^b R(v, v_1, x)\, dW(v_1)\right].$$

Furthermore, equation (23) provides the initial condition

$$R(v, z, 0) = 0. \tag{28}$$

It is to be noted that the function R is determined by the initial-value problem contained in equations (27) and (28). The plan is to express $J(x, x, z)$ in terms of $R(v, z, x)$, using equation (22). Then $J(\tau, x, z)$ is computed from the differential equation (19), employing the representation in equation (17) for $\Phi(\tau, x)$ in terms of the function J. This completes the discussion of the first factor on the right-hand side of equation (13).

Next the second factor, $\varphi(x, x)$, is considered. From equation (8) it follows that

$$\varphi(x, x) = g(x) + c(x)\int_0^x k(x - y)\varphi(y, x)\, dy, \tag{29}$$

which may be rewritten as

$$\varphi(x, x) = g(x) + c(x)\int_0^x \int_a^b e^{-(x-y)z}\, dW(z)\varphi(y, x)\, dy. \tag{30}$$

This suggests introducing the new function

$$e(z, x) = \int_0^x e^{-(x-y)z}\varphi(y, x)\, dy, \tag{31}$$

so that

$$\varphi(x, x) = g(x) + c(x)\int_a^b e(z, x)\, dW(z). \tag{32}$$

The differential equation for the function $e(z, x)$ is

$$e_x(z, x) = -ze(z, x) + \int_0^x e^{-(x-y)z}\Phi(y, x)\varphi(x, x)\, dy + \varphi(x, x), \tag{33}$$

where equation (13) has been utilized. Using the representation for Φ in terms of J and the definition of the function R, this becomes

$$e_x(z, x) = -ze(z, x) + \varphi(x, x)\left[1 + \int_a^b R(z, v, x)\, dW(v)\right]. \tag{34}$$

The initial condition for the function $e(z, x)$ is

$$e(z, 0) = 0, \qquad a \le z \le b. \tag{35}$$

The function $e(z, x)$ is determined by the initial condition in the above equation and the differential equation

$$\begin{aligned} e_x(z, x) = -ze(z, x) + &\left[g(x) + c(x)\int_a^b e(z_1, x)\, dW(z_1)\right] \\ &\times \left[1 + \int_a^b R(z, v, x)\, dW(v)\right], \qquad x > 0. \end{aligned} \tag{36}$$

A recapitulation is now in order. On the interval $0 \le x \le \tau$ the functions $R(v, z, x)$ and $e(z, x)$ are determined by the differential equations and initial conditions

$$\begin{aligned} R_x(v, z, x) = -(v + z)R(v, z, x) + c(x)&\left[1 + \int_a^b R(v_1, z, x)\, dW(v_1)\right] \\ &\times \left[1 + \int_a^b R(v, z_1, x)\, dW(z_1)\right], \end{aligned} \tag{37}$$

$$R(v, z, 0) = 0, \qquad a \le v, z \le b, \tag{38}$$

$$\begin{aligned} e_x(z, x) = -ze(v, x) + &\left[g(x) + c(x)\int_a^b e(v, x)\, dW(v)\right] \\ &\times \left[1 + \int_a^b R(z, v, x)\, dW(v)\right], \end{aligned} \tag{39}$$

$$e(z, 0) = 0, \qquad a \le z \le b. \tag{40}$$

At $x = \tau$, J and φ have the initial conditions

$$J(\tau, \tau, z) = c(\tau)\left[1 + \int_a^b R(v, z, \tau)\, dW(v)\right], \tag{41}$$

and

$$\varphi(\tau, \tau) = g(\tau) + c(\tau)\int_a^b e(z, \tau)\, dW(z). \tag{42}$$

For $x > \tau$, the functions J and φ are determined by the differential equations

(43)
$$J_x(\tau, x, z) = -zJ(\tau, x, z) + c(x)\left[1 + \int_a^b R(v, z, x)\, dW(v)\right] \times \int_a^b J(\tau, x, v)\, dW(v),$$

and

$$\varphi_x(\tau, x) = \left[g(x) + c(x)\int_a^b e(v, x)\, dW(v)\right]\int_a^b J(\tau, x, v)\, dW(v). \tag{44}$$

In the computational procedure the integrals are approximated as sums according to a quadrature formula. The result is a system of ordinary differential equations for the dependent variables, subject, of course, to initial conditions.

These results will now be specialized to the case of multiple-scattering in a finite slab. The basic integral equation for radiative transfer in a slab containing emitting sources is

$$\varphi(\tau, x) = g(\tau) + \frac{c(\tau)}{2}\int_0^x E_1(|\tau - y|)\varphi(y, x)\, dy, \tag{45}$$

where x is the optical thickness of the medium, τ is the optical altitude, $c(\tau)$ is the albedo for single scattering, and $g(\tau)$ is the distribution of internal sources. As before, $E_1(r)$ is the first exponential integral function.

The auxiliary functions are the source function due to monodirectional illumination at $\tau = x$ with direction cosine μ_0,

$$J(\tau, x, \mu_0) = \frac{c(\tau)}{4} e^{-(x-\tau)/\mu_0} + \frac{c(\tau)}{2}\int_0^x E_1(|\tau - y|)J(y, x, \mu_0)\, dy, \tag{46}$$

the reflection function

$$R(\mu, \mu_0, x) = 4\int_0^x e^{-(x-y)/\mu} J(y, x, \mu_0)\, dy, \tag{47}$$

and the emergent intensity due to the internal sources

$$e(\mu, x) = \frac{1}{\mu}\int_0^x e^{-(x-y)/\mu}\varphi(y, x)\, dy. \tag{48}$$

The system of differential-integral equations is

$$\begin{aligned} R_x(\mu, \mu_0, x) = -\left(\frac{1}{\mu} + \frac{1}{\mu_0}\right)R + c(x)\left[1 + \frac{1}{2}\int_0^1 R(\mu', \mu_0, x)\frac{d\mu'}{\mu'}\right] \\ \times \left[1 + \frac{1}{2}\int_0^1 R(\mu, \mu'', x)\frac{d\mu''}{\mu''}\right], \end{aligned} \tag{49}$$

$$\begin{aligned} e_x(\mu, x) = -\frac{1}{z} e(\mu, x) + \frac{1}{\mu}\left[g(x) + \frac{c(x)}{2}\int_0^1 e(\mu', x)\, d\mu'\right] \\ \times \left[1 + \frac{1}{2}\int_0^1 R(\mu, \mu'', x)\frac{d\mu''}{\mu''}\right], \end{aligned} \tag{50}$$

(51)
$$J_x(\tau, x, \mu_0) = -\frac{1}{\mu_0} J(\tau, x, \mu_0) + \frac{c(x)}{2}\left[1 + \frac{1}{2}\int_0^1 R(\mu, \mu_0, x)\frac{d\mu}{\mu}\right] \times \int_0^1 J(\tau, x, \mu')\frac{d\mu'}{\mu'},$$

(52)
$$\varphi_x(\tau, x) = 2\left[g(x) + \frac{c(x)}{2}\int_0^1 e(\mu, x)\,d\mu\right]\int_0^1 J(\tau, x, \mu')\frac{d\mu'}{\mu'},$$

and the initial conditions are

(53)
$$R(\mu, \mu_0, 0) = 0,$$

(54)
$$e(\mu, 0) = 0,$$

(55)
$$J(\tau, x, \mu_0)\big|_{x=\tau} = \frac{c(\tau)}{4}\left[1 + \frac{1}{2}\int_0^1 R(\mu, \mu_0, \tau)\frac{d\mu}{\mu}\right],$$

(56)
$$\varphi(\tau, x)\big|_{x=\tau} = g(\tau) + \frac{c(\tau)}{2}\int_0^1 e(\mu, \tau)\,d\mu.$$

Let us now discuss in detail the numerical determination of the reflection function [3]. A medium of finite optical thickness x bounded by two horizontal parallel, infinite plates absorbs radiation and it scatters radiation, each scattering being assumed to be isotropic. The probability of reemission of absorbed energy, also called the albedo for single scattering, is denoted by c. The scattering properties of the medium are vertically stratified, so that c depends on the height τ above the bottom. The lower boundary is a completely absorbing barrier so that no radiation is reflected from it. (This, however, is not an essential requirement.) The medium is uniformly illuminated on its upper surface by monodirectional radiation of net flux π with area normal to the direction of propagation. The direction of the rays is denoted by the direction cosine μ_0 with respect to an inward drawn normal to the upper surface.

After multiple scatterings, radiation emerges from the top of the slab with an angular distribution of intensity described by the reflection coefficient $r(\mu, \mu_0, x)$:

(57) $r(\mu, \mu_0, x) =$ the intensity of radiation emerging from the top of a slab of thickness x due to parallel rays of radiation of net flux π incident at the top with direction cosine μ_0, the direction cosine of the emergent radiation being μ.

The reflection function $R(\mu, \mu_0, x)$ is introduced

(58)
$$r(\mu, \mu_0, x) = R(\mu, \mu_0, x)/4\mu.$$

It is symmetric in μ_0 and μ, i.e.

(59)
$$R(\mu, \mu_0, x) = R(\mu_0, \mu, x).$$

From the preceding theory, or directly on physical grounds, one obtains the differential-integral equation which $R(\mu, \mu_0, x)$ satisfies

$$(60)\qquad R_x(\mu, \mu_0, x) = -\left(\frac{1}{\mu} + \frac{1}{\mu_0}\right)R + c(x)\left[1 + \frac{1}{2}\int_0^1 R(\mu, \mu', x)\frac{d\mu'}{\mu'}\right] \times \left[1 + \frac{1}{2}\int_0^1 R(\mu'', \mu_0, x)\frac{d\mu''}{\mu''}\right].$$

This equation for inhomogeneous media was first given in **[9]**. Analysis of the slab of thickness zero leads to the initial condition

$$(61)\qquad R(\mu, \mu_0, 0) = 0.$$

This condition depends on the assumption of the scattering properties of the lower boundary. Note that in equation (60) the partial derivative is with respect to x, the thickness. The integrals are on the angular parameters.

It is convenient to use the method of discrete ordinates **[10]**. The integrals are approximated as sums. The reflection function, originally a function of thickness and two angles, is reduced to a matrix function of thickness only. The reflection matrix is approximately the reflection function evaluated at discrete values of the direction cosines of incident and emergent rays,

$$(62)\qquad R_{ij}(x) \simeq R(z_i, z_j, x).$$

The basic equations are the system of ordinary differential equations and the initial conditions

$$(63)\qquad R'_{ij}(x) = -\left(\frac{1}{z_i} + \frac{1}{z_j}\right)R_{ij}(x) + c(x)\left[1 + \frac{1}{2}\sum_{k=1}^{N} R_{ik}(x)\frac{w_k}{z_k}\right] \times \left[1 + \frac{1}{2}\sum_{k=1}^{N} R_{kj}(x)\frac{w_k}{z_k}\right], \qquad x > 0,$$

$$(64)\qquad R_{ij}(0) = 0, \qquad i = 1, 2, \ldots, N; \quad j = 1, 2, \ldots, N.$$

If a gaussian quadrature method of order N is employed, the direction cosines $\{z_i\}$ are the roots of the shifted Legendre polynomial of degree N, $P_N(1 - 2x)$, and the weights w_i are the Christoffel weights; see **[10]**.

This problem is readily solved by numerical integration on a modern computer. The integration is carried from $x = 0$ with the known initial conditions to $x =$ the desired thickness. A parameter study is simultaneously made of all thicknesses between these two limits. While $N \sim 3$ may be sufficient, both a higher accuracy and a more detailed knowledge of the angular distribution of intensity are obtained with larger values of N, at the cost of more computing. The choice of $N = 7$ is a satisfactory compromise in some instances. With $N = 7$, a step length of integration of 0.01, and an Adams-Moulton fourth-order integration formula, the computing time is several minutes for all slab thicknesses up to 100 optical units.

These analytical and computational procedures are readily extended to more complex cases such as time dependence **[10]**, energy dependence **[11]**, anisotropy **[12]**, and spherical geometry **[13]**.

Numerical results for the reflection function are given in **[3]**, **[14]**, and **[15]**.

Values of the source function J are provided in **[16]** and **[17]**. Some numerical results for the emission function e are given in **[18]**. Some results for energy-dependent transport are given in **[11]**. Values of Φ are available in **[19]** and **[20]**.

III. **Inverse problems.** In recent years much work has been done on the numerical solution of nonlinear two-point and multi-point boundary value problems for systems of ordinary differential equations [7], **[21]**, **[22]**. Coupled with the great advances in the speed and high-speed storage capabilities of current computing machines, this means it is possible to think in terms of nonlinear boundary value problems for systems in the order of several hundred or several thousand. Furthermore, this number is rapidly growing.

This implies, as will be seen, that many "system identification," "inverse," or "black-box" problems are well within the reach of current computational capability. Basically, it will be shown that it is fruitful to consider many system identification problems as the problem of fitting a differential equation to a set of observations **[6]**, **[23]**.

In transport theory, inverse problems take the form: determine the structure of a medium and the nature of the sources, given information about the radiation fields.

In radiative transfer theory alone there are many inverse problems. For example, on the basis of emitted intensities, estimate the distribution of internal sources **[18]**. Or, on the basis of reflection and transmission measurements of multiple scattering, deduce the phase function for anisotropic single scattering **[24]**. Still another problem: estimate the source function given reflection data **[25]**.

In the following section an inverse problem is formulated. This is viewed as a nonlinear boundary value problem, amenable to treatment by the quasilinearization method. Numerical results obtained are then presented. (A more detailed discussion is given in **[26]**.)

Consider an absorbing and scattering medium which is illuminated by parallel rays of the sun. The calculation of the reflected intensities has been described above.

Here, we treat the inverse problem in which the albedo as a function of height, $c(x)$, is an unknown function which is to be estimated based on measurements of multiply scattered reflected energy

$$\beta_{ij} \simeq r(z_i, z_j, c_0). \tag{65}$$

The thickness of the medium, c_0, is also unknown. We limit ourselves to the case in which $c(x)$ has a known form, but contains unknown parameters

$$c(x) = 0.5 + ax + bx^2, \qquad 0 \le x \le c_0, \tag{66}$$

where a, b, and c_0 are the unknown parameters.

The statement of the inverse problem is given by the equations

$$\begin{aligned} R_x(\mu, \mu_0, x) = &-\left(\frac{1}{\mu_0} + \frac{1}{\mu}\right) R(\mu, \mu_0, x) + c(x)\left[1 + \frac{1}{2}\int_0^1 R(\mu, \mu', x)\frac{d\mu'}{\mu'}\right] \\ &\times \left[1 + \frac{1}{2}\int_0^1 R(\mu'', \mu_0, x)\frac{d\mu''}{\mu''}\right], \qquad 0 \le x \le c_0, \end{aligned} \tag{67}$$

$$R(\mu, \mu_0, 0) = 0, \tag{68}$$

$$c(x) = 0.5 + ax + bx^2, \tag{69}$$

$$\min_{a,b,c_0} \sum_{i,j} [R(z_i, z_j, c_0) - 4z_i\beta_{ij}]^2. \tag{70}$$

The last expression requires the determination of a, b, and c_0 to make the sum of squares of deviations between theoretical and observed values a minimum. Of course, other criteria for the goodness of fit could be used.

To fix the end point of integration, the thickness of the medium is normalized to unity by a change of independent variable,

$$s = x/c_0, \qquad 0 \leq s \leq 1. \tag{71}$$

The equations become, using the same symbol R to denote the reflection function,

$$\begin{aligned} R_s(\mu, \mu_0, s) = c_0\Big\{ -\left(\frac{1}{\mu_0} + \frac{1}{\mu}\right) R(\mu, \mu_0, s) + c(s)\left[1 + \frac{1}{2}\int_0^1 R(\mu, \mu', s)\frac{d\mu'}{\mu'}\right] \\ \times \left[1 + \frac{1}{2}\int_0^1 R(\mu'', \mu_0, s)\frac{d\mu''}{\mu''}\right]\Big\}, \qquad 0 \leq s \leq 1, \end{aligned} \tag{72}$$

$$R(\mu, \mu_0, 0) = 0, \tag{73}$$

$$c(s) = 0.5 + ac_0 s + bc_0^2 s^2, \tag{74}$$

$$\min_{a,b,c_0} \sum_{i,j} [R(z_i, z_j, 1) - 4z_i\beta_{ij}]^2. \tag{75}$$

In finite-ordinate form, the basic equations become

$$\begin{aligned} \dot{R}_{ij}(s) = c_0\Big\{ -\left(\frac{1}{z_i} + \frac{1}{z_j}\right) R_{ij}(s) + c(s)\left[1 + \frac{1}{2}\sum_{k=1}^N R_{ik}(s)\frac{w_k}{z_k}\right] \\ \times \left[1 + \frac{1}{2}\sum_{k=1}^N R_{kj}(s)\frac{w_k}{z_k}\right]\Big\}, \qquad 0 \leq s \leq 1, \end{aligned} \tag{76}$$

$$R_{ij}(0) = 0, \tag{77}$$

$$c(s) = 0.5 + ac_0 s + bc_0^2 s^2, \tag{78}$$

$$\min_{a,b,c_0} \sum_{i,j} [R_{ij}(1) - 4z_i\beta_{ij}]^2. \tag{79}$$

Here the dot indicates differentiation with respect to s. The quantity under observation, $R_{ij}(1)$, satisfies a system of ordinary differential equations containing three unknown constants. Conditions are given on R_{ij} at $s = 0$, and additional conditions are given at $s = 1$ for the determination of a, b, and c_0. The problem is one of fitting a system of differential equations to data, a conceptual extension of polynomial fitting or regression analysis. Problems of this type are referred to as system identification problems, and various methods of approach are available. We now describe one [7], **[26]**.

Consider an $M + N$ dimensional vector x that is a solution of the differential equation

$$\dot{x} = f(x), \qquad 0 < t. \tag{80}$$

The first M components of $x(0)$ are unknown:

$$x_i(0) = c_i = \text{unknown}, \qquad i = 1, 2, \ldots, M, \tag{81}$$

and the last N components of $x(0)$ are known,

$$x_i(0) = c_i = \text{known}, \qquad i = M + 1, M + 2, \ldots, M + N. \tag{82}$$

At various times t_i, $i = 1, 2, \ldots, R$, where

$$t_i \leq t_{i+1}, \qquad i = 1, 2, \ldots, R - 1, \tag{83}$$

certain linear combinations of the components of $x(t_i)$ are "observed" so that

$$(x(t_i), \alpha_i) \simeq b_i, \qquad i = 1, 2, \ldots, R. \tag{84}$$

Our aim is to select the unknown initial components of $x(0)$, $x_1(0)$, $x_2(0), \ldots,$ $x_M(0)$, so as to minimize S, where

$$S = \sum_{i=1}^{R} [(x(t_i), \alpha_i) - b_i]^2. \tag{85}$$

An effective numerical method for carrying out this minimization is this. First an initial estimate of the minimizing initial vector $x(0)$ is obtained, using all available a priori physical information. Then equation (80) is integrated numerically from $t = 0$ to $t = t_R$ using these initial conditions. Call the solution so obtained $x = z(t)$, $0 \leq t \leq t_R$. Introduce the linearized vector equation

$$\dot{w} = f(z) + J(z)(w - z), \qquad 0 < t \leq t_R, \tag{86}$$

where the components of the Jacobian matrix J are given by the relationship

$$J_{ij} = \partial f_i(z)/\partial x_j, \qquad i, j = 1, 2, \ldots, M + N. \tag{87}$$

It can now be assumed that the matrix J is known computationally on the interval $0 \leq t \leq t_R$. Use is made of the method of complementary functions to represent the general solution of equation (86) in the form

$$w(t) = p(t) + \sum_{j=1}^{M} c_j h_j(t), \qquad 0 \leq t \leq t_R. \tag{88}$$

The function $p(t)$ is a particular solution of equation (86) for which the initial value of the ith component is

$$p_i(0) = \begin{cases} 0, & i = 1, 2, \ldots, M \\ c_i, & i = M + 1, M + 2, \ldots, M + N. \end{cases} \tag{89}$$

The vector functions $h_j(t)$ are solutions of the homogeneous form of equation (86). The ith component of $h_i(0)$, $i = 1, 2, \ldots, M$, is unity; the remaining components are zero. It is assumed that the $M + 1$ vector functions, $p(t)$ and $h_i(t)$, $i = 1, 2, \ldots, M$, are produced numerically as the solution of initial-value problems on the interval $0 \le t \le t_R$. The constants $c_1, c_2, \ldots, c_M$ are then chosen so as to minimize the sum S_1, where

$$S_1 = \sum_{i=1}^{R}\left[\left(p(t_i) + \sum_{j=1}^{M} c_j h_j(t_i), \alpha_i\right) - b_i\right]^2. \tag{90}$$

This leads to the set of linear algebraic equations

$$\partial S_1/\partial c_j = 0, \qquad j = 1, 2, \ldots, M, \tag{91}$$

for the determination of the unknown constants $c_1, c_2, \ldots, c_M$. It is assumed that these linear equations can be solved numerically. Their solution provides the new approximation to the initial conditions which minimize the sum S. The new approximation to the minimizing vector function $x(t)$, $0 \le t \le t_R$ is given by equation (88), since now the entire right-hand side is known numerically for $0 \le t \le t_R$. The calculation outlined is then repeated over and over until the percent change in passing from one approximation to the next is sufficiently small, or until a fixed number of iterations passes.

If the initial approximation is sufficiently good, quadratic convergence will result; otherwise no convergence takes place. For a partial discussion see [7].

A series of investigations of the inverse problem for the estimation of c and the thickness was carried out with the aid of a digital computer and the method of quasilinearization.

The system of linearized equations in the $(n + 1)$st approximation is

$$\begin{aligned}\dot{R}_{ij}^{n+1} = {} & c_0^n\left\{-\left(\frac{1}{z_i} + \frac{1}{z_j}\right)R_{ij}^n + c^n f_i(R^n) f_j(R^n)\right\} + c_0^n\left\{-\left(\frac{1}{z_i} + \frac{1}{z_j}\right)(R_{ij}^{n+1} - R_{ij}^n)\right. \\ & + \tfrac{1}{2}c^n\left[f_i \sum_k (R_{kj}^{n+1} - R_{kj}^n)\frac{w_k}{z_k} + f_j \sum_l (R_{il}^{n+1} - R_{il}^n)\frac{w_l}{z_l}\right]\Bigg\} \\ & + (a^{n+1} - a^n)c_0^n(c_0^n s) f_i(R^n) f_j(R^n) + (b^{n+1} - b^n)c_0^n(c_0^n s)^2 f_i(R^n) f_j(R^n) \\ & + (c_0^{n+1} - c_0^n)\left\{-\left(\frac{1}{z_i} + \frac{1}{z_j}\right)R_{ij}^n + c^n f_i(R^n) f_j(R^n)\right. \\ & \left. + [a^n c_0^n s + 2b^n (c_0^n s)^2] f_i(R^n) f_j(R^n)\right\}\end{aligned} \tag{92}$$

$$\dot{a}^{n+1} = 0, \tag{93}$$

$$\dot{b}^{n+1} = 0, \tag{94}$$

$$\dot{c}_0^{n+1} = 0, \tag{95}$$

where

$$c^n = \tfrac{1}{2} + a^n c_0^n s + b^n (c_0^n s)^2 \tag{96}$$

$$f_i(R^n) = 1 + \frac{1}{2}\sum_j R_{ij}^n \frac{w_j}{z_j}. \tag{97}$$

The dot over a letter indicates a derivative with respect to s. The initial conditions are

$$R_{ij}^{n+1}(0) = 0, \tag{98}$$

and the terminal conditions are

$$\partial Q/\partial a^{n+1} = 0, \tag{99}$$

$$\partial Q/\partial b^{n+1} = 0, \tag{100}$$

$$\partial Q/\partial c_0^{n+1} = 0, \tag{101}$$

where

$$Q = \sum_{i,j} [R_{ij}^{n+1}(1) - 4z_i\beta_{ij}]^2. \tag{102}$$

Some numerical results are next given [**26**]. First, it should be mentioned that the data were generated computationally by solving for the reflection matrix as described earlier, using as the true parameter values

$$a = 2.0, \qquad b = -2.0, \qquad c_0 = 1.0. \tag{103}$$

A gaussian quadrature formula with $N = 7$ is used. The differential equations are integrated, using a fourth-order Adams-Moulton integration method and a step length of 0.01.

In the first system identification trial 49 accurate values of the reflection matrix are assumed given. The initial estimates of the parameters are 10 to 50 percent in error,

$$a \cong 2.2, \qquad b \cong -1.8, \qquad c_0 \cong 1.5. \tag{104}$$

The quasilinearization method successfully refines the estimates to

$$a = 1.9990, \qquad b = -1.9982, \qquad c_0 = 1.004, \tag{105}$$

in the fourth approximation. The error is at most 0.4 percent, and the $c(t)$ profile is very accurately determined. The calculation consumes about 5 minutes of IBM 7044 time, no attempt having been made to streamline the calculations. The feasibility of the calculation was our sole concern.

After this successful trial, another trial is made with initial estimates (Trial 2)

$$a \cong 2.2, \qquad b \cong -1.8, \qquad c_0 \cong 0.5. \tag{106}$$

This time the solution diverges because this set of initial estimates is so poor.

In the next two trials, errors are deliberately introduced into the observations according to the formulas

$$n_i = (1 \pm p)t_i, \tag{107}$$

where t_i is the true observation, p is a decimal fraction, and n_i is the "noisy" observation. In Trial 3, $p = 0.01 = 1\%$, and in Trial 4, $p = 0.05 = 5\%$. The final estimated values are, for Trial 3

$$(108)\qquad a = 1.99, \qquad b = -1.98, \qquad c_0 = 1.003,$$

and for Trial 4

$$(109)\qquad a = 1.91, \qquad b = -1.85, \qquad c_0 = 1.02,$$

showing the corruption of the estimates by the noise in the observations. (In these trials, the initial estimates are the correct values of the parameters. Because of the errors in the data, these estimates deteriorate.)

In Trial 5, random numbers g_i drawn from a gaussian distribution with zero mean and standard deviation unity are added to corrupt the data

$$(110)\qquad m_i = (1 + pg_i)t_i.$$

The results of a trial in which $p = 0.01$ are

$$(111)\qquad a = 1.95, \qquad b = -1.93, \qquad c_0 = 1.01,$$

slightly poorer than the results of Trial 3.

That the angle of incidence of the illumination has an effect on the estimation is seen from the results (not tabulated here) of experiments in which only seven reflection measurements for one fixed input angle are employed. It is noted that the gaussian errors lead to consistently poorer results. The estimation results are poorer when based on reflection measurements with incident rays near grazing.

Finally, the effect that the criterion has on the estimates of parameters is tested. The criterion to be used is

$$(112)\qquad \min_{a,b,c_0}\left\{\max_{i,j}\left|\frac{R_{ij}(1) - 4z_i\beta_{ij}}{4z_i\beta_{ij}}\right|\right\},$$

i.e. minimize the maximum percent deviation. This is a linear programming problem for the determination of the parameters at each stage of the quasilinearization procedure **[27]**. The data are produced for input angle 60 degrees and with ± 2 percent errors (Trial 7) and with 2 percent gaussian errors (Trial 8). The results are

$$(113)\qquad a = 1.9995, \qquad b = -1.9996, \qquad c_0 = 1.00001,$$

and

$$(114)\qquad a = 1.76, \qquad b = -1.67, \qquad c_0 = 1.04,$$

respectively. The maximum deviations are 0.02 for Trial 6, precisely as expected, and 0.029 for Trial 7.

From numerical studies such as these, it is seen that the capability to convert noisy data into satisfactory estimates of unknown structures is potentially available. This opens the way to the planning of experiments to obtain measurements in sufficient quantity, and with sufficient accuracy, to determine the structure of various media **[28]**, **[29]**.

IV. **Discussion.** We now indicate some of the lines along which the previous discussions may be extended.

Initial-value methods for determining resolvents and Green's functions are given in **[30]**, **[31]**, **[32]**. Kernels more general than those considered in this paper are dealt with in **[32]**. Differential equations for the eigenvalues as functions of the interval length are given in **[33]**. The smallest eigenvalue as a function of the interval length can be studied by integrating the equation for R until R becomes sufficiently large **[34]**, **[35]**. Bellman **[36]** and Krein **[37]** have suggested other approaches to the study of Fredholm integral equations in which the interval length is varied. It would be of interest to study the interconnections. Still another initial-value method has been proposed by Bellman **[38]**. In **[19]** two new functions b and h were introduced. When $k = k(|\tau - y|)$, it is known that b and h satisfy initial-value problems and that Φ, J, and R may all be expressed *algebraically* in terms of these basic functions. It remains to be seen if the resolvent itself may be so expressed. It can be established directly that $\varphi(y, x)$ computed as indicated does indeed satisfy the integral equation.

The inverse problem of determining the local scattering diagram based on measurements of multiply scattered reflected radiation is discussed in **[24]**. In **[18]** and **[39]** the position of internal sources is estimated on the basis of observations of multiply scattered radiation escaping from the medium. The positions of interfaces in a vertically stratified medium are determined on the basis of reflection measurements **[15]**. A start on the study of inverse problems in wave propagation has been made **[26]**, **[40]**.

In keeping with our rapidly expanding ability to solve ordinary differential equations numerically, we may expect initial-value methods to grow in importance in the theory and applications of integral equations.

References

1. V. V. Sobolev, *A treatise on radiative transfer*, Van Nostrand, New York, 1963.

2. I. Busbridge, *The mathematics of radiative transfer*, Cambridge Univ. Press, Cambridge, 1960.

3. R. Bellman, R. Kalaba and M. Prestrud, *Invariant imbedding and radiative transfer in slabs of finite thickness*, American Elsevier, New York, 1963.

4. L. Collatz, *The numerical treatment of differential equations*, Springer-Verlag, Berlin, 1960.

5. H. Kagiwada and R. Kalaba, *An initial value method for Fredholm integral equations of convolution type*, The RAND Corporation, RM-5186-PR, November 1966.

6. R. Bellman, H. Kagiwada and R. Kalaba, *Quasilinearization, system identification and prediction*, Internat. J. Engrg. Sci. **3** (1965), 327–334.

7. R. Bellman and R. Kalaba, *Quasilinearization and nonlinear boundary value problems*, American Elsevier, New York, 1965.

8. H. Kagiwada, R. Kalaba and A. Schumitzky, *An initial-value method for Fredholm integral equations with displacement kernels: Reflection functions*, The RAND Corporation, RM-5306-PR, April 1967.

9. R. Bellman and R. Kalaba, *On the principle of invariant imbedding and propagation through inhomogeneous media*, Proc. Nat. Acad. Sci. U.S.A. **42** (1956), 629–632.

10. R. Bellman, H. Kagiwada, R. Kalaba and M. Prestrud, *Invariant imbedding and time-dependent transport processes*, American Elsevier, New York, 1964.

11. D. Mathews, K. Hansen and E. Mason, *Deep penetration of radiation by the method of invariant imbedding*, Nuclear Sci. and Engrg. **27** (1967), 263–270.

12. R. Bellman, R. Kalaba and S. Ueno, *Invariant imbedding and diffuse reflection from a two-dimensional flat layer*, Icarus (4) **1** (1963), 297–303.

13. R. Bellman, H. Kagiwada and R. Kalaba, *Invariant imbedding and radiative transfer in spherical shells*, J. Comp. Phys. (2) **1** (1966), 245–256.

14. R. Kalaba, R. Bellman, H. Kagiwada and S. Ueno, *Computational results for diffuse transmission and reflection for homogeneous finite slabs with isotropic scattering*, The RAND Corporation, RM-4775-NASA, October 1965.

15. ———, *Inverse problems in radiative transfer: Layered media*, Icarus (2) **4** (1965), 119–126.

16. ———, *Invariant imbedding and the computation of internal fields for transport processes*, J. Math. Anal. Appl. (3) **12** (1965), 541–548.

17. R. Bellman, H. Kagiwada and R. Kalaba, *Numerical results for the auxiliary equation of radiative transfer*, J. Quant. Spectrosc. Radiat. Transfer **6** (1966), 291–310.

18. R. Bellman, H. Kagiwada, R. Kalaba and S. Ueno, *Numerical results for the estimation of source distributions from external field measurements*, J. Comp. Phys. (4) **1** (1967).

19. H. Kagiwada and R. Kalaba, *A new initial-value method for internal intensities in radiative transfer*, Astrophys. J. (1) **147** (1967), 301–309.

20. ———, *Numerical results for internal intensities in atmospheres illuminated by isotropic sources*, The RAND Corporation, RM-4958-PR, April 1966.

21. R. Bellman, H. Kagiwada and R. Kalaba, *Invariant imbedding and the numerical integration of boundary-value problems for unstable linear systems of ordinary differential equations*, Comm. ACM. (2) **10** (1967), 100–102.

22. ———, *Numerical studies of a two-point nonlinear boundary value problem using dynamic programming, invariant imbedding, and quasilinearization*, The RAND Corporation, RM-4069-PR, March 1964.

23. R. Bellman et al., *Estimation of heart parameters using skin potential measurements*, Comm. ACM. (11) **7** (1964), 666–668.

24. H. Kagiwada and R. Kalaba, *Estimation of local anisotropic scattering properties using measurements of multiply scattered radiation*, J. Quant. Spectrosc. Radiat. Transfer **7** (1967), 295–303.

25. R. Bellman, H. Kagiwada and R. Kalaba, *Numerical inversion of Laplace transforms and some inverse problems in radiative transfer*, J. Atmos. Sci. **23** (1966), 555–559.

26. H. Kagiwada, *Computational aspects of inverse problems in analytical mechanics, transport theory, and wave propagation*, Ph.D. thesis, University of Kyoto, 1965.

27. R. Bellman, H. Kagiwada and R. Kalaba, *Quasilinearization, boundary-value problems and linear programming*, IEEE Trans. on Automatic Control (2) AC-**10** (1965), 199.

28. Z. Sekera, *Determination of atmospheric parameters from measurement of polarization of upward radiation by satellite or space probe*, The RAND Corporation, RM-5158-PR, October 1966.

29. L. Kaplan, *Inference of atmospheric structure from remote radiation measurements*, J. Opt. Soc. Amer. **49** (1959), 1004–1007.

30. H. Kagiwada and R. Kalaba, *An initial-value method suitable for the computation of certain Fredholm resolvents*, J. Math. Phys. Sci. (1) **1** (1967).

31. ———, *A practical method for determining Green's functions using Hadamard's variational formula*, J. Optimiz. Theory and Appl. (1) **1** (1967).

32. H. Kagiwada, R. Kalaba and A. Schumitzky, *An initial value method for Fredholm integral equations*, J. Math. Anal. Appl. (1) **19** (1967).

33. ———, *Differential systems for eigenvalues of Fredholm integral equations*, The RAND Corporation, RM-5361-PR, June 1967.

34. R. Bellman, H. Kagiwada and R. Kalaba, *Quasilinearization and the calculation of eigenvalues*, Comm. ACM. (7) **9** (1966), 522–524.

35. G. M. Wing, *An introduction to transport theory*, Wiley, New York, 1962.

36. R. Bellman, *Functional equations in the theory of dynamic programming*—VII: *An integro-differential equation for the Fredholm resolvent*, Proc. Amer. Math. Soc. **8** (1957), 435–440.

37. M. G. Kreĭn, *A new method for solving linear integral equations of the first and second kind*, Dokl. Akad. Nauk SSSR. (3) **100** (1955), 413–416. (Russian)

38. R. Bellman, *A new approach to the numerical solution of a class of linear and nonlinear integral equations of Fredholm type*, Proc. Nat. Acad. Sci. (6) **54** (1965), 1501–1503.

39. R. Bellman, H. Kagiwada and R. Kalaba, *Estimation of internal source distributions using external field measurements in radiative transfer*, Icarus (3) **5** (1966), 274–278.

40. R. Bellman, D. Detchmendy, H. Kagiwada and R. Kalaba, *On the identification of systems and the unscrambling of data*—III: *One-dimensional wave and diffusion processes*, The RAND Corporation, RM-5173-PR, October 1966.

RAND Corporation

Solution of the discrete ordinate equations in one and two dimensions

E. M. Gelbard, James A. Davis, L. A. Hageman

1. **Discrete ordinate P_L and double-P_L approximations in slab geometry.**

1.1. In reactor design and development work one rarely encounters a transport problem which can be solved analytically. It doesn't follow that analytic techniques are unimportant: this is far from true. Today, as in the past, analytic methods have two essential functions. First it is, generally, through a study of analytic results that one gains some understanding of neutron transport processes. Secondly, analytic calculations serve as computational standards, benchmarks which one can use to test the accuracy of more practical approximate methods. Still, in all reactor laboratories we know of, computer programs based on approximate numerical methods carry almost all of the computational work load.

Many different numerical techniques have been devised to treat the transport equation and it is not feasible to discuss them all here. Instead we shall focus our attention on one single method, the discrete ordinate method. No other method will be dealt with *explicitly*. But it should be noted that, in slab and spherical geometries, the P_L and double-P_L equations can be cast into discrete ordinate form [**1**], [**2**]. Thus any algorithm which may be used to solve the discrete ordinate equations is applicable, also, to the P_L and double-P_L equations for slabs and spheres. There is a relation between the P_L and discrete ordinate equations in cylindrical geometry [**3**], but it is a relation which has not yet been fully explored.

In multidimensional geometries all links between spherical harmonics and discrete ordinate approximations seem to disappear.[1] We shall see, in addition,

[1] See discussion on page 150.

that the numerical problems which confront us in one-dimensional geometries, and those we encounter in multidimensional geometries, are somewhat different. Therefore it will be convenient to divide this paper sharply into halves. In the first half we discuss numerical techniques which have been used to treat one-dimensional problems and only later, in the second half, will we turn our attention to two-dimensional configurations. All the methods we discuss, when slightly modified, are applicable to the multigroup transport equation. However, to simplify our exposition, we shall set aside the multigroup problem and deal exclusively, with the monoenergetic neutron transport equation.

1.2. In slab geometry the neutron transport equation takes the form

$$\mu \frac{\partial \varphi(x, \mu)}{\partial x} + \Sigma_T(x)\varphi(x, \mu) = \int_{-1}^{1} \Sigma_S(x, \mu, \mu')\varphi(x, \mu')\, d\mu' + S(x, \mu). \tag{1.2.1}$$

Here

μ is the cosine of the angle between the neutron's velocity and the x axis:
$\varphi(x, \mu)$ is the directional flux:
$\Sigma_T(x)$ is the total cross section:
$\Sigma_S(x, \mu, \mu')$ is the differential scattering cross section from μ' to μ:
$S(x, \mu)$ is the neutron source density.

Equation (1.2.1) is a function of two continuous variables and, before it can be solved numerically, both variables shall have to be discretized. We deal later with the discretization of the spatial variable: only the angular variable concerns us here. Let us define a set of ordinates μ_i ($i = 1, 2, \ldots, N$), and weights w_i, such that

$$\int_{-1}^{1} \Sigma_S(x, \mu_i, \mu')\varphi(x, \mu')\, d\mu' \approx \sum_{j=1}^{N} \Sigma_S(x, \mu_i, \mu_j) w_j \varphi(x, \mu_j). \tag{1.2.2}$$

Given equation (1.2.2) we may write

$$\mu_i \frac{d\varphi(x, \mu_i)}{dx} + \Sigma_T(x)\varphi(x, \mu_i) \approx \sum_{j=1}^{N} \Sigma_{Sij}(x) w_j \varphi(x, \mu_j) + S_i(x).$$

Here $\Sigma_{sij}(x) \equiv \Sigma_S(x, \mu_i, \mu_j)$, $S_i(x) \equiv S(x, \mu_i)$. If we now introduce functions, $\phi_i(x)$, which identically satisfy equations (1.2.3),

$$\mu_i \frac{d\varphi_i(x)}{dx} + \Sigma_T(x)\varphi_i(x) = \sum_{j=1}^{N} \Sigma_{Sij}(x) w_j \varphi_j(x) + S_i(x), \tag{1.2.3}$$

we have reason to hope that $\varphi_i(x) \approx \varphi(x, \mu_i)$.

Equations (1.2.3) are one-dimensional discrete ordinate equations in their most general form. In deriving equations (1.2.3) we have approximated an integral via a quadrature formula but we have not insisted on any *particular* quadrature formula. To simplify the work that follows it will, however, be convenient to put some weak restrictions on both the weights and ordinates. We shall assume

1. that all the weights are positive,
2. that the ordinates are symmetrically placed about $\mu = 0$, and
3. that the weights are symmetric about $\mu = 0$.

Thus if μ_i is an ordinate there must be some other ordinate, say μ_j, such that $\mu_i = -\mu_j$. Furthermore $w_i = w_j$.

We've already pointed out that both the slab P_L and the slab double-P_L equations can be thrown into discrete ordinate form by suitable linear transformations. In the resulting discrete ordinate equations the weights and ordinates satisfy all the above conditions. One finds, for example, that the slab P_L weights and ordinates are, simply, Gauss quadrature weights and ordinates, and that these weights and ordinates have all the properties we require. Thus the slab P_L and double-P_L equations belong to the general class of equations which we propose to deal with in this paper.

If all cross sections are continuous functions of x the discrete ordinate equations, supplemented by appropriate boundary conditions, will completely determine the φ_i's. Many different boundary conditions occur in practice, and the construction of boundary conditions is often no simple matter. But it is not a matter which concerns us here. In this paper we will discuss only the form of the boundary conditions and not the procedures used to derive them.

In formulating boundary conditions it is usually convenient to make sharp distinctions between rightward- and leftward-moving neutrons. For this and other reasons we now introduce a new notation which displays, explicitly, the direction of the neutron's motion. If there is an ordinate at $\mu = 0$, this ordinate will be designated as μ_0: w_0 and φ_0 are, respectively, the weight and flux at μ_0. Suppose that there are $2n$ nonzero ordinates. Positive ordinates will be labeled $\mu_1^+, \mu_2^+, \ldots, \mu_n^+$, with ordinates numbered so that $\mu_1^+ > \mu_2^+ > \cdots \mu_n^+$. The weights and fluxes corresponding to these positive ordinates will be denoted by w_i^+ and ψ_i. Negative ordinates will be labeled $\mu_1^-, \mu_2^-, \ldots, \mu_n^-$ with corresponding weights w_i^- and fluxes χ_i. We assume that $\mu_i^- = -\mu_i^+$ and $w_i^- = w_i^+$. Suppose that the left-hand boundary of the problem configuration is at x_L. Then the left-hand boundary condition takes the form

$$\boldsymbol{\psi}(x_L) = \begin{pmatrix} \psi_1(x_L) \\ \psi_2(x_L) \\ \cdot \\ \cdot \\ \cdot \\ \psi_n(x_L) \end{pmatrix} = B_L \boldsymbol{\chi}(x_L).$$

Here B_L is an n by n matrix. Similarly at a right-hand boundary $\boldsymbol{\chi}(x_R) = B_R \boldsymbol{\psi}(x_R)$.

Probably the most commonly used boundary conditions are the reflecting and vacuum conditions. If there is a reflecting boundary (whether a right-hand or left-hand boundary) at $x = x_0$, then $\boldsymbol{\psi}(x_0) = \boldsymbol{\chi}(x_0)$: it follows that $B = I$. Unfortunately no such trivial argument leads us to satisfactory vacuum conditions.

In fact the formulation of vacuum conditions involves substantial theoretical difficulties, and vacuum conditions of various types have been proposed. The simplest of these (though often not the most accurate[2]) is the Mark condition [**4**], i.e. $B = 0$. Because of its simplicity we adopt Mark's condition in this paper, and set $B = 0$ at vacuum boundaries.

At points where any cross section is discontinuous, equations (1.2.3) have no clear meaning. We refer to such points as "interfaces" or "interface points." If we are to use equations (1.2.3) to treat problems with interfaces we must set down conditions ("interface conditions") which will govern the flux at interface points. Now, the true flux is a continuous function of x at all μ except $\mu = 0$. It is natural, then, to require that ψ_i and χ_i should be continuous across interfaces. On the other hand φ_0 is simply a linear combination of the other approximate fluxes: in fact, it follows from (1.2.3) that

$$(1.2.4)\qquad \begin{aligned}&[\Sigma_T(x) - w_0\Sigma_{00}(x)]\varphi_0(x)\\ &\quad = \sum_{j\neq 0}^{n} [\Sigma_S(x, \mu_0, \mu_j^+)w_j^+\psi_j(x) + \Sigma_S(x, \mu_0, \mu_j^-)w_j^-\chi_j(x)] + S_0(x).\end{aligned}$$

Therefore no interface (or boundary) conditions need be applied to φ_0. Since φ_0 is not truly an independent variable the ordinate at $\mu = 0$ contributes little to the accuracy of the discrete ordinate equations. For this reason approximations with odd numbers of ordinates have rarely been used, although it has recently been shown that they have some valuable properties [**5**]. In all the work which follows we assume, for simplicity, that N, the number of ordinates, is even.

2. **Iterative methods for slabs.**

2.1. In matrix notation equations (1.2.3) take the form[3]

$$(2.1.1)\qquad D(d\boldsymbol{\phi}/dx) + E\boldsymbol{\phi} = S.$$

Here $\boldsymbol{\phi}$ and S are vectors,

$$\boldsymbol{\phi} \equiv \begin{pmatrix} \varphi_1 \\ \varphi_2 \\ \cdot \\ \cdot \\ \cdot \\ \varphi_N \end{pmatrix}, \qquad S \equiv \begin{pmatrix} S_1 \\ S_2 \\ \cdot \\ \cdot \\ \cdot \\ S_N \end{pmatrix},$$

while D and E are matrices:

$$D \equiv \text{Diag}\,\{\mu_1, \mu_2, \ldots, \mu_N\},$$
$$E \equiv (E_{ij}),\ E_{ij} \equiv \Sigma_T\,\delta_{ij} - \Sigma_{Sij}, \quad \text{where } \delta_{ij} \text{ is the Kronecker delta.}$$

---

[2] Mark's condition is perfectly satisfactory in double-P_L approximations but not in P_L approximations.

[3] In the argument which follows, the neutron's direction of motion is not relevant. For the moment, therefore, we revert to a notation which does not distinguish leftward- from rightward-moving fluxes.

To determine the flux we append, to equation (2.1.1), boundary conditions on $\boldsymbol{\phi}$, conditions which we apply at $x = 0$ and $x = x_R$.

Now, it may appear that equation (2.1.1) will yield to the most simple and straightforward attack. In fact it may be solved formally without difficulty. To achieve a formal solution we introduce a matrix, M, such that

(2.1.2) $$D(dM/dx) + EM = 0, \qquad M(0) = I.$$

This matrix is called the "matrizer" or "matricant" [**6**] of the homogeneous equation (2.1.1). We define, also, a vector, $\boldsymbol{v}$, which satisfies the discrete ordinate equations, but which vanish at the left boundary:

(2.1.3) $$D(d\boldsymbol{v}/dx) + E\boldsymbol{v} = \boldsymbol{S}, \qquad \boldsymbol{v}(0) = 0.$$

It follows from the linearity of the discrete ordinate equations that any solution vector, $\boldsymbol{u}(x)$, can be written in the form $\boldsymbol{u}(x) = M(x)\boldsymbol{u}(0) + \boldsymbol{v}(x)$ and therefore

(2.1.4) $$\boldsymbol{\phi}(x) = M(x)\boldsymbol{\phi}(0) + \boldsymbol{v}(x).$$

Of course we are not, generally, given $\boldsymbol{\phi}(0)$. But the initial conditions (reflecting or vacuum conditions, for example) normally impose $N/2$ linear constraints on $\boldsymbol{\phi}(0)$. Similarly the outer boundary conditions impose $N/2$ linear constraints on $\boldsymbol{\phi}(x_R)$. Thus boundary conditions, jointly with equation (2.1.4), give us N equations which uniquely determine the flux at the origin. Knowing that flux we can construct the flux at any point.

It seems easy enough to make practical use of our formal solution. To do this we introduce a mesh grid and difference approximations transforming equations (2.1.2) and (2.1.3) into difference equations. We might write, for example,[4]

(2.1.5a) $$D\left(\frac{M_{n+1} - M_n}{x_{n+1} - x_n}\right) + E_n M_n = 0, \qquad M_0 = I,$$

(2.1.5b) $$D\left(\frac{\boldsymbol{v}_{n+1} - \boldsymbol{v}_n}{x_{n+1} - x_n}\right) + E_n \boldsymbol{v}_n = \boldsymbol{S}_n, \qquad \boldsymbol{v}_0 = 0.$$

Equations (2.1.5) could be used as step-ahead equations to determine, successively, all the M_n's and $\boldsymbol{v}_n$'s. Unfortunately recursion relations based on equation (2.1.5) are unstable against the accumulation of roundoff error, as we shall see below.

Let us suppose that, in solving equation (2.1.5b), we commit a roundoff error at point m and at no other point. For $n \geqq m$ our computation does not give us $\boldsymbol{v}_n$ but a vector $\tilde{\boldsymbol{v}}_n$ instead. Clearly $\tilde{\boldsymbol{v}}_n$ will not satisfy equation (2.1.5b). It will, however, satisfy a very similar equation:

$$D\left[\frac{\tilde{\boldsymbol{v}}_{n+1} - \tilde{\boldsymbol{v}}_n}{x_{n+1} - x_n}\right] + E_n\tilde{\boldsymbol{v}}_n = \boldsymbol{S}_n + \delta_{n+1,m}\left[\frac{D}{x_{n+1} - x_n}\right]\boldsymbol{\epsilon}, \qquad \tilde{\boldsymbol{v}}_0 = \boldsymbol{v}_0.$$

Here $\delta_{n,m}$ is the Kronecker delta, while $\boldsymbol{\epsilon}$ is a vector whose magnitude depends on the magnitude of the roundoff error. If $\boldsymbol{\epsilon}_n = \tilde{\boldsymbol{v}}_n - \boldsymbol{v}_n$, then $\boldsymbol{\epsilon}_n = 0$ for $n < m$,

[4] Equations (2.1.5) would not be very accurate. We cite them only for illustrative purposes.

$\boldsymbol{\epsilon}_m = \boldsymbol{\epsilon}$ and

$$D\left[\frac{\boldsymbol{\epsilon}_{n+1} - \boldsymbol{\epsilon}_n}{x_{n+1} - x_n}\right] + E_n\boldsymbol{\epsilon}_n = 0, \qquad n \geqq m + 1. \tag{2.1.6}$$

Now we define a vector function, $\boldsymbol{\epsilon}(x)$, such that

$$D(d\boldsymbol{\epsilon}(x)/dx) + E(x)\boldsymbol{\epsilon}(x) = 0 \tag{2.1.7}$$

for $x > x_m$, and $\boldsymbol{\epsilon}(x_m) = \boldsymbol{\epsilon}$. The reader will observe that equation (2.1.6) is a difference equation based on (2.1.7). Therefore we may expect that $\boldsymbol{\epsilon}_n$ will be equal, approximately, to $\boldsymbol{\epsilon}(x_n)$[5] and we can use equation (2.1.7) in place of (2.1.6) to study roundoff errors. Such a procedure is useful because the form of (2.1.7) is not dependent on our method of differencing.[6] In most approximations to equation (2.1.3) the propagation of roundoff will be governed by (2.1.7).

At this point it is convenient to multiply (2.1.7) by D^{-1} and to write

$$d\boldsymbol{\epsilon}(x)/dx = -K(x)\boldsymbol{\epsilon}(x). \tag{2.1.8}$$

Here $K(x) = D^{-1}E(x)$. Before we can proceed it will be necessary to know something about the properties of $K(x)$. To simplify our analysis we assume that the material properties of the diffusing medium are position independent. In this case the eigenvalues of K have special significance. For suppose that, in equation (2.1.1), we set $S = 0$ and search for exponential solutions, $\boldsymbol{f}(x)$, of the resulting homogeneous equations. Let $\boldsymbol{f}(x) = \boldsymbol{q}e^{-kx}$. Here $\boldsymbol{q}$ is a vector and k a scalar. We find that

$$[kI - D^{-1}E]\boldsymbol{q} = [kI - K]\boldsymbol{q} = 0.$$

Thus the eigenvalues, λ_ν, of K are the permissible values of the decay constant, k. As a result of some recent work **[8]**, **[9]** there is a good deal we can say about these eigenvalues. It is known that, when all the Σ_{Sij} are positive and the number of ordinates is even,[7] the eigenvalues of K are all real, and the λ_ν occur in positive and negative pairs. Further it is known that the eigenvectors, $\boldsymbol{q}_\nu$, of K form a complete set. Expanding $\boldsymbol{\epsilon}(x)$ in terms of these eigenvectors,

$$\boldsymbol{\epsilon}(x) = \sum_{\nu=1}^{N} c_\nu(x)\boldsymbol{q}_\nu,$$

we see from equation (2.1.8) that $dc_\nu/dx = -\lambda_\nu c_\nu$. Consequently

$$\boldsymbol{\epsilon}(x) = 0, \qquad x < x_m$$

$$\boldsymbol{\epsilon}(x) = \sum_{\nu=1}^{N} c_\nu(x_m)\boldsymbol{q}_\nu \exp(-\lambda_\nu(x - x_m)), \qquad x \geq x_m.$$

[5] This sort of argument is presented more rigorously in Henrici's book on difference equations [7].

[6] Of course this procedure has its own disadvantages. It tells us nothing about the effect of mesh width on the behavior of roundoff. Furthermore it is true that, if one elects (as many authors do) to study difference equations instead of differential equations, one has all the tools of matrix theory at one's disposal. We've chosen our approach because of its simplicity and generality.

[7] The reader will recall that we have already assumed N to be even. See page 133.

Since half the λ_ν's are negative it follows that $\boldsymbol{\epsilon}(x)$ increases without bound as x goes to infinity. We rule out as unlikely the possibility that the $c_\nu(x_m)$ vanish for all ν such that λ_ν is negative.

What can we conclude from this analysis? In any practical problem x lies in a finite range. It cannot become infinite and neither can the error, $\boldsymbol{\epsilon}(x)$. But if the problem configuration contains a region many mean free paths in width, the roundoff error deep in that region can become very large. Experience shows that roundoff amplification is a serious problem even if no single region is wide. Roundoff errors will become large if the problem configuration as a whole is wide in mean free paths.

We do not mean to imply that the method just described is useless. It can be used, and it has been used by Daitch **[10]**,[8] to solve problems of certain types. But use of this method cannot be recommended as a general practice. We shall have to look for alternatives, to seek some way to mitigate the effects of roundoff. It is roundoff which forces us into the complex strategies used to solve one-dimensional transport problems.

2.2. We consider, now, the oldest iterative method devised to control roundoff.[9] For the sake of simplicity we assume the scattering and source are isotropic. In this case the discrete ordinate equations (1.2.3) take the form

$$\nu(d\boldsymbol{\psi}(x)/dx) + \Sigma_T(x)\boldsymbol{\psi}(x) = [\tfrac{1}{2}\Sigma_S(x)\Phi(x) + \tfrac{1}{2}s(x)]\boldsymbol{e}$$

$$-\nu(d\boldsymbol{\chi}(x)/dx) + \Sigma_T(x)\boldsymbol{\chi}(x) = [\tfrac{1}{2}\Sigma_S(x)\Phi(x) + \tfrac{1}{2}s(x)]\boldsymbol{e},$$

where

$$\nu \equiv \text{Diag}\,\{\mu_1^+, \mu_2^+, \ldots, \mu_n^+\}, \qquad \Sigma_S(x) = 2\Sigma_{Sij},$$

$$\boldsymbol{e} \equiv \begin{pmatrix} 1 \\ 1 \\ \cdot \\ \cdot \\ \cdot \\ 1 \end{pmatrix} \qquad s(x) = 2S_i(x),$$

and

$$\Phi \text{ is the scalar flux, i.e. } \quad \Phi \equiv \sum_j [w_j^+\psi_j + w_j^-\chi_j].$$

Again we impose the boundary conditions

$$\boldsymbol{\psi}(0) = B_L\boldsymbol{\chi}(0), \qquad \boldsymbol{\chi}(x_R) = B_R\boldsymbol{\psi}(x_R).$$

To initiate iteration we guess the scalar flux, $\Phi(x) \approx \Phi^{[0]}(x)$, and the incoming flux at the left-hand boundary, $\boldsymbol{\psi}^{[1]}(0) = \boldsymbol{t}$. We then proceed to solve the equation

$$\nu\frac{d\boldsymbol{\psi}^{[1]}(x)}{dx} + \Sigma_T(x)\boldsymbol{\psi}^{[1]}(x) = [\tfrac{1}{2}\Sigma_S(x)\Phi^{[0]}(x) + \tfrac{1}{2}s(x)]\boldsymbol{e} \tag{2.2.1}$$

---

[8] Daitch introduces special techniques to minimize roundoff errors.

[9] Variants of this method have been used to solve the P_L, double-P_L, and S_n equations **[22]**.

with the boundary condition

$$\psi^{[1]}(0) = \boldsymbol{t}. \tag{2.2.2}$$

Equations (2.2.1) and (2.2.2) determine the rightward-moving flux at all points and, in particular, at the right-hand boundary. To compute the leftward-moving flux we utilize the right-hand boundary conditions

$$\chi^{[1]}(x_R) = B_R\psi^{[1]}(x_R)$$

and solve the equations

$$-v\frac{d\chi^{[1]}(x)}{dx} + \Sigma_T(x)\chi^{[1]}(x) = [\tfrac{1}{2}\Sigma_S(x)\Phi^{[0]}(x) + \tfrac{1}{2}s(x)]\boldsymbol{e}. \tag{2.2.3}$$

At this point we have completed one iteration and prepare for the next, computing

$$\Phi^{[1]}(x) = \sum_j [w_j^+\psi_j^{[1]} + w_j^-\chi_j^{[1]}], \qquad \psi^{[2]}(0) = B_L\chi^{[1]}(0),$$

etc. The iterative process may be continued until some convergence criterion is met.

We are assuming, of course, that the process does converge though we cannot prove, in all cases of practical interest, that it will. One can prove convergence, however, in certain simple situations. Suppose, for example, that $B_L = 0$, or $B_L = I$, and that $B_R = 0$, or $B_R = I$. Suppose, further, that the absorption cross section Σ_a, where $\Sigma_a = \Sigma_T - \Sigma_S$, is nonnegative and is positive somewhere. Let the scattering cross section be everywhere nonnegative. Then it can be shown that the principle of neutron conservation guarantees convergence.

In practice equation (2.2.1) will be solved by differencing and stepping recursively (or "marching") from left to right. Thus, in marching, we follow the neutron's motion. It is in this way that we ensure roundoff stability. For suppose that a roundoff error is committed at x_0 in solving equation (2.2.1). Let ϵ_i be the resulting error in $\psi_i^{[1]}$. One can show that, if the difference approximation is stable and the mesh is fine enough, then

$$\epsilon_i(x) = \epsilon_i(x_0)\exp[-(x - x_0)\Sigma_T/\mu_i^+], \qquad x > x_0, \tag{2.2.4}$$

while $\epsilon_i(x) = 0$ for $x < x_0$. We see that the cumulative roundoff error, $\epsilon_i(x)$, decays exponentially as we march to the right. Similarly we find that the effect of a roundoff error decays exponentially as we march to the *left* computing *leftward-moving* fluxes.

The iterative process just described is inefficient in one respect. In the course of the computation one does not update the scattering rate so as to make immediate use of latest values of the directional flux. A modified process in which latest values are used is discussed, by L. Hageman [**12**]. Hageman, in the same paper, also investigates the utility of overrelaxation as applied to this modified process.

3. **Stable noniterative methods for slabs.**

3.1. In §1.1 we described a noniterative computational method which we quickly abandoned. It was a somewhat unsatisfactory method, unstable against the accumulation of roundoff errors. To circumvent roundoff difficulties we turned our attention to iterative methods, but this change in approach was not absolutely necessary. There are many satisfactory noniterative techniques which may be used to solve the discrete ordinate equations. We now discuss one such method, developed by E. Schmidt and E. Gelbard [**13**].

We have noted earlier that, if the source and scattering cross sections are isotropic, the discrete ordinate equations take the form

$$v(d\boldsymbol{\psi}/dx) + \Sigma_T\boldsymbol{\psi} = [\tfrac{1}{2}\Sigma_S\Phi + \tfrac{1}{2}s]\boldsymbol{e}, \tag{3.1.1}$$

$$\begin{aligned} -v(d\boldsymbol{\chi}/dx) + \Sigma_T\boldsymbol{\chi} = [\tfrac{1}{2}\Sigma_S\Phi + \tfrac{1}{2}s]\boldsymbol{e}, \qquad \boldsymbol{\psi}(0) = B_L\boldsymbol{\chi}(0),\\ \boldsymbol{\chi}(x_R) = B_R\boldsymbol{\psi}(x_R). \end{aligned} \tag{3.1.2}$$

Define $\boldsymbol{g} = \boldsymbol{\psi} + \boldsymbol{\chi}$, $\boldsymbol{u} = \boldsymbol{\psi} - \boldsymbol{\chi}$. From equations (3.1.1) and (3.1.2) it is easy to show that

$$v\boldsymbol{u}' + \Sigma_T\boldsymbol{g} = \boldsymbol{e}\left[\Sigma_S\sum_{j=1}^{n} w_j^+ g_j + s\right], \tag{3.1.3}$$

$$v\boldsymbol{g}' + \Sigma_T\boldsymbol{u} = 0, \tag{3.1.4}$$

where the prime indicates differentiation with respect to x. For the sake of brevity we write

$$\boldsymbol{u}' + \Sigma_0\boldsymbol{g} = \boldsymbol{S}, \tag{3.1.5}$$

$$\begin{aligned} \boldsymbol{g}' + \Sigma_1\boldsymbol{u} = 0, \qquad \boldsymbol{u}(0) = -(1 + B_L)^{-1}(1 - B_L)\boldsymbol{g}(0),\\ \boldsymbol{u}(x_R) = (1 + B_R)^{-1}(1 - B_R)\boldsymbol{g}(x_R). \end{aligned} \tag{3.1.6}$$

Here Σ_0 and Σ_1 are matrices which we need not display explicitly. It is assumed, in equation (3.1.6), that $1 + B_L$ and $1 + B_R$ are nonsingular matrices and, in practice, we have always found that they are.

Now we make the substitution

$$\boldsymbol{u}(x) = M(x)\boldsymbol{g}(x) + \boldsymbol{N}(x), \qquad M(x) \equiv (m_{ij}(x)), \qquad \boldsymbol{N}(x) \equiv \begin{pmatrix} N_1(x)\\ N_2(x)\\ \cdot\\ \cdot\\ \cdot\\ N_n(x) \end{pmatrix}, \tag{3.1.7}$$

and seek to define M and $\boldsymbol{N}$ in such a way as to guarantee the validity of the discrete ordinate equations. We find that

$$\boldsymbol{g}' + \Sigma_1 M\boldsymbol{g} = -\Sigma_1\boldsymbol{N}, \tag{3.1.8}$$

and

$$\boldsymbol{u}' = (M' - M\Sigma_1 M)\boldsymbol{g} - M\Sigma_1\boldsymbol{N} + \boldsymbol{N}'. \tag{3.1.9}$$

Inserting (3.1.9) into (3.1.5) we get (3.1.10),

$$(M' - M\Sigma_1 M + \Sigma_0)\boldsymbol{g} - M\Sigma_1\boldsymbol{N} + \boldsymbol{N}' = \boldsymbol{S}. \tag{3.1.10}$$

Clearly equation (3.1.10) will be satisfied if

$$M' - M\Sigma_1 M = -\Sigma_0, \tag{3.1.11}$$

$$N' - M\Sigma_1 N = S. \tag{3.1.12}$$

If, in addition,

$$M(0) = -(1 + B_L)^{-1}(1 - B_L), \qquad N(0) = 0, \tag{3.1.13}$$

and

$$g(x_R) = [(1 + B_R)^{-1}(1 - B_R) - M(x_R)]^{-1}N(x_R), \tag{3.1.14}$$

then it can be shown that equations (3.1.5) and (3.1.6) will also be satisfied. Thus, to solve the discrete ordinate equations we

(A) solve equations (3.1.11), (3.1.12) and (3.1.13), stepping rightward from the origin:
(B) determine $g(x_R)$ from (3.1.14):
(C) compute $g(x)$ from (3.1.8), stepping leftward from the right-hand boundary and, finally
(D) compute $u(x)$ from (3.1.7).

The behavior of roundoff in the course of the computation just described is analyzed in **[14]**. There it is shown that the computation is stable against the accumulations of roundoff errors.[10]

Thus, to use Schmidt's method, we must solve equations (3.1.8), (3.1.11) and (3.1.12): so far we have not specified how this should be accomplished. Of course it is possible to substitute difference approximations into these equations[11] and to solve the resulting difference equations by a marching process. It is also possible, however, to solve (3.1.8), (3.1.11) and (3.1.12) analytically and, thus, as E. Schmidt has pointed out, to generate *exact* difference equations. Again, the exact difference equations can be solved by marching. Schmidt has done a great deal to make both methods practical and effective, and has applied both in computer programs.

4. **Discrete ordinate equations for spheres.**

4.1. Certainly much remains to be said about the discrete ordinate equations in slab geometry, but an exhaustive treatment of this subject alone would fill a large monograph. It is necessary that we divert our attention, now, to problems which arise in other geometries. There are two other one-dimensional geometries which are important in reactor physics, namely, spherical and cylindrical geometry. Unfortunately we cannot consider both, and we have chosen to deal with the simpler of the two. Through our discussion of spherical geometry we hope to bring to light some characteristic properties of discrete ordinate equations in curvilinear coordinate systems.

---

[10] In **[14]** the reader will also find a discussion of the close connection between the above method and invariant imbedding.

[11] Some differencing techniques will be described in §5.

We shall not derive the spherical transport equation: it is enough, for our purposes, to note that this equation has the form

$$\mu \frac{\partial\varphi(r,\mu)}{\partial r} + \frac{(1-\mu^2)}{r}\frac{\partial\varphi(r,\mu)}{\partial\mu} + \Sigma_T(r)\varphi(r,\mu) = \int_{-1}^{1} \Sigma_S(r,\mu,\mu')\varphi(r,\mu')\,d\mu' + S(r,\mu). \tag{4.1.1}$$

Here $r = |\boldsymbol{r}|$, where $\boldsymbol{r}$ is the radius vector from the center of the sphere to a field point. As in §1.2 we introduce a set of weights and ordinates such that

$$\int_{-1}^{1} \Sigma_S(r,\mu_i,\mu')\varphi(r,\mu')\,d\mu' \approx \sum_{j=1}^{N} \Sigma_{Sij}(r) w_j \varphi(r,\mu_j) \tag{4.1.2}$$

but, in spherical geometry, equation (4.1.2) does not completely characterize a discrete ordinate approximation. We need, in addition, to approximate the angular derivative, $\partial\varphi(r,\mu)/\partial\mu$, through some differencing technique. Let us assume that

$$\left.\frac{\partial\varphi(r,\mu)}{\partial\mu}\right|_{\mu_i} \approx \sum_{j=1}^{N} A_{ij}\varphi(r,\mu_j) \tag{4.1.3}$$

and introduce functions, $\varphi_i(r)$, such that

$$\mu_i \frac{d\varphi_i(r)}{dr} + \frac{(1-\mu_i^2)}{r}\sum_{j=1}^{N} A_{ij}\varphi_j(r) + \Sigma_T(r)\varphi_i(r) = \sum_{j=1}^{T}\Sigma_{Sij}(r) w_j\varphi_j(r) + S_i(r). \tag{4.1.4}$$

Again (as in §1.1) we have reason to hope that $\varphi_i(r) \approx \varphi(r,\mu_i)$. In matrix notation

$$d\boldsymbol{\phi}/dr + (1/r)B\boldsymbol{\phi} + E\boldsymbol{\phi} = \boldsymbol{S}. \tag{4.1.5}$$

It is possible, of course, to define the weights and ordinates, and the matrix elements A_{ij}, in many different ways. So far as we know, however, the only spherical discrete ordinate equations which have ever been used are the P_L and double-P_L equations **[15]**. Goertzel has shown **[16]** that the spherical P_L equations can be thrown into discrete ordinate form.[12] When this is done one finds, as in slab geometry, that the P_L weights and ordinates are, simply, Gauss-quadrature weights and ordinates. As for the matrix B, it can be shown that, in a P_L approximation

$$B_{ij} = \frac{(1-\mu_i^2)}{\mu_i}\sum_{l=0}^{L}\left(\frac{2l+1}{2}\right) P_l'(\mu_i) P_l(\mu_j) w_j,$$

where

$$P_l'(\mu_i) \equiv \left.\frac{d}{d\mu} P_l(\mu)\right|_{\mu=\mu_i}$$

We shall make use of this explicit form of B in §5.1.

[12] The S_n equations for spheres **[17]** differ, in form, from equation (4.1.4) and therefore, by one definition, they are not discrete ordinate equations.

5. **Spatial differencing techniques in spherical geometry.**

5.1. In discussing the slab discrete ordinate equations we have said very little about spatial differencing methods. Actually the derivation of spatial difference equations in slab geometry is a perfectly straightforward process, while, in spherical geometry, the singularity at the origin gives rise to some interesting problems. It is for this reason that we have, so far, passed over all questions concerning spatial difference approximations and, for the same reason, we choose to deal with such questions here.

Both the iterative and noniterative methods described in earlier sections can be adapted for use in spherical geometry. Invariably in iterative computations[13] the discrete ordinate equations are differenced centrally, so that the difference equations take the form

(5.1.1)

$$\frac{\boldsymbol{\phi}(r_{n+1}) - \boldsymbol{\phi}(r_n)}{(r_{n+1} - r_n)} + \frac{1}{(r_{n+1} + r_n)} B[\boldsymbol{\phi}(r_{n+1}) + \boldsymbol{\phi}(r_n)] + \tfrac{1}{2}E_n[\boldsymbol{\phi}(r_{n+1}) + \boldsymbol{\phi}(r_n)]$$
$$= \tfrac{1}{2}[\boldsymbol{S}(r_{n+1}) + \boldsymbol{S}(r_n)], \qquad n = 0, 1, \ldots, M.$$

Here the r_n are mesh points, while the E_n are cross section matrices appropriate to the n'th mesh interval. More sophisticated differencing techniques are difficult to incorporate into the iteration equations. Now, in high order discrete ordinate approximations the angular flux may vary very rapidly with r. Therefore it seems worthwhile to difference such equations very accurately if this can be done without great inconvenience. It happens that high-precision difference equations can easily be put into noniterative computations. Suppose, for example, that we expand the flux in a Taylor series about the n'th mesh point. We may truncate this series wherever we like, and use the discrete ordinate equations to evaluate all the coefficients we have chosen to retain. By such a procedure we are led to difference equations of the form

$$\boldsymbol{\phi}(r_{n+1}) = G_n\boldsymbol{\phi}(r_n) + \boldsymbol{H}_n. \tag{5.1.2}$$

Here $\boldsymbol{H}_n$ is a complicated vector function of the source vector, $\boldsymbol{S}(r)$, and of its derivatives, while G is an N by N matrix. On introducing functions $\boldsymbol{u}$ and $\boldsymbol{g}$, defined as in §3.1, we find that

$$\boldsymbol{u}(r_{n+1}) = \boldsymbol{u}_{n+1} = P_n^{(1)}\boldsymbol{u}_n + P_n^{(2)}\boldsymbol{g}_n + \boldsymbol{P}_n^{(3)}, \tag{5.1.3}$$

$$\boldsymbol{g}_{n+1} = Q_n^{(1)}\boldsymbol{u}_n + Q_n^{(2)}\boldsymbol{g}_n + \boldsymbol{Q}_n^{(3)}, \tag{5.1.4}$$

where all the P's and Q's are known functions of n. Again we postulate that

$$\boldsymbol{u}_n = M_n\boldsymbol{g}_n + \boldsymbol{N}_n, \tag{5.1.5}$$

[13] We refer, here, specifically to computations based on the method of §2.

and conclude that

$$u_{n+1} = (P_n^{(1)} M_n + P_n^{(2)}) g_n + P_n^{(1)} N_n + P_n^{(3)}, \tag{5.1.6}$$

$$g_{n+1} = (Q_n^{(1)} M_n + Q_n^{(2)}) g_n + Q_n^{(1)} N_n + Q_n^{(3)}. \tag{5.1.7}$$

But

$$u_{n+1} = M_{n+1} g_{n+1} + N_{n+1}. \tag{5.1.8}$$

Therefore

$$\begin{aligned} M_{n+1}(Q_n^{(1)} M_n + Q_n^{(2)}) g_n + M_{n+1}(Q_n^{(1)} N_n + Q_n^{(3)}) + N_{n+1} \\ = (P_n^{(1)} M_n + P_n^{(2)}) g_n + P_n^{(1)} N_n + P_n^{(3)}. \end{aligned} \tag{5.1.9}$$

If equation (5.1.9) is to be identically satisfied, then

$$M_{n+1} = (Q_n^{(1)} M_n + Q_n^{(2)})^{-1} (P_n^{(1)} M_n + P_n^{(2)}), \tag{5.1.10}$$

$$N_{n+1} = (P_n^{(1)} - M_{n+1} Q_n^{(1)}) N_n + P_n^{(3)} - M_{n+1} Q_n^{(3)}. \tag{5.1.11}$$

Equations (5.1.10) and (5.1.11) are recursion relations for M and N, and experience shows that they are stable against roundoff accumulation.

To compute the quantities M and N we must establish starting values for both quantities from one of the boundary conditions, then use (5.1.10) and (5.1.11) to step away from the boundary. It is precisely at this point that the $1/r$ singularity becomes troublesome. Because of the presence of this singularity it is generally impossible to calculate G_n and H_n at $r_n = 0$. Suppose, for example, that we expand $\phi(r)$ and $S(r)$ in Taylor series about $r = 0$,

$$\phi(r) = \sum_{n=0}^{\infty} \mathscr{A}_n r^n, \qquad S(r) = \sum_{n=0}^{\infty} S_n r^n \tag{5.1.12}$$

and substitute these series into equation (4.1.5). We find that[14]

$$(nI + B)\mathscr{A}_n + E\mathscr{A}_{n-1} = S_{n-1}, \qquad n = 1, 2, \ldots, \infty. \tag{5.1.13}$$

Here, of course, the $\mathscr{A}_n$ and S_n are N dimensional vectors. Now, it is easy to show that, in any P_L approximation beyond P_1, -2 is an eigenvalue of B. In fact

$$\begin{aligned} \sum_j B_{ij}(1 - \mu_j^2) &= \frac{1 - \mu_i^2}{\mu_i} \sum_{l=0}^{L} \left(\frac{2l+1}{2}\right) \sum_{j=1}^{L+1} P_l(\mu_i) P_l'(\mu_j) w_j (1 - \mu_j^2) \\ &= \frac{1 - \mu_i^2}{\mu_i} \sum_{l=0}^{L} \left(\frac{2l+1}{2}\right) P_l(\mu_i) \int_{-1}^{1} P_l'(\mu)(1 - \mu^2)\, d\mu, \quad L > 1, \\ &= -\frac{1 - \mu_i^2}{\mu_i} \sum_{l=0}^{L} \left(\frac{2l+1}{2}\right) P_l(\mu_i) \int_{-1}^{1} 2\mu P_l(\mu)\, d\mu \\ &= -2(1 - \mu_i^2). \end{aligned} \tag{5.1.14}$$

---

[14] We have assumed, here, that the cross sections are constant in some neighborhood of the origin.

Therefore the matrix $nI + B$ is singular when $n = 2$ and we cannot determine $\mathscr{A}_2$. This difficulty is not peculiar to the P_L approximations, but resides in the transport equation itself. Since the flux at the origin is isotropic it is completely characterized by one single number, and the value of this number alone does not determine the flux in the neighborhood of the origin.

It is necessary, then, that we avoid the use of equation (5.1.2) at $r_n = 0$. Thus we cannot define the P's or Q's at $r_n = 0$, and we have no way to march outward from the origin. We can, however, take r_0 to lie on the *outer* boundary of the problem configuration, determine M_0 and N_0 from the outer boundary condition. In practice one finds that this procedure fails in very high order P_L (and in high order double-P_L) approximations: it fails because one encounters severe roundoff problems in computing M_n and N_n for $r_n \approx 0$. A somewhat better approach is described below.

We shall try to construct solutions of the discrete ordinate equations (solutions valid for r small) by combining $N/2$ homogeneous solutions with one solution of the inhomogeneous equation. The resulting linear combination of solutions will contain $N/2$ arbitrary constants which can be adjusted, ultimately, to fit $N/2$ conditions at the outer boundary of the problem configuration. We assume here, as we have earlier, that there are precisely $N/2$ outer boundary conditions. To facilitate obtaining solutions to the inhomogeneous equation, we postulate that the source is isotropic and that the source and cross sections are position independent in some neighborhood of the origin. In other words for some $r_0 > 0$, $S_i(r) = S$, $\Sigma_T(r) = \Sigma_T$, and $\Sigma_{Sij}(r) = \Sigma_{Sij}$ when $r \leq r_0$. We assume, further, that $\sum_{j=1}^N A_{ij} = 0$ so that, in equation (4.1.4), $\sum_{j=1}^N A_{ij}\varphi_j = 0$ when the flux is isotropic.

Suppose that $\boldsymbol{y}_\nu(r)$ satisfies the homogeneous discrete ordinate equations, and that it can be expanded in powers of r:

$$\boldsymbol{y}_\nu = r^\nu \sum_{n=0}^{\infty} \boldsymbol{b}_{\nu n} r^n. \tag{5.1.15}$$

Substituting (5.1.15) into (4.1.5) and assuming that $\boldsymbol{S} = 0$, we see that

$$(\nu I + B)\boldsymbol{b}_{\nu 0} = 0. \tag{5.1.16}$$

Thus $-\nu$ must be an eigenvalue of B, and $\boldsymbol{b}_{\nu 0}$ must be a corresponding eigenvector. Further

$$[(n + \nu)I + B]\boldsymbol{b}_{\nu n} + E\boldsymbol{b}_{\nu(n-1)} = 0, \qquad n = 1, 2, \ldots, \infty. \tag{5.1.17}$$

If $(n + \nu)I + B$ is nonsingular for all $n \geq 1$, then clearly equations (5.1.16) and (5.1.17), together, determine at least one set of series coefficients. On the other hand suppose that for some integer m, greater than or equal to one, $(m + \nu)I + B = V$ is singular, but that $E\boldsymbol{b}_{\nu(m-1)}$ is orthogonal to the null space of V transpose.[15] Then, again, it is possible to solve equations (5.1.17) though $b_{\nu m}$ is not uniquely determined.

---

[15] This occurs in the P_L approximations.

Finally, if $Eb_{\nu(m-1)}$ is not orthogonal to the null space of V transpose, then solutions of the homogeneous equations can still be found, but only by methods more powerful than the one we have discussed. In this last case a peculiar situation arises: it may happen that m is very large. Strictly speaking, then, equations (5.1.17) cannot be solved, though one can compute $b_{\nu 0}, b_{\nu 1}, \ldots, b_{\nu(m-1)}$. We find, in such cases that the truncated series, $y_\nu = r^\nu \sum_{n=0}^{m-1} b_{\nu n} r^n$, can be treated as if it were a homogeneous solution, and that it can be used in constructing an accurate approximate solution of the discrete ordinate equation.

Granting the assumptions we have made, it is easy to construct a particular solution (valid for $r \leq r_0$) to the discrete ordinate equations. In fact, as can be seen from equation (4.1.5), the vector $Z = E^{-1}S$ is such a particular solution.

In all cases we have studied, the matrix B has exactly $N/2$ linearly independent eigenvectors belonging to nonpositive eigenvalues. We shall assume this to be true here[16] and let $-\nu_j$, $j = 1, \ldots, N/2$, be the nonpositive eigenvalues corresponding to the set of $N/2$ independent eigenvectors α_j, $j = 1, \ldots, N/2$, If y_{ν_j} is the homogeneous solution (5.1.15) obtained by using ν_j and α_j in (5.1.16), then the unique finite solution of the discrete ordinate equations can be cast in the form

(5.1.18) $$\phi(r) = Z + \sum_{j=1}^{N/2} t_j y_{\nu_j}(r), \quad r \leq r_0.$$

In notation which distinguishes forward and backward directions we may write

(5.1.19) $$\psi(r_0) = Z^+(r_0) + \sum_{j=1}^{N/2} t_j y^+_{\nu_j}(r_0).$$

(5.1.20) $$\chi(r_0) = Z^-(r_0) + \sum_{j=1}^{N/2} t_j y^-_{\nu_j}(r_0).$$

Combining equations (5.1.19) and (5.1.20) one can generally eliminate the t_j, thus obtaining a linear relation between $\psi(r_0)$ and $\chi(r_0)$. This relation can be treated as a boundary condition on ϕ at r_0, from which one determines starting values M_0 and N_0. Given M_0 and N_0, M_n and N_n can be computed at all spatial mesh points with the aid of equations (5.1.10) and (5.1.11).[17]

6. Discrete ordinate equations in *X*-*Y* geometry.

6.1. We shift our attention now to the solution of the neutron transport equation in two-dimensional geometries. From a practical point of view the one-dimensional and two-dimensional neutron transport problems are vastly different. In a typical one-dimensional problem we may be called upon to compute 10 fluxes

---

[16] It is true in both the P_L and the double-P_L approximations. It seems likely that a finite solution to the discrete ordinate equations does not exist or is not unique if B does not have exactly $N/2$ independent eigenvectors belonging to nonpositive eigenvalues.

[17] It is difficult to attribute the above methods to any single person. The use of series expansions about the origin was suggested to the authors by C. Maynard. Computational algorithms incorporating such expansions were developed, independently, by E. Schmidt and E. Gelbard at Bettis, and by Chi Chung Wang at the University of Wisconsin. Wang's work is discussed in a Doctoral Dissertation to be published by the University of Wisconsin.

at each of 100 mesh points, but in two dimensions it is often necessary to compute *64* fluxes at all nodes of a mesh grid containing thousands of points. Because, in multidimensional geometries the number of unknowns is very large, direct inversion methods do not seem feasible and only iterative methods of solution sppear to be practical. Usually, in iterative computations, the effect of any single roundoff error is suppressed, just as other errors are suppressed, by the iterative process itself. Thus, in more than one dimension, it is not roundoff propagation which makes the transport problem difficult. Instead we have to face more serious difficulties of many sorts which seem to stand arrayed against us.

With the tools we have at hand today, one-dimensional discrete ordinate equations may be solved quite easily, and by many different methods. Given the capabilities of modern computers, and of modern numerical techniques, one-dimensional problems no longer challenge our ingenuity.[18] On the other hand, in many respects, the methods which have been used to attack two-dimensional problems are still inadequate. Two-dimensional computations are still extremely expensive. Consider, for example, a two-dimensional discrete ordinate computation run on the CDC-6600. The problem configuration contained 2500 spatial mesh points and 64 ordinates per point. All neutrons were taken to be at the same energy. Using a fairly typical S_n method it took 15 minutes to solve this problem. While 15 minutes is not a prohibitively large computing time, the number of points in the mesh grid was relatively small. In reactor design work even a coarse mesh grid may often contain more than 10,000 points. Furthermore in a practical problem there are usually at least four energy groups. At present the reactor designer at many installations cannot really afford to run practical problems on existing S_n computer programs. Clearly it is worthwhile, in two dimensions, to try to make our discrete ordinate computations more efficient.

To improve two-dimensional computations it seems necessary that we thoroughly understand the methods used to solve two-dimensional problems. Unfortunately, however, the theory underlying these methods is still primitive. Here again one sees a difference, and a most important difference, between transport computations in one and two dimensions. In one dimension, for example, we find many reliable spatial differencing techniques available to us, classical techniques which have been carefully investigated. But difference approximations to the two-dimensional discrete ordinate equations are relatively unfamiliar, and we shall soon see that they sometimes exhibit unexpected peculiarities. Iterative methods are much more important in two dimensions than in one. Yet little theoretical work has yet been done on iterative methods applicable to two-dimensional discrete ordinate calculations, and still less on acceleration techniques.

We propose, in the remainder of this paper, not to suggest any radically new computational methods, but to analyze some old methods. For the sake of brevity we will focus our attention on a few methods which seem amenable to analysis, and

[18] Of course great ingenuity is still required in the construction of *analytic*, as opposed to numerical, solutions.

pass over a good deal of work which, while it may be important, is purely empirical.

6.2. In this paper we shall deal with a single two-dimensional geometry, namely, X-Y geometry. If we assume (as we will from now on) that the source and scattering are isotropic, the X-Y discrete ordinate equations take the following form:

$$(6.2.1)\quad \mathbf{\Omega}^i \cdot \nabla \varphi^i(\mathbf{r}) + \Sigma_T(\mathbf{r})\varphi^i(\mathbf{r}) = \frac{1}{4\pi}\Sigma_S(\mathbf{r})\Phi(\mathbf{r}) + \frac{1}{4\pi}s(\mathbf{r}), \quad i = 1, 2, \ldots, N.$$

Here our notation is essentially the same as in earlier sections. The reader should notice, however, that the indices which label the various ordinates are now superscripts rather than subscripts. As for interface and boundary conditions, these two are essentially the same as in one dimension.

Throughout the work which follows we shall always require that equations (6.2.1) be invariant under 90° rotations about the Z axis. It follows that N will be a multiple of 4,[19] i.e. $N = 4\tau$. Furthermore there will be τ ordinates, $\mathbf{\Omega}^k = (\Omega_x^k, \Omega_y^k)$, $k = 1, 2, \ldots, \tau$, such that $\Omega_x^k, \Omega_y^k > 0$. These τ ordinates will be called "principal ordinates," or ordinates in the "principal directions." With each principal ordinate, $\mathbf{\Omega}^k$, we associate three others, designated $\mathbf{\Omega}^{k_1}$, $\mathbf{\Omega}^{k_2}$, and $\mathbf{\Omega}^{k_3}$, such that

$$\mathbf{\Omega}^{k_1} = (-\Omega_x^k, \Omega_y^k), \quad \mathbf{\Omega}^{k_2} = (\Omega_x^k, -\Omega_y^k), \quad \text{and} \quad \mathbf{\Omega}^{k_3} \equiv (-\Omega_x^k, -\Omega_y^k).$$

If $\mathbf{\Omega}^k$ is an ordinate, then $\mathbf{\Omega}^{k_1}$, $\mathbf{\Omega}^{k_2}$, and $\mathbf{\Omega}^{k_3}$ must also be ordinates, and it must be true that $w^k = w^{k_1} = w^{k_2} = w^{k_3}$. So far as we know all X-Y discrete ordinate equations described in published literature have such 90° symmetry.

By now a good deal of information is available to guide us in formulation of *one-dimensional* discrete ordinate equations. It is known, for example, that certain Gaussian weights and ordinates are useful in problems of certain types. In contrast there is little one can say about weights and ordinates in two dimensions. We do not know how weights and ordinates should be selected, though many alternatives have been proposed. And it is difficult to predict, in a given situation, how many ordinates will be needed to attain some prescribed degree of accuracy.

There is, however, one awkward feature common to all X-Y discrete ordinate equations, regardless of one's choice of weights and ordinates. In pure absorbers, and in X-Y geometry, the solutions of discrete ordinate equations always exhibit ray effects. Consider, for example, the flux produced by a line source in an infinite absorbing medium, i.e. a medium in which the scattering cross section is zero. In such a case it is easy to show that the φ^i vanish, except along certain lines, or "rays." These rays emanate from the source and extend in the directions of the various ordinates.

To see what ray effects may do in practice we examine a more complicated problem configuration, depicted in Figure 1. Here all the boundaries are reflecting boundaries[20] and the source is confined to region I. Suppose, initially, that there

[19] We assume here that no ordinates are vertical, or horizontal. Further, if there is an ordinate at $\mathbf{\Omega} = (\Omega_x, \Omega_y)$, then there must also be an ordinate at $\mathbf{\Omega} = (\Omega_y, \Omega_x)$.

[20] Reflecting boundary conditions are defined explicitly by equations (7.1.1) and (7.1.2).

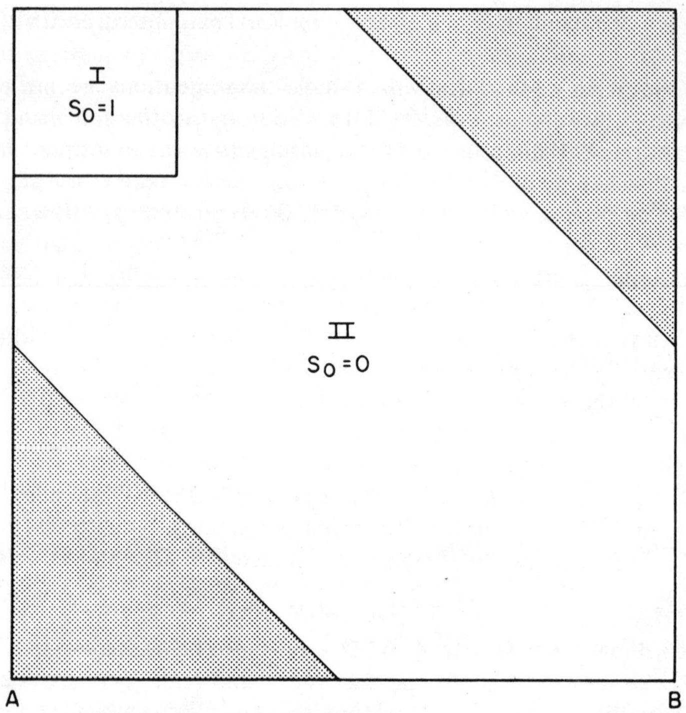

FIGURE 1. Simple cell problem.

is only one ordinate per octant and that Ω_x and Ω_y are equal in magnitude. If $\Sigma_S = 0$ it is easy to show that the φ^i and the scalar flux will vanish in the dotted regions. But it follows from the transport equation that the scalar flux is nowhere zero and is, in fact, greater at the point marked "A" than it is at point B. If $\Sigma_S \neq 0$ the ray effects become blurred, but vestiges of ray effects tend to remain. Thus, even when Σ_S is fairly large, we may still find an anomalous flux dip produced, at point A, by the discrete ordinate approximation. As we increase the number of ordinates the ray effects become weaker, but more complicated. As a result it is sometimes difficult to tell whether a peculiar dip, or peak, in the flux is spurious or real.

Since the spherical harmonics equations have full rotational symmetry it is clear that their solutions show *no* ray effects. It follows that, in X-Y geometry (or X-Y-Z geometry), no discrete ordinate equations are equivalent to the spherical harmonics equations. There is some confusion in the literature of transport theory regarding the connection between discrete-ordinate and spherical harmonics equations. To avoid compounding this confusion we shall try to make our language more precise. We insist, in this paper, that all allowable X-Y discrete ordinate equations be formally identical with equation (6.2.1). In (6.2.1) all cross sections and sources are to be taken directly from the given problem configuration. One is not permitted to manipulate the solution vector by tampering with these given parameters. If our definition of the X-Y discrete ordinate equations is adopted, then the discrete ordinate and spherical harmonics equations are not equivalent. It is impossible, in other words, to find a linear transformation, independent of position, which takes spherical harmonics fluxes into discrete ordinate fluxes. Ray effects, as we've described them, do not exist in other multidimensional geometries, though other anomalies may take their place. It is natural to ask whether there is a relation between discrete ordinate and spherical harmonics equations in other multidimensional geometries but, apparently, this question has not yet been answered.

7. Spatial differencing.

7.1. As in earlier sections, it will sometimes be helpful to make use of a notation which displays, explicitly, certain information about the neutron's direction of motion. Before proceeding further we define such a notation. At every point, (x, y), within the problem configuration, there are τ principal ordinates. Correspondingly there are, at (x, y), τ "principal fluxes," i.e. τ fluxes in the principal directions. Let $\boldsymbol{\psi}(x, y)$ be a vector whose components are these principal fluxes, i.e. let $\psi^k(x, y) = \varphi^k(x, y)$. Similarly, define vectors $\boldsymbol{\chi}(x, y)$, $\boldsymbol{\eta}(x, y)$, and $\boldsymbol{\xi}(x, y)$ such that $\chi^k(x, y) = \varphi^{k_1}(x, y)$, $\eta^k(x, y) = \varphi^{k_2}(x, y)$, $\xi^k(x, y) = \varphi^{k_3}(x, y)$. Thus, for example, $\psi^k(x, y)$ is the flux, at (x, y), in the direction (Ω_x^k, Ω_y^k): $\eta^k(x, y)$ lies in the direction $(\Omega_x^k, -\Omega_y^k)$, etc.

In such notation reflecting (or "symmetric") and Mark vacuum conditions take a very simple form. We shall assume, in the remainder of this paper, that the problem configuration is rectangular, so that $0 \leq x \leq X$, $0 \leq y \leq Y$. Further, we shall always impose reflecting conditions at the upper and right-hand boundaries. It follows that

(7.1.1) $$\boldsymbol{\chi}(X, y) = \boldsymbol{\psi}(X, y), \qquad \boldsymbol{\xi}(X, y) = \boldsymbol{\eta}(X, y),$$

(7.1.2) $$\boldsymbol{\chi}(x, Y) = \boldsymbol{\xi}(x, Y), \qquad \boldsymbol{\psi}(x, Y) = \boldsymbol{\eta}(x, Y).$$

At the lower boundary we permit either a symmetry condition,

(7.1.3) $$\boldsymbol{\psi}(x, 0) = \boldsymbol{\eta}(x, 0), \qquad \boldsymbol{\chi}(x, 0) = \boldsymbol{\xi}(x, 0),$$

or the Mark condition,

(7.1.4) $$\boldsymbol{\psi}(x, 0) = \boldsymbol{\chi}(x, 0) = 0.$$

Similarly, at the left-hand boundary, either

(7.1.5) $$\boldsymbol{\psi}(0, y) = \boldsymbol{\chi}(0, y), \qquad \boldsymbol{\eta}(0, y) = \boldsymbol{\xi}(0, y),$$

or

(7.1.6) $$\boldsymbol{\psi}(0, y) = \boldsymbol{\eta}(0, y) = 0.$$

7.2. Having formulated boundary conditions we return, now, to our discussion of the discrete ordinate equations themselves. In equations (6.2.1) we have already discretized the angular parameter, $\mathbf{\Omega}$, but the neutron's position is still specified by continuous spatial coordinates. If the discrete ordinate equations are to be solved numerically one has to discretize x and y as well. Now, it has been demonstrated, above, that the introduction of discrete $\mathbf{\Omega}$'s may have peculiar side effects which disappear in one dimension. In discretizing the spatial variables we shall, very soon, encounter a similar situation.

Consider, for example, the spatial difference equations which are generated by an ordinary central difference approximation. This is the spatial difference approximation most often used in one-dimensional discrete-ordinate calculations, and it seems as reasonable to work with central differences in two dimensions as in one.

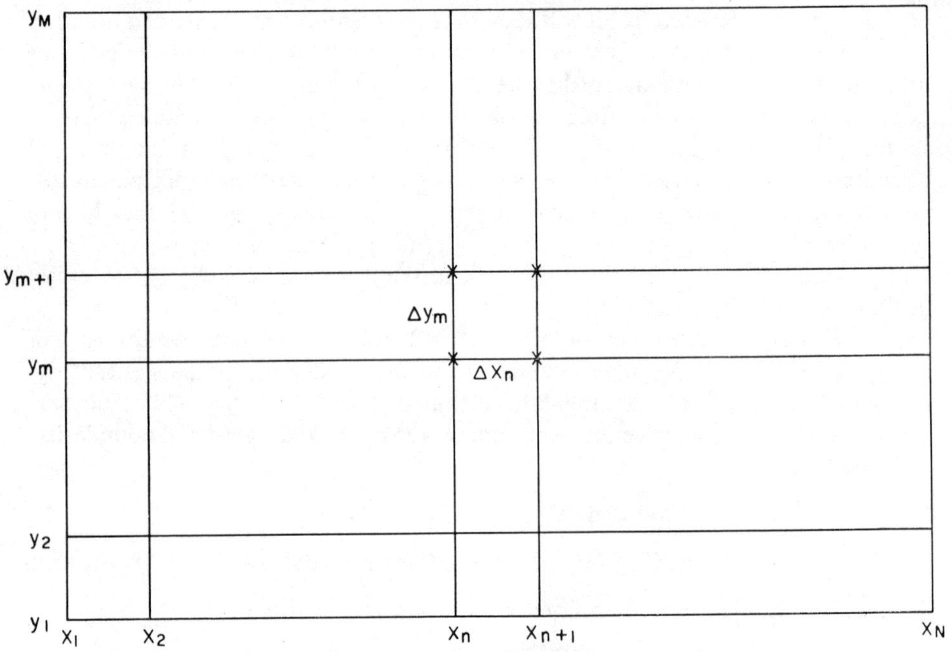

FIGURE 2. Mesh grid.

To derive central difference equations we first overlay, on the problem configuration, a rectangular mesh grid (as in Figure 2). We then assume that the fluxes, within each mesh box, are linear in x when y is constant and linear in y when x is constant. Thus in each mesh box each flux has the form

$$\varphi^i(x, y) = a^i + b^i x + c^i y + d^i xy. \tag{7.2.1}$$

Clearly, in every direction $\mathbf{\Omega}^i$, four fluxes are associated with any given mesh box. It is easy to solve for the four coefficients a^i, b^i, c^i, and d^i, in terms of these four fluxes. We see, then, that all the coefficients in equation (7.2.1) may be written as linear combinations of the fluxes at four mesh points.

If we insert equation (7.2.1) into (6.2.1), then integrate the resulting equations over mesh box (n, m),[21] we find that

$$\begin{aligned}(7.2.2)\quad & \tfrac{1}{2}\Omega_x^i\,\Delta y_m(\varphi^i_{n+1,m+1} + \varphi^i_{n+1,m} - \varphi^i_{n,m+1} - \varphi^i_{n,m}) \\ & + \tfrac{1}{2}\Omega_y^i\,\Delta x_n(\varphi^i_{n+1,m+1} + \varphi^i_{n,m+1} - \varphi^i_{n+1,m} - \varphi^i_{n,m}) + \Sigma_{T n,m}\,\Delta x_n\,\Delta y_m\bar{\varphi}^i_{n,m} \\ & = \frac{1}{4\pi}\Delta x_n\,\Delta y_m\,\Sigma_{S n,m}\overline{\Phi}_{n,m} + \frac{1}{4\pi}\Delta x_n\,\Delta y_m\bar{s}_{n,m}.\end{aligned}$$

Here the subscripts designate mesh points so that, for example, $\varphi^i_{n,m}$ is a flux at the point (n, m). Further

$$\begin{aligned}\bar{\varphi}_{n,m} &\equiv \tfrac{1}{4}[\varphi^i_{n+1,m+1} + \varphi^i_{n+1,m} + \varphi^i_{n,m+1} + \varphi^i_{n,m}],\\ \overline{\Phi}_{n,m} &\equiv \tfrac{1}{4}[\Phi_{n+1,m+1} + \Phi_{n+1,m} + \Phi_{n,m+1} + \Phi_{n,m}],\\ \bar{s}_{n,m} &\equiv \tfrac{1}{4}[s_{n+1,m+1} + s_{n+1,m} + s_{n,m+1} + s_{n,m}].\end{aligned}$$

In matrix notation equations (7.2.2) take the form

$$(7.2.3)\qquad A\boldsymbol{\varphi} = \boldsymbol{S},$$

where the components of $\boldsymbol{\varphi}$ are flux values at the various mesh points. In (7.2.3) we have used the boundary conditions (7.1.1)–(7.1.6) to eliminate the inward directed fluxes along the boundaries. Thus, boundary conditions are implicit in A.

So far as we know John Bennett was the first to use the central difference scheme in X-Y geometry. Apparently he was also the first to discover that, if all four boundaries are reflecting boundaries, the matrix A is singular. Bennett pointed out that the vectors $\boldsymbol{v}(k)$, $k = 1, 2, \ldots, \tau$, with components

$$(7.2.4)\qquad \begin{aligned} v^i_{n,m}(k) &= (-1)^{n+m}, \quad i = k, k_1, k_2, k_3,\\ &= 0 \quad \text{otherwise,}\end{aligned}$$

all lie in the null space of A, i.e. $A\boldsymbol{v}(k) = \boldsymbol{0}$. In fact, it can be shown **[18]** that, in general, the equation $A\boldsymbol{\varphi} = \boldsymbol{S}$ has no solution.

That such a respectable differencing technique should fail seems somewhat surprising. Given that the central difference equations are singular, one might suspect that other reasonable difference equations also have hidden peculiarities. In fact the central difference equations and the diamond equations, discussed below, are closely related, and it is natural to suppose that both have similar properties.

In differencing the two-dimensional discrete-ordinate equations Carlson's diamond scheme **[17]** is probably more often used than any other. Like the central differencing equations, the diamond equations are obtained by integrating the discrete ordinate equations over each mesh box. Integration over mesh box (n, m)

[21] The point (n, m) is the mesh point $x = x_n$, $y = y_m$. In writing (7.2.2) we assume that cross sections are constant over mesh boxes. The box numbered n, m has the point (n, m) at its lower left-hand corner.

gives

(7.2.5)
$$\Omega_x^i\,\Delta y_m[V^i_{n+1,m} - V^i_{n,m}] + \Omega_y^i\,\Delta x_n[H^i_{n,m+1} - H^i_{n,m}] + \Delta x_n\,\Delta y_m\left[\Sigma_{T n,m} N^i_{n,m} - \sum_{j=1}^{4r} \frac{\Sigma_{S n,m}}{4\pi} w^j N^j_{n,m}\right] = \Delta x_n\,\Delta y_m \frac{S_{n,m}}{4\pi}.$$

Here the variables $N^i_{m,n}$ represent average fluxes within mesh boxes. The $V^i_{n,m}$ are flux averages taken over *vertical* sides of mesh boxes, while the $H^i_{n,m}$ are averages over *horizontal* sides. If we regard all variables in (7.2.5) as independent variables we have, at this point, more unknowns than equations. To eliminate the extra variables it is necessary to postulate some additional relations among the N's, V's, and H's. Carlson assumes that, in each box,

(7.2.6)
$$N^i_{n,m} = \tfrac{1}{2}[V^i_{n,m} + V^i_{n+1,m}] = \tfrac{1}{2}[H^i_{n,m} + H^i_{n,m+1}].$$

If we make these assumptions the diamond equations, in matrix notation, take the form

(7.2.7)
$$\begin{pmatrix} \sigma & \delta_x & \delta_y \\ I & -P & 0 \\ I & 0 & -Q \end{pmatrix} \begin{pmatrix} \boldsymbol{N} \\ \boldsymbol{V} \\ \boldsymbol{H} \end{pmatrix} = \begin{pmatrix} \boldsymbol{S} \\ 0 \\ 0 \end{pmatrix}$$

Here the coefficient matrix, A, has been partitioned to show its structure. The submatrix, P, relates $\boldsymbol{N}$ and $\boldsymbol{V}$, while Q relates $\boldsymbol{N}$ and $\boldsymbol{H}$. Boundary conditions are incorporated, implicitly, in the coefficient matrix. If all boundaries are reflecting, one can demonstrate that δ_x and δ_y, as well as P and Q, are singular and that, in fact, the matrix A is singular. Yet, because of the form of the source term, the matrix equation has solutions and, for a given $\boldsymbol{S}$, there is only one solution vector $\boldsymbol{N}$ [**18**]. This $\boldsymbol{N}$ is unique, though $\boldsymbol{V}$ and $\boldsymbol{H}$ are not. It follows that there is a unique matrix, B, such that $\boldsymbol{N} = B\boldsymbol{S}$.

It would be reassuring if one could demonstrate that the diamond equations, in some sense, are valid difference equations. Probably, as all the mesh widths go to zero, $\boldsymbol{N}$ will approach the solution of the discrete ordinate equation; but apparently no one has shown that this is true. At least one would like to know that, in differencing the differential equations, some of the most fundamental properties of the differential equations are preserved. Now it is easy to prove that if the source is positive then, under reasonable conditions, the discrete ordinate fluxes will be nonnegative. It seems desirable that the solutions of the diamond equations should behave in the same way. In other words the vector $\boldsymbol{N}$ should be nonnegative when $\boldsymbol{S}$ is nonnegative and, since $\boldsymbol{N} = B\boldsymbol{S}$, B should be a nonnegative matrix. Unfortunately this is not always true. One can construct problem configurations such that B has negative elements and *retains* negative elements even as the maximum mesh width goes to zero.

In practice the diamond equations seem to work quite well. Still, because of their theoretical peculiarities one doesn't know how far to trust them. For this reason it may be worthwhile to look for alternatives, for equations whose coefficient matrices are nonsingular matrices with nonnegative inverses. It is not difficult to construct difference equations which have these properties. Consider, for example, a set of equations which we'll call the "simple three-point equations." These are very similar to equations based on Carlson's "step model." To derive the simple three-point equations it is necessary to replace all derivatives, in (6.2.1) by backward differences. One finds, in this way, that

$$\begin{aligned}(7.2.8)\qquad &\Omega_x^i \frac{(\psi^i_{n+1,m+1} - \psi^i_{n,m+1})}{\Delta x_n} + \Omega_y^i \frac{(\psi^i_{n+1,m+1} - \psi^i_{n+1,m})}{\Delta y_m} + \Sigma_{Tn,m}\psi^i_{n+1,m+1} \\ &\quad = \frac{1}{4\pi}\Sigma_{Sn,m}\Phi_{n+1,m+1} + \frac{1}{4\pi} s_{n+1,m+1}, \quad i = 1, 2, \ldots, \tau,\end{aligned}$$

$$\begin{aligned}(7.2.9)\qquad &\Omega_x^i \frac{(\chi^i_{n,m+1} - \chi^i_{n+1,m+1})}{\Delta x_n} + \Omega_y^i \frac{(\chi^i_{n,m+1} - \chi^i_{n,m})}{\Delta y_m} + \Sigma_{Tn,m}\chi^i_{n,m+1} \\ &\quad = \frac{1}{4\pi}\Sigma_{Sn,m}\Phi_{n,m+1} + \frac{1}{4\pi} s_{n,m+1}, \quad i = 1, 2, \ldots, \tau,\end{aligned}$$

$$\begin{aligned}(7.2.10)\qquad &\Omega_x^i \frac{(\xi^i_{n,m} - \xi^i_{n+1,m})}{\Delta x_n} + \Omega_y^i \frac{(\xi^i_{n,m} - \xi^i_{n,m+1})}{\Delta y_m} + \Sigma_{Tn,m}\xi^i_{n,m} \\ &\quad = \frac{1}{4\pi}\Sigma_{Sn,m}\Phi_{n,m} + \frac{1}{4\pi} s_{n,m}, \quad i = 1, 2, \ldots, \tau,\end{aligned}$$

$$\begin{aligned}(7.2.11)\qquad &\Omega_x^i \frac{(\eta^i_{n+1,m} - \eta^i_{n,m})}{\Delta x_n} + \Omega_y^i \frac{(\eta^i_{n+1,m} - \eta^i_{n+1,m+1})}{\Delta y_m} + \Sigma_{Tn,m}\eta^i_{n+1,m} \\ &\quad = \frac{1}{4\pi}\Sigma_{Sn,m}\Phi_{n+1,m} + \frac{1}{4\pi} s_{n+1,m}, \quad i = 1, 2, \ldots, \tau.\end{aligned}$$

Note that these difference equations, like equations (6.2.1), are invariant under 90° rotations about Z.

If we gather the coefficients of the fluxes into a coefficient matrix A, then the form of A will depend on the ordering of the equations and unknowns. We choose an ordering such that all terms involving total cross sections appear on the diagonal of A. A simple argument reveals that A is, then, an irreducibly diagonally dominant M matrix **[19]**. Therefore, regardless of the mesh structure, A has a positive inverse.

Nevertheless the simple three-point equations have an important disadvantage. Analysis in one dimension indicates that, in spite of their attractive features, the simple three-point equations are often less accurate than the diamond equations.[22] Thus we have, at this stage, inaccurate equations which we feel we understand and seemingly accurate equations which one cannot always trust. We need, instead, equations which are, at the same time, theoretically simple and reasonably accurate. There *are* difference equations which seem to fit this description, equations which

[22] Note that, in one dimension, the diamond equations are identical with the central difference equations, while the simple three-point equations are backward difference equations.

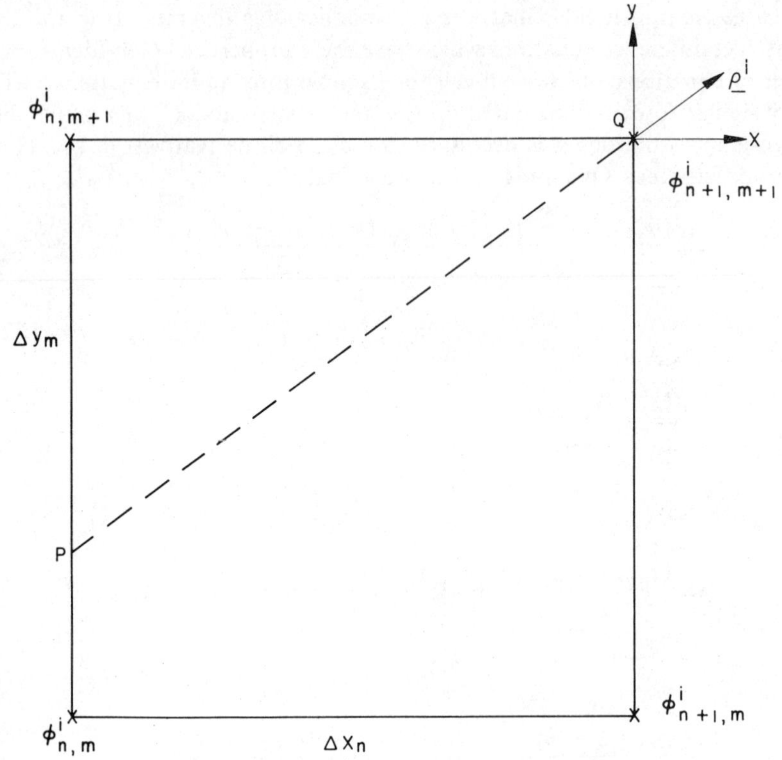

FIGURE 3. Mesh box (n, m).

we'll call the "variable three-point equations." [23] These are derived, below, by central differencing along characteristic lines.

Let $\mathbf{\Omega}^i$ be a principal ordinate and $\boldsymbol{\rho}^i$ its projection in the x-y plane. Further, as in Figure 3, let PQ be a line segment through the point Q and parallel to $\boldsymbol{\rho}^i$; let ρ be the length of the line segment, PQ. In equation (6.2.1), set $\mathbf{\Omega} = \mathbf{\Omega}^i$. Now, integrating (6.2.1) along PQ, and assuming that φ^i and Φ are linear in the variable of integration, one finds that

$$[1 - (\Omega_z^i)^2]^{1/2}[\varphi^i_{n+1,m+1} - \varphi^i(P)] + \tfrac{1}{2}\rho\Sigma_{Tn,m}[\varphi^i_{n+1,m+1} + \varphi^i(P)] = \tfrac{1}{2}\rho\,\frac{\Sigma_{Sn,m}}{4\pi}[\Phi_{n+1,m+1} + \Phi(P)] + \tfrac{1}{2}\rho\,\frac{S_{n,m}}{4\pi}. \tag{7.2.12}$$

Finally one arrives at the variable three point equation by expressing fluxes at P, via linear interpolation, in terms of fluxes at (n, m) and $(n, m + 1)$. Had the direction $\mathbf{\Omega}^i$ been such that the point P were on a horizontal mesh line, then the flux at P would be expressed in terms of the fluxes at (n, m) and $(n + 1, m)$. In deriving equation (7.2.12) it was assumed that $\mathbf{\Omega}^i$ is a principal ordinate. We will, however, postulate that the difference equations are again invariant with respect

[23] These equations were used in a program, written at BAPL, by J. Bennett [20].

to 90° rotations. It follows that one can deduce, from equations (7.2.12), three other sets of equations involving ordinates which are not principal ordinates.

Let $t^i_{n,m}$ be equal to $2\,|\Omega^i_y/\Delta y_m|$, or to $2\,|\Omega^i_x/\Delta x_n|$, whichever is larger. Suppose that

$$t^i_{n,m} > [\Sigma_{Tn,m} - w^i\Sigma_{Sn,m}/4\pi] \tag{7.2.13}$$

for all n, m and i. Suppose, again, that the coefficients of the variable three-point equations are formed into a coefficient matrix A. Then one can show that A (if it is appropriately ordered) is a diagonally dominant M matrix. Hence when the above inequality is satisfied, A is nonsingular and has a positive inverse. It seems likely that the variable three-point equations are *more* accurate than the simple three-point equations, but often *less* accurate than the diamond equations.

8. Methods of solution.

8.1. Difference equations of various types have been derived in §7 and it is natural to ask, now, how these equations can be solved. We've pointed out already that all methods of solution which seem feasible, today, are iterative. There are, however, many iterative techniques which have been used, or might be used, in discrete ordinate computations. Since we cannot discuss them all here we shall be forced, again, to pass over much interesting and useful work and to focus our attention on a few selected topics. Below, in this section, we deal with the general Jacobi and Gauss-Seidel block methods. Later, in §9, we consider a particular Jacobi iteration process and show that one can speed convergence of this process through use of cyclic Chebyshev polynomial techniques.

8.2. Before describing the Jacobi and Gauss-Seidel procedures we first write equation (7.2.3) in the following partitioned form:

$$\begin{pmatrix} A_{1,1} & A_{1,2} & \cdots & A_{1,Q} \\ A_{2,1} & A_{2,2} & & A_{2,Q} \\ \cdot & \cdot & \cdot & \cdot \\ \cdot & \cdot & \cdot & \cdot \\ \cdot & \cdot & \cdot & \cdot \\ A_{Q,1} & A_{Q,2} & \cdots & A_{Q,Q} \end{pmatrix} \begin{pmatrix} \boldsymbol{\varphi}_1 \\ \boldsymbol{\varphi}_2 \\ \cdot \\ \cdot \\ \cdot \\ \boldsymbol{\varphi}_Q \end{pmatrix} = \begin{pmatrix} s_1 \\ s_2 \\ \cdot \\ \cdot \\ \cdot \\ s_3 \end{pmatrix}. \tag{8.2.1}$$

Here all diagonal submatrices are square. We assume now, and in the remainder of our work, that A is an irreducible diagonally dominant M matrix.[24] It follows [**19**, p. 85] that the $A_{i,i}$ are also M matrices and, hence, nonsingular. Now let $A = D - (L + U)$, where $D \equiv \text{Diag}\,(A_{1,1} \cdots A_{Q,Q})$, while L and U are, respectively, strictly lower and upper triangular matrices. Given D, L, and U the block Jacobi method takes the form

$$D\boldsymbol{\varphi}(l+1) = (L+U)\boldsymbol{\varphi}(l) + \boldsymbol{s}, \tag{8.2.2}$$

[24] This is true, for example, if the difference equations are simple three-point equations. It is also true if they are variable three-point equations, provided that A is appropriately ordered and that inequality (7.2.13) is satisfied. See discussion in §7.2.

while, for the Gauss-Seidel method

$$D\boldsymbol{\varphi}(l+1) = L\boldsymbol{\varphi}(l+1) + U\boldsymbol{\varphi}(l) + \mathbf{s}. \tag{8.2.3}$$

In both above equations the index, l, is the iteration index. We refer to the matrix $J \equiv D^{-1}(L+U)$ as the Jacobi iteration matrix and to the matrix $\mathscr{L}_1 \equiv (D-L)^{-1}U$ as the Gauss-Seidel iteration matrix.

Throughout our work we shall use the asymptotic convergence rate, R_∞ (see **[19]**), as a measure of the efficiency of iterative processes. Given a convergent matrix iterative process, $\mathbf{x}(l+1) = C\mathbf{x}(l) + \mathbf{g}$, the number of iterations required to reduce the initial error by some specified factor is roughly proportional to $[R_\infty(C)]^{-1}$. Here $R_\infty(C) = -\ln \rho(C)$, and $\rho(C)$ is the spectral radius of C.[25]

From the fact that A is an M matrix it follows **[19**, p. 90] that the Jacobi and Gauss-Seidel processes, as defined above, must converge. Moreover, since $D^{-1}(L+U)$ is nonnegative, we know **[21]** that $R_\infty(\mathscr{L}_1) > R_\infty(J)$. Thus, for any partitioning (8.2.1) of A, the associated Gauss-Seidel process converges faster than the corresponding Jacobi process. Of course the convergence rate of either process depends on the structure of the partitioned coefficient matrix. Generally, as D is enlarged to include more nonzero elements, the convergence rate will increase. Unfortunately the work required to invert D will also increase. Therefore it is difficult to settle on an optimum partitioning procedure, or even to evaluate alternative procedures. Without evaluating alternatives we shall consider only one partitioning of A, a two-cyclic partitioning **[19]** suggested by J. Bennett **[22]**. We show next that the Jacobi process based on this partitioning can be accelerated by the cyclic Chebyshev semi-iterative method **[23]**.[26]

9. The Chebyshev acceleration method. Let $\boldsymbol{\alpha}$ be a vector made up of the components of $\boldsymbol{\psi}$ and $\boldsymbol{\chi}$. In other words let all the components of $\boldsymbol{\psi}$ and $\boldsymbol{\chi}$ now be regarded as components of a single vector, $\boldsymbol{\alpha}$. Similarly, define a vector, $\boldsymbol{\beta}$, whose components are those of $\boldsymbol{\eta}$ and $\boldsymbol{\xi}$. When the new unknown vectors are introduced into equation (7.2.3) this basic matrix equation takes on the form

$$\begin{pmatrix} A_1 & -B_1 \\ -B_2 & A_2 \end{pmatrix} \begin{pmatrix} \boldsymbol{\alpha} \\ \boldsymbol{\beta} \end{pmatrix} = \begin{pmatrix} \mathbf{s}_1 \\ \mathbf{s}_2 \end{pmatrix}. \tag{9.1.1}$$

The coefficient matrix, A, as it is partitioned in equation (9.1.1), is a two-cyclic matrix. The cyclic Chebyshev method, when applied to (9.1.1), is defined by the following iteration equations **[19**, p. 132]:

$$\boldsymbol{\alpha}(l+1) = a_{l+1}[\hat{\boldsymbol{\alpha}}(l+1) - \boldsymbol{\alpha}(l)] + \boldsymbol{\alpha}(l), \tag{9.1.2}$$

where

$$A_1\hat{\boldsymbol{\alpha}}(l+1) = B_1\boldsymbol{\beta}(l) + \mathbf{s}_1; \tag{9.1.3}$$

$$\boldsymbol{\beta}(l+1) = b_{l+1}[\hat{\boldsymbol{\beta}}(l+1) - \boldsymbol{\beta}(l)] + \boldsymbol{\beta}(l) \tag{9.1.4}$$

[25] If λ_i, $i = 1, 2, \ldots$, are all the eigenvalues of the matrix, C, then $\rho(C) = \max_i |\lambda_i|$.

[26] Chebyshev polynomials could also be used to accelerate the Gauss-Seidel process but we feel (for reasons we won't discuss here) that it is advantageous to accelerate the Jacobi process instead.

where

$$A_2\hat{\boldsymbol{\beta}}(l+1) = B_2\boldsymbol{\alpha}(l+1) + \mathbf{s}_2. \tag{9.1.5}$$

In the above equations the quantities a_{l+1} and b_{l+1} are Chebyshev acceleration parameters which have yet to be specified.

Given the partitioning (9.1.1) we can (at least in principle) construct the corresponding Jacobi iteration matrix J. To make full use of the power of the Chebyshev acceleration method one needs some information about the eigenvalues of J and, more precisely, about the domain in which the eigenvalues lie. When the domain is known it may be possible to optimize the sequence of acceleration parameters. Thus, for example

(A) if all the eigenvalues of J lie in or on an ellipse, as in Figure 4,

(B) if, in Figure 4, $\epsilon \leq \rho < 1$,

(C) and if J has a diagonal Jordan form,

then the optimum parameters have the following values [**24**]:

$$a_1 = 1.0, \qquad b_1 = 1/(1 - \tfrac{1}{2}\Lambda^2) \tag{9.1.6}$$

$$a_{l+1} = 1/(1 - \tfrac{1}{4}\Lambda^2 b_l), \qquad b_{l+1} = 1/(1 - \tfrac{1}{4}\Lambda^2 a_{l+1}), \qquad l = 1, 2, \ldots .$$

Here $\Lambda^2 = \rho^2 - \epsilon^2$. Actually one can show that there are infinitely many ellipses with properties A and B. To define the Chebyshev parameters we select the ellipse, oriented as in Figure 4, with smallest possible ϵ and ρ. From the fact that A is an irreducible diagonally dominant M matrix, it follows that λ_0, the maximum modulus eigenvalue of J, is positive and less than unity. Therefore, the smallest possible ρ is λ_0 and the smallest possible ϵ is no larger than λ_0. Furthermore, numerical experiments suggest that J is indeed similar to a diagonal matrix. Thus it seems likely that condition C, like A and B, is satisfied in practice.

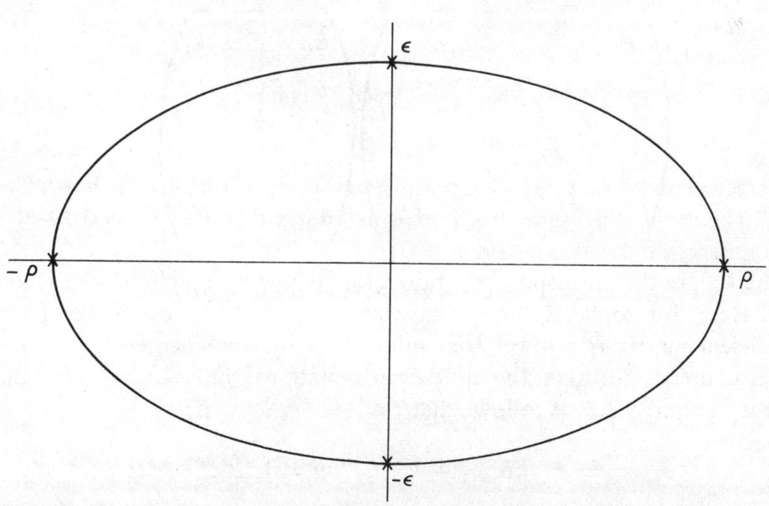

FIGURE 4. Eigenvalue domain of J.

It is known [**25**] that the effectiveness of the cyclic Chebyshev method decreases as the ratio ϵ/ρ increases. In fact as ϵ/ρ approaches unity the convergence rate of the accelerated Jacobi process approaches the convergence rate of the Gauss-Seidel process. Acceleration methods are important, however, when ρ is close to one. Some analysis, joined with practical experience, leads us to believe that ϵ/ρ is generally small when ρ is close to one. Thus, Chebyshev acceleration should often be useful.

Of course one rarely forms the matrix J explicitly and one has, a priori, little knowledge of its eigenvalues. It is possible, however, to guess ϵ and ρ, to compute tentative Chebyshev parameters from these guesses, and by observing successive accelerated iterates, to deduce improved estimates of ϵ and ρ. Algorithms for reestimating ϵ and ρ are discussed in [**25**].

9.2. Above we have described one specific iterative procedure which may be used to solve the discrete ordinate equations. During the course of each iteration it is necessary to solve for $\hat{\boldsymbol{\alpha}}(l+1)$ and $\hat{\boldsymbol{\beta}}(l+1)$ in the equations of the form

$$A_1\hat{\boldsymbol{\alpha}}(l+1) = B_1\boldsymbol{\beta}(l) + s_1, \tag{9.2.1}$$

$$A_2\hat{\boldsymbol{\beta}}(l+1) = B_2\boldsymbol{\alpha}(l+1) + s_2. \tag{9.2.2}$$

Fortunately this task is not as difficult as it may seem, since A_1 and A_2 are very sparse matrices.

To solve the equations (9.2.1), for example, we first partition $\hat{\boldsymbol{\alpha}}$ into subvectors. With each component of $\hat{\boldsymbol{\alpha}}$ we associate a mesh point and a direction. Certain components of $\hat{\boldsymbol{\alpha}}$ are associated with the various points of the mth mesh row. Let these components constitute an unknown subvector $\hat{\boldsymbol{\alpha}}_m$. In terms of the new unknown vectors equation (9.2.1) takes on the form

$$\begin{pmatrix} E_1 & & & \\ -F_1 & E_2 & & \\ & \ddots & \ddots & \\ & & -F_{M-1} & E_M \end{pmatrix} \begin{pmatrix} \hat{\boldsymbol{\alpha}}_1 \\ \hat{\boldsymbol{\alpha}}_2 \\ \vdots \\ \hat{\boldsymbol{\alpha}}_M \end{pmatrix} = \begin{pmatrix} \boldsymbol{k}_1 \\ \boldsymbol{k}_2 \\ \vdots \\ \boldsymbol{k}_M \end{pmatrix}, \tag{9.2.3}$$

where the $\boldsymbol{k}_m$ are the corresponding subvectors of the right side of (9.2.1) and are known. Here, for convenience, we have dropped the iteration index. In (9.2.3) the coefficient matrix is a lower triangular block matrix since each row of mesh points is coupled, through the difference equations, only to the row directly beneath it. From (9.2.3) it follows that

$$E_m\hat{\boldsymbol{\alpha}}_m = F_{m-1}\hat{\boldsymbol{\alpha}}_{m-1} + \boldsymbol{k}_m, \qquad m = 1, 2, \ldots, M, \tag{9.2.4}$$

where $F_0 = 0$. Clearly (9.2.4) can be solved recursively for $\hat{\boldsymbol{\alpha}}_1, \hat{\boldsymbol{\alpha}}_2, \ldots, \hat{\boldsymbol{\alpha}}_m$.

In effect, when solving (9.2.4) recursively, we are, first, computing all upward-moving fluxes at points of the first mesh row. Next we are computing upward-moving fluxes at the points of the second row, etc. Of course one can write equations, analogous to (9.2.4), for the downward-moving fluxes. When all upward-moving fluxes are computed, we calculate from these analogous equations the downward-moving fluxes, $\hat{\boldsymbol{\beta}}_{M-1}$, at the points of row $M - 1$. Next we calculate $\hat{\boldsymbol{\beta}}_{M-2}$, etc. The downward-moving fluxes $\hat{\boldsymbol{\beta}}_M$ are obtained from $\hat{\boldsymbol{\alpha}}_M$ using the boundary condition (7.1.2).

For any given m, equations (9.2.4) are similar to the one-dimensional discrete ordinate equations, and they can be solved by methods similar to methods used in one dimension. Thus they might be solved by direct inversion or by iterative techniques. If they are solved iteratively the process by which they are solved is often called an "inner" iterative process, since the "inner" iterations are contained, so to speak, in the main, or "outer" iterative cycle. Since the convergence rate of the outer iteration is affected by the degree of convergence achieved by the inner iterations, there is an optimum number of inner iterations one should do. Of course, it is not known how to determine this optimum number. However, the criterion of terminating the inner iterative process when the initial error has been reduced by a factor of 10 seems to work pretty well.

10. Conclusion.

10.1. It is impossible, in any short paper, to give an exhaustive account of all important work that has been done on the discrete ordinate method. We regret, in particular, that we have had so little opportunity to discuss the work of Carlson and Lee.[27] These two authors have developed many of the numerical procedures, and many of the S_n computer programs, now in use. They have carried out a good deal of experimentation with various difference approximations, iterative techniques and strategems for accelerating convergence. But since their work has been mainly empirical it does not fall within the scope of this paper. One of the accelerative devices used in S_n programs is particularly interesting. We refer, here, to a technique which is usually called "normalization" and which has been introduced in many different guises, into all sorts of reactor computations. But, in spite of its wide range of utility, normalization (like so many valuable computational devices) has not been subjected to any rigorous analysis.

Another interesting acceleration technique is Kopp's "synthetic method" **[26]**. Kopp has used this method extensively to solve one-dimensional and two-dimensional problems, but theoretical work on Kopp's method is too primitive to be very helpful in practice.

Thus, we find ourselves, again and again, in a very awkward situation. The two-dimensional transport problem is an important problem to a great many reactor physicists and engineers. Yet many of the methods of solution available to us have no firm mathematical foundation, and the work of the applied mathematician in the field has, in a sense, just begun.

---

[27] See, for example, **[17]**, **[27]**, and **[28]**.

REFERENCES

1. B. Davison and J. B. Sykes, *Neutron Transport Theory*, Clarendon Press, Oxford, 1957.

2. J. Yvon, *La diffusion macroscopique des neutrone une méthods d'approximation*, J. Nucl. Energy I, **4** (1957), 305–318.

3. H. Waldinger et al., *Numerical integration of the spherical harmonics equations*, Nucl. Sci. and Eng. **18** (1964).

4. A. Weinberg and E. Wigner, *The physical theory of neutron chain reactors*, Univ. of Chicago Press, Chicago, Ill., 1958 (see also [**1**]).

5. J. Mingle, *The even-order spherical harmonics method in cylindrical geometry*, Nucl. Sci. and Eng. **20** (1964).

6. F. R. Gantmacher, *Matrix theory*, Vol. 2, Chelsea, New York, 1959.

7. P. Henrici, *Discrete variable methods in ordinary differential equations*, Wiley, New York, 1962.

8. E. M. Gelbard, *Properties of the P_L and double-P_L relaxation lengths in the presence of anisotropic scattering*, Nucl. Sci. and Eng. **24** (1966).

9. E. M. Gelbard and S. Kaplan, *Reality of relaxation lengths in various approximate forms of the slab transport equation*, Nucl. Sci. and Eng. **26** (1966).

10. P. Daitch et al., *CEPTR, an* IBM-704 *code to solve the P_3 approximation to the one velocity transport equation in cylindrical geometry*, MPC-20, 1959; for discussion of boundary conditions see also W. W. Clendenin, *Effect of zero gradient boundary conditions in cylindrical geometry*, Nucl. Sci. and Eng. **14** (1962).

11. C. E. Lee, *The discrete S_n approximation to transport theory*, Los Alamos Scientific Laboratory Report LA-2595, March, 1962.

12. L. Hageman, *On a numerical approximation to the solution of the one-dimensional, one-velocity transport equation with isotropic scattering*, WAPD-T-706, 1959.

13. E. Schmidt, *A stable and noniterative solution to the discrete ordinate equations*, Trans. ANS **6** (1963).

14. S. Kaplan and E. M. Gelbard, *Invariant imbedding and the integration techniques of reactor theory*, J. Math. Anal. Appl. **11** (1965), 538–547.

15. D. Drawbaugh and L. Noderer, *The double spherical harmonics method for cylinders and spheres*, Nucl. Sci. and Eng. **6** (1959); see also E. Schmidt and E. M. Gelbard, *A double-P_N method for spheres and cylinders*, WAPD-T-1945, 1966.

16. G. Goertzel, *The method of discrete ordinates*, Nucl. Sci. and Eng. **4** (1958).

17. B. G. Carlson, *The numerical theory of neutron transport*, Vol. I, *Statistical Physics*, Academic Press, New York, 1963.

18. J. A. Davis, L. A. Hageman, and R. B. Kellogg, *Singular difference approximations for the discrete ordinate equations in x-y geometry*, Nucl. Sci. and Eng. **29** (1967), 237–243.

19. R. S. Varga, *Matrix iterative analysis*, Prentice-Hall, Englewood Cliffs, N.J., 1962.

20. J. H. Bennett, TXY01: *A one-group transport program for x-y geometry*, WAPD-TM-482, October, 1964.

21. P. Stein and R. L. Rosenberg, *On the solution of linear simultaneous equations by iteration*, J. London Math. Soc. **23** (1948).

22. J. H. Bennett, *Accelerating convergence of discrete ordinate methods*, Nucl. Sci. and Eng. **19** (1964), 209.

23. G. H. Golub and R. S. Varga, *Chebyshev semi-iterative methods, successive overrelaxation iterative methods and second order Richardson iterative methods*, Numer. Math. **3** (1961).

24. A. J. Clayton, *Further results on polynomials having least maximum modulus over an ellipse in the complex plane*, U.K.A.E.A. Memorandum-AEEW-M. 348, 1963.

25. L. A. Hageman, *The Chebyshev polynomial method of iteration*, WAPD-TM-537, 1967.

26. H. J. Kopp, *Synthetic method solution of the transport equation*, Nucl. Sci. and Eng. **17** (1963).

27. B. G. Carlson and C. E. Lee, *Mechanical quadrature and the transport equation*, Los Alamos Scientific Laboratory Report LAMS-2573 (June 1961).

28. B. G. Carlson, *Numerical formulation and solution of neutron transport problems*, LA-2996, 1964.

WESTINGHOUSE ELECTRIC CORPORATION BETTIS ATOMIC POWER LABORATORY
WEST MIFFLIN, PENNSYLVANIA

Mathematical methods suggested by transport theory

G. M. Wing[1]

I. **Introduction.** In the last several years my personal research interests have shifted from transport problems *per se* to an investigation of mathematical methods that have either been originally developed for the study of transport phenomena, or that have experienced a considerable growth because of their application to such phenomena. My studies have included consideration of new areas of mathematics, numerical analysis, and physics to which these concepts may be applied. I hope then that I may digress from the main topics of this symposium and talk more about the tools of the trade than of the trade itself.

It is a truism, of course, that the interaction between mathematics and the physical sciences has often provided a very considerable impetus to both. That studies in transport theory should be contributing new ideas to mathematics is not at all surprising. Any discussion of this interaction is certain to leave out many important aspects and to slight many interesting examples. I apologize in advance for these oversights and omissions, as well as for the fact that most of the topics that are included and emphasized are those which have attracted my own interests.

Perhaps the most striking contribution to mathematical analysis that can be traced quite directly to a transport question is the method of Wiener-Hopf, originally applied to the Milne problem **[22]**. Surely it is not necessary to dwell at length on the beauty and depth of this theory, nor to attempt a list of applications, extensions, and generalizations associated with it. It seems regrettable that an exposition of the Wiener-Hopf theory is not made a regular part of all good complex variable theory courses.

The requirements of the atomic energy program have provided tremendous stimulus to the study of transport phenomena. Often these needs have demanded

[1] Work supported in part by National Science Foundation Grants GP 4180, 5965, and 5967.

extensive numerical calculations. Specific numerical devices have been invented for solution of such problems; computing machines have sometimes been developed with the realization that these questions would account for a significant amount of their running time. Monte Carlo evolved in part for the investigation of these problems. It would be extremely difficult to evaluate the impact that transport theory has had on the science of numerical analysis and computing in the last generation. Perhaps everyone who uses a modern computer owes some debt to the transport theorists.

The singular function method of Case provides another example [**5**]. While it must be admitted that the first application of the idea was to a problem in plasma physics [**21**], the technique has certainly come to fruition in its applications to neutron transport questions, and will doubtless find numerous applications to other fields. In fact, the singular integral equation approach of Mullikin—inspired, I believe, by Case's work—has already been employed in investigations of hydrodynamics and signal detection [**11**]–[**14**], [**16**].

There is also the close relationship with random walk and branching processes which have been investigated by researchers in probability theory [**8**]. It seems certain that such researches have already made specific contributions to the general theory of probability, stochastic processes, and statistics. It may even be possible to find results in abstract functional analysis that are directly traceable to the existence of transport theory, since numerous abstract and semiabstract approaches to transport problems have been made (see, for example [**4**]).

Obviously one could go on at length with these general remarks and speculations but it is now appropriate to become more specific. I have chosen three topics to discuss in some detail today. The first stems from some work of T. W. Mullikin. In [**15**] he used a very clever trick to employ the Rayleigh-Ritz method to obtain *upper* bounds on the eigenvalues of the integral equations arising in one-velocity neutron transport problems in slabs and spheres. Such information provides estimates on the critical size of such configurations. In §II I shall discuss some generalizations of these ideas to rather wide classes of integral equations. This research has led not only to consideration of bounds on eigenvalues but also to a study of their "shapes" in a sense to be described later.

Several speakers at this conference have discussed the method of invariant imbedding. While it is possible to trace the basic ideas involved back at least to Stokes [**19**] it was the work of Ambarzumian [**2**] which first attracted significant attention. The use made by Chandrasekhar [**6**] of this method of invariance in solving numerous outstanding problems in astrophysics is another beautiful chapter in the history of transport theory—or radiative transfer, as the astronomers prefer. It was Richard Bellman who first suggested about twelve years ago that the basic ideas of Ambarzumian and Chandrasekhar might lie quite deep and might be widely applicable. In §III the imbedding ideas will be applied to a class of integral equations which very considerably generalize those arising from transport problems. The possibility of using these methods in numerical studies seems good.

§IV also represents in a sense an application of the imbedding method, but so far removed from its original source that the technique is barely recognizable. We shall see in that section that the basic ideas of Stokes and Ambarzumian—when well understood—can be used to provide a practical method of studying and computing the eigenvalues of Schrödinger-like equations. There are strong indications that in some cases the device may have distinct numerical advantages over those more commonly used.

II. **Eigenvalues of integral equations.** The integral equation probably most familiar to workers in transport theory is

$$\varphi(x) = \frac{c}{2}\int_{-a}^{a} E_1(|x - x'|)\varphi(x')\,dx'. \tag{2.1}$$

This equation arises from the study of time independent, one-velocity, isotropic transport in a uniform slab. Here c is the average number of secondaries per collision, $2a$ is the slab thickness measured in units of mean free path, and

$$E_1(|u|) = \int_1^\infty \frac{e^{-s|u|}}{s}\,ds. \tag{2.2}$$

The physical system is just critical when (2.1) has a nonnegative solution φ; hence when $2/c = \lambda_1$, where λ_1 is the largest eigenvalue of the E_1 kernel.

The Rayleigh-Ritz principle basically provides lower bounds for λ_1,

$$\lambda_1 = \sup_{\|\psi\|=1} (E_1\psi, \psi), \tag{2.3}$$

where the notation and the norm used are the classical ones:

$$\|\psi\|^2 = \int_{-a}^{a} \psi^2(x)\,dx, \tag{2.4}$$

$$(K\psi, \psi) = \int_{-a}^{a}\int_{-a}^{a} K(x, y)\psi(x)\psi(y)\,dx\,dy. \tag{2.5}$$

(We confine our study to the real case for convenience.)

In [15] Mullikin used the following argument to obtain upper bounds on λ_1. Observe that

$$(E_1\psi, \psi) = \int_1^\infty \frac{ds}{s}\int_{-a}^{a}\int_{-a}^{a} e^{-s|x-y|}\psi(x)\psi(y)\,dx\,dy. \tag{2.6}$$

But, by the Rayleigh-Ritz principle again, if $\|\psi\| = 1$,

$$\int_{-a}^{a}\int_{-a}^{a} e^{-s|x-y|}\psi(x)\psi(y)\,dx\,dy \leqq \tilde{\lambda}_1(s) \tag{2.7}$$

where $\tilde{\lambda}_1(s)$ is the largest (positive) eigenvalue of

$$\tilde{\lambda}_1(s)\tilde{\varphi}_1(x, s) = \int_{-a}^{a} e^{-s|x-y|}\tilde{\varphi}_1(y, s)\,dy. \tag{2.8}$$

Using (2.7) in (2.6) one finds from (2.3) that

$$\lambda_1 \leqq \int_1^\infty \frac{\tilde{\lambda}_1(s)}{s}\,ds. \tag{2.9}$$

Now (2.8) can be solved explicitly and an analytic expression for $\tilde{\lambda}_1(s)$ is known. Thus (2.9) may be evaluated. Excellent upper bounds for λ_1 are thus found.

To obtain a considerable generalization we cite a result due to J. Hersch [**9**].

THEOREM (HERSCH). *Let $K(x, y)$ be real, symmetric, and in L_2 on $-a \leqq x, y \leqq a$ and suppose its first n eigenvalues are positive. Let S_j, $j \leqq n$, be any set of orthonormal functions $\psi_1, \psi_2, \psi_3, \ldots, \psi_j$ on $-a \leqq x \leqq a$. Then*

$$\lambda_1 + \lambda_2 + \cdots + \lambda_j = \max_{S_j} \{(K\psi_1, \psi_1) + (K\psi_2, \psi_2) + \cdots + (K\psi_j, \psi_j)\}. \tag{2.10}$$

The maximum is assumed when the ψ_i are the corresponding eigenfunctions of K.

With this we prove

THEOREM 2.1. *Let $k(x, y, s)$ be defined and real for each fixed s, $-\infty < s < \infty$, symmetric in x and y, and in L_2 on $-a \leqq x, y \leqq a$. Let $\tilde{\lambda}_j(s)$ be the eigenvalues corresponding to k and suppose $\tilde{\lambda}_j(s) \geqq 0$ for all s and for $1 \leqq j \leqq n$. Further, assume $K(x, y)$ is in L_2 for $-a \leqq x, y \leqq a$ and such that for some $F(s) \geqq 0$*

$$K(x, y) = \int_{-\infty}^{\infty} k(x, y, s) F(s)\,ds. \tag{2.11}$$

Finally, suppose

$$\mathscr{K}(x, y) = \int_{-\infty}^{\infty} |k(x, y, s)|\, F(s)\,ds \tag{2.12}$$

is in L_2 on $-a \leqq x, y \leqq a$. Then if $\lambda_1, \lambda_2, \ldots, \lambda_n$ are the first n eigenvalues of K

(2.13)

$$\lambda_1 + \lambda_2 + \cdots + \lambda_j \leqq \int_{-\infty}^{\infty} \{\tilde{\lambda}_1(s) + \tilde{\lambda}_2(s) + \cdots + \tilde{\lambda}_j(s)\} F(s)\,ds, \quad j = 1, 2, \ldots, n.$$

PROOF. If $\varphi_1, \varphi_2, \ldots, \varphi_n$ are the orthonormalized eigenfunctions of K then for $j \leqq n$,

$$\begin{aligned}
\lambda_1 + \cdots + \lambda_j &= (K\varphi_1, \varphi_1) + (K\varphi_2, \varphi_2) + \cdots + (K\varphi_j, \varphi_j) \\
&= \int_{-a}^{a}\int_{-a}^{a}\left\{\int_{-\infty}^{\infty} k(x, y, s) F(s)\,ds\right\} \\
&\quad \times \{\varphi_1(x)\varphi_1(y) + \cdots + \varphi_j(x)\varphi_j(y)\}\,dx\,dy \qquad (2.14) \\
&= \int_{-\infty}^{\infty} F(s)\,ds \left\{\int_{-a}^{a}\int_{-a}^{a} k(x, y, s)[\varphi_1(x)\varphi_1(y) + \cdots \right. \\
&\quad \left. + \varphi_j(x)\varphi_j(y)]\,dx\,dy\right\},
\end{aligned}$$

the interchange of order being legitimate because $\mathscr{K}(x, y)$ is assumed in L_2. But by the theorem of Hersch

$$(2.15)\quad \int_{-a}^{a}\int_{-a}^{a} k(x, y, s)[\varphi_1(x)\varphi_1(y) + \cdots + \varphi_j(x)\varphi_j(y)]\,dx\,dy \leqq \tilde{\lambda}_1(s) + \cdots + \tilde{\lambda}_j(s), \quad j \leqq n.$$

The result is now immediate.

Direct use of the Hersch result provides lower bounds on the sums of eigenvalues, while the above theorem gives upper bounds. Obvious manipulations then provide both upper and lower bounds on individual eigenvalues, although these become in general less satisfactory as j increases due to loss of information introduced by the subtractions required.

Theorem 2.1 has been used to obtain excellent bounds on λ_2 for the E_1-kernel. The applicability of the result is severely limited by the almost complete lack of analytical and numerical information concerning eigenvalues of nontrivial kernels $k(x, y, s)$. This problem has been partially overcome by a numerical scheme for calculating such eigenvalues. The device is particularly appropriate when k is a difference kernel which is highly oscillatory for large s. The details of this method are found in [**17**]. With the information thus obtained bounds on the eigenvalues of a large number of interesting kernels K have been found (see [**23**]).

An examination of some of these eigenvalues considered as functions of the interval length $2a$ has suggested that the "shape" of these functions may be describable in terms of the "shape" of the kernels involved. Out of these considerations have emerged theorems of the following sort.

THEOREM 2.2. *Let $K(u)$ be nonnegative and not identically zero for $0 \leqq u \leqq 2T$. Suppose $g(t)$ is defined on $0 < t < t_0$, $0 < g(t) < T$, and is such that $K(g(t)u)$ is a convex function of t on $0 < t < t_0$ for each fixed u, $0 \leqq u \leqq 2$. Let*

$$(2.16)\quad \lambda_1(g(t))\varphi_1(x, g(t)) = \int_{-1}^{1} K(g(t)\,|x - y|)\varphi_1(y, g(t))\,dy.$$

Then $\lambda_1(g(t))$ is a convex function of t, $0 < t < t_0$.

PROOF. The following proof is due to W. R. Boland. Let α be any number in $[0, 1]$ and choose t_1 and t_2 in $(0, t_0)$. By hypothesis

$$(2.17)\quad K(g(\alpha t_1 + (1 - \alpha)t_2)u) \leqq \alpha K(g(t_1)u) + (1 - \alpha)K(g(t_2)u).$$

Call $t = \alpha t_1 + (1 - \alpha)t_2$ and observe that $\varphi_1(x, g(t))$ can be chosen to be nonnegative since K is nonnegative. Then from (2.16) and (2.17)

$$(2.18)\quad \begin{aligned} \lambda_1(g(t)) &= \int_{-1}^{+1}\int_{-1}^{+1} K(g(t)\,|x - y|)\varphi_1(x, g(t))\varphi_1(y, g(t))\,dx\,dy \\ &\leqq \alpha\int_{-1}^{1}\int_{-1}^{1} K(g(t_1)\,|x - y|)\varphi_1(x, g(t))\varphi_1(y, g(t))\,dx\,dy \\ &\quad + (1 - \alpha)\int_{-1}^{1}\int_{-1}^{1} K(g(t_2)\,|x - y|)\varphi_1(x, g(t))\varphi_1(y, g(t))\,dx\,dy. \end{aligned}$$

The classical Rayleigh-Ritz principle now yields

$$\lambda_1(g(t)) \leqq \alpha\lambda_1(g(t_1)) + (1 - \alpha)\lambda_1(g(t_2)), \tag{2.19}$$

the desired result.

Attempts to establish results for sums of eigenvalues require rather restrictive hypotheses on K and are not entirely satisfactory (see [**24**]). This whole class of theorems is being investigated by Mr. Boland and seems likely to produce new results regarding eigenvalues of Sturm Liouville problems, partial differential equations, etc.

Sight should not be lost of the fact that convexity theorems such as Theorem 2.2 provide further information on upper bounds. Thus if the right hand side of (2.19) is known or can be estimated from above then upper bounds on $\lambda_1(g(t))$ are available for all t in $[t_1, t_2]$. (The fact that the interval of integration runs from -1 to 1 is obviously of no consequence since an appropriate transformation may always be made.) Ideally one would attempt to choose a g so that the hypotheses are satisfied and so that the curve $\lambda_1(g(t))$ considered as a function of g is almost a straight line. Iterative schemes for obtaining this type of behavior are a possibility. In these an admissible g might be "improved" in such a way as to remain admissible and cause the function $\lambda_1(g(t))$ to become flatter.

III. **A class of Fredholm integral equations.** In this section we shall extend the classical work of Chandrasekhar on the X and Y functions associated with certain integral equations of radiative transfer [**6**]. In essence these functions convert a Fredholm equation into an initial value problem.

Two points should be mentioned. First, Chandrasekhar himself noted that a considerable generalization of his work was possible (see [**6**]), but he did not pursue the investigation because of its lack of immediate transport significance. Secondly, results similar to those obtained here have been established [**10**] without resorting to the transport theory arguments which we shall use. It is the author's feeling, however, that the transport analogies are of great help in understanding and extending the method.

Let us consider

$$\varphi(z) = g(z) + \int_{\tau_0}^{\tau} c(z')K(|z - z'|)\varphi(z')\,dz' \tag{3.1}$$

where K has an integral representation

$$K(u) = \int_0^\infty k(s')e^{-a(s')u}\,ds'. \tag{3.2}$$

We assume

(H1) $a(s)$ and $k(s)$ are piecewise continuous even functions on $-\infty < s < \infty$.
(H2) $c(z) \geqq 0$ is continuous in z for $\tau_0 \leqq z \leqq \tau$.
(H3) $\operatorname{Re} a(s) \geqq 0$, $-\infty < s < \infty$.
(H4) $\int_0^\infty |k(s')|\,ds' < \infty$.

It should be noted that at once that for $K(u) = E_1(u)$ one has $a(s) = s$, $s > 0$, and $k(s) = 0$ for $0 < s < 1$, $k(s) = 1/s$ for $s > 1$. Thus (H4) is violated and the theory to be presented here does not include the case of classical interest. This omission in part is for the sake of simplifying the arguments. The hypotheses (H1)–(H4) can be considerably relaxed though just how much is a matter currently under study.

We next introduce the "Boltzmann-like" equation

$$\text{(3.3a)} \quad \operatorname{sgn} s \frac{\partial N}{\partial z}(z, s) + a(s)N(z, s) = k(s)c(z)\int_{-\infty}^{\infty} N(z, s')\, ds', \quad \tau_0 \leq z \leq \tau$$

$$\text{(3.3b)} \qquad N(\tau_0, s) = 0, \qquad 0 < s < \infty,$$

$$\text{(3.3c)} \qquad N(\tau, s) = f(s), \qquad -\infty < s < 0$$

and add the hypothesis

$$\text{(H5)} \qquad \int_{-\infty}^{0} |f(s')|\, ds' < \infty.$$

Equation (3.3) has many pseudo-physical interpretations. If one multiplies by $|s|$ and considers s to play the rôle of a generalized "direction" (analogous to the classical μ) then $|s|\, a(s)$ can be thought of as a cross section[2] (dependent now on "direction") while $2k(s)c(z)/a(s)$ is analogous to the average number of secondaries per collision. This last term depends on both position and "direction" and is the term that is most unreasonable from a physical standpoint (see, however, **[18]** for some recent comments). There is no reason to require $f(s) \geqq 0$ or to expect that $N(z, s) \geqq 0$.

Again one may interpret (3.3) as representing transport on a line in which case s plays the role of a frequency or energy **[25]**. However, no physical interpretation is really necessary. Using completely standard arguments (see **[26]**) the following result may be established.

THEOREM 3.1. *Let $N(z, s)$ be continuously differentiable in z, $\tau_0 \leqq z \leqq \tau$, uniformly integrable in s, $-\infty < s < \infty$, and let N satisfy* (3.3). *Then*

$$\text{(3.4)} \qquad \varphi(z) = \int_{-\infty}^{\infty} N(z, s')\, ds'$$

satisfies (3.1) *provided*

$$\text{(3.5)} \qquad g(z) = \int_{-\infty}^{0} e^{-a(s')(\tau - z)} f(s')\, ds'.$$

Conversely, if φ is the solution to (3.1) *with g representable by* (3.5) *and if the operator* ***K*** *defined by*

$$\text{(3.6)} \qquad \boldsymbol{K}\cdot = \int_{\tau_0}^{\tau} c(z')K(|z - z'|)\cdot dz'$$

[2] $|s|\, a(s)$ corresponds to the cross section σ used by other authors in this volume.

does not have unity as an eigenvalue, then the function N defined by

$$\text{(3.7a)}\qquad N(z, s) = k(s)\int_{\tau_0}^{z} c(z')\varphi(z')e^{-a(s)(z-z')}\,dz', \qquad s > 0$$

$$\text{(3.7b)}\quad N(z, s) = k(s)\int_{z}^{\tau} c(z')\varphi(z')e^{-a(s)(z'-z)}\,dz' + e^{-a(s)(\tau-z)}f(s), \qquad s < 0$$

is a solution of (3.1). *Finally the correspondence provided between the N and* φ *functions is unique.*

Thus any Fredholm equation of the type (3.3) is equivalent to a Boltzmann-like problem and this relationship may be exploited. From the theory of integral equations one may now obtain an expression for φ in terms of a resolvent kernel. Equation (3.7) may be used then to obtain a similar type of representation for N. In order to emphasize the importance of the parameters τ and τ_0 in the problem we henceforth write

$$\varphi = \varphi(z, \tau, \tau_0), \qquad N = N(z, \tau, \tau_0, s).$$

THEOREM 3.2. *The functions* φ *and N have integral representations*

$$\text{(3.8)}\qquad \varphi(z, \tau, \tau_0) = \int_{-\infty}^{0} R(z, \tau, \tau_0, s')f(s')\,ds',$$

and

$$\text{(3.9)}\qquad N(z, \tau, \tau_0, s) = e^{a(s)(z-\tau)}f(s)H(s) + \int_{-\infty}^{0} \mathscr{R}(z, \tau, \tau_0, s, s')f(s')\,ds',$$

$$H(s) = \begin{cases} 0, & s \geqq 0, \\ 1, & s < 0, \end{cases}$$

where R and $\mathscr{R}$ *are continuously differentiable with respect to* τ, τ_0, *and z in any interval* $\tilde{\tau}_0 \leqq \tau_0 \leqq z \leqq \tau \leqq \tilde{\tau}$ *in which* **K** *does not have unity as an eigenvalue. Moreover, R and* $\mathscr{R}$ *are bounded and piecewise continuous in s' and the only possible discontinuities in s' appear at discontinuities of k and a. The same remarks hold for s in the case of* $\mathscr{R}$.

Theorem 3.2 may be regarded as simply a statement of Duhamel's principle for the functions φ and N.

Returning now to equation (3.3) we find that differentiation with respect to τ is legitimate:

$$\text{(3.10)}\qquad \begin{gathered} \operatorname{sgn} s\,\frac{\partial N_2}{\partial z}(z, \tau, \tau_0, s) + a(s)N_2 = k(s)c(z)\int_{-\infty}^{\infty} N_2(z, \tau, \tau_0, s')\,ds', \\ N_2(\tau_0, \tau, \tau_0, s) = 0, \qquad s > 0, \qquad N_2(\tau, \tau, \tau_0, s) = -N_1(\tau, \tau, \tau_0, s), \qquad s < 0. \end{gathered}$$

This problem is resolved by Theorem 3.2 and the use of (3.3):

$$\text{(3.11)}\qquad \begin{aligned} \varphi_2(z, \tau, \tau_0) &= \int_{-\infty}^{0} R(z, \tau, \tau_0, s')\{-N_1(\tau, \tau, \tau_0, s')\}\,ds' \\ &= \int_{-\infty}^{0} R(z, \tau, \tau_0, s')\{-a(s)N(\tau, \tau, \tau_0, s') + k(s')c(\tau)\varphi(\tau, \tau, \tau_0)\}\,ds'. \end{aligned}$$

Again using (3.8) we obtain

$$\varphi_2(z, \tau, \tau_0) = \int_{-\infty}^{0} f(s')R_2(z, \tau, \tau_0, s')\, ds'. \tag{3.12}$$

Since $N(\tau, \tau, \tau_0, s) = f(s)$ for $s < 0$ we may combine equations (3.11) and (3.12) to obtain

$$\begin{aligned} 0 = \int_{-\infty}^{0} f(s')\, ds' \Big\{ &-R_2(z, \tau, \tau_0, s') - a(s')R(z, \tau, \tau_0, s') \\ &+ c(\tau)R(\tau, \tau, \tau_0, s') \int_{-\infty}^{0} R(z, \tau, \tau_0, s'')k(s'')\, ds'' \Big\}. \end{aligned} \tag{3.13}$$

This last equation must hold for quite arbitrary f and we conclude that

$$\begin{aligned} \frac{\partial R}{\partial \tau}(z, \tau, \tau_0, s) = &-a(s)R(z, \tau, \tau_0, s) \\ &+ c(\tau)R(\tau, \tau, \tau_0, s) \int_{-\infty}^{0} R(z, \tau, \tau_0, s')k(s')\, ds' \end{aligned} \tag{3.14}$$

at all values of s where R is continuous.

If we choose $z = \tau = \tau_0$ then from (3.8)

$$\varphi(\tau, \tau, \tau) = \int_{-\infty}^{0} R(\tau, \tau, \tau, s')f(s')\, ds', \tag{3.15}$$

while from (3.3) and (3.4)

$$\varphi(\tau, \tau, \tau) = \int_{-\infty}^{0} f(s')\, ds'. \tag{3.16}$$

Hence

$$R(\tau, \tau, \tau, s) = 1, \tag{3.17}$$

at all values of s for which R is continuous.

To understand this result it is perhaps helpful to think of $f(s)$ *formally* as a delta function:

$$f(s) = \delta(s - s_1), \qquad -\infty < s_1 < 0. \tag{3.18}$$

Then $\varphi(z, \tau, \tau_0) = R(z, \tau, \tau_0, s_1)$. As a notational matter we write $R(z, \tau, \tau_0, s) = \bar{\varphi}(z, \tau, \tau_0, s)$ and conclude that

$$\begin{aligned} (\partial\bar{\varphi}/\partial\tau)(z, \tau, \tau_0, s) = &-a(s)\bar{\varphi}(z, \tau, \tau_0, s) \\ &+ c(\tau)\bar{\varphi}(\tau, \tau, \tau_0, s) \int_{-\infty}^{0} \bar{\varphi}(z, \tau, \tau_0, s')k(s')\, ds', \\ \tilde{\tau}_0 \leq \tau_0 \leq z \leq \tau \leq \tilde{\tau}, &\qquad \bar{\varphi}(\tau, \tau, \tau, s) = 1. \end{aligned} \tag{3.19}$$

Pursuing the physical analogy begun earlier we observe that $\bar{\varphi}(z, \tau, \tau_0, s_1)$ is the "integrated flux" at z in a "slab" extending from τ_0 to τ when the input at the "top," $z = \tau$, is $\delta(s - s_1)$.

Equation (3.19) is not adequate to determine $\bar{\varphi}$. It would be sufficient if $\bar{\varphi}(\tau, \tau, \tau_0, s)$ were known for $\tilde{\tau}_0 \leqq \tau \leqq \tilde{\tau}$ since then one could integrate on τ

starting with $\tau = z$ as the initial value. Clearly another equation must be found. To do this we consider a *new* "transport" problem in which (3.3a) holds and the boundary conditions (3.3b) and (3.3c) are replaced by

$$\tilde{N}(\tau_0, \tau, \tau_0, s) = \tilde{f}(s), \quad s > 0, \tag{3.20b}$$

$$\tilde{N}(\tau, \tau, \tau_0, s) = 0, \quad s < 0. \tag{3.20c}$$

(Here the tilde is used to distinguish this problem from the preceding.) This corresponds to an input at the "bottom" of the slab. Using Theorem 3.2 we conclude easily that there exists a function $\tilde{R}$ such that

$$\tilde{\varphi}(z, \tau, \tau_0) = \int_0^\infty \tilde{f}(s')\tilde{R}(z, \tau, \tau_0, s')\, ds'.$$

Differentiating (3.3a) with respect to τ_0 now yields

$$\operatorname{sgn} s \frac{\partial N_3}{\partial z}(z, \tau, \tau_0, s) + a(s)N_3 = k(s)c(z)\int_{-\infty}^\infty N_3(z, \tau, \tau_0, s')\, ds', \tag{3.21}$$

$$N_3(\tau_0, \tau, \tau_0, s) = -N_1(\tau_0, \tau, \tau_0, s), \quad s > 0, \qquad N_3(\tau, \tau, \tau_0, s) = 0, \quad s < 0.$$

Thus

$$\begin{aligned} \varphi_3(z, \tau, \tau_0) &= \int_0^\infty \tilde{R}(z, \tau, \tau_0, s')\{-N_1(\tau_0, \tau, \tau_0, s')\}\, ds' \\ &= \int_0^\infty \tilde{R}(z, \tau, \tau_0, s')\{a(s')N(\tau_0, \tau, \tau_0, s') \\ &\qquad - c(\tau_0)k(s')\varphi(\tau_0,\tau,\tau_0)\}\, ds' \\ &= -c(\tau_0)\int_0^\infty k(s')\tilde{R}(z, \tau, \tau_0, s')\, ds' \int_{-\infty}^0 f(s'')R(\tau_0, \tau, \tau_0, s'')\, ds''. \end{aligned} \tag{3.22}$$

Again, from (3.8)

$$\varphi_3(z, \tau, \tau_0) = \int_{-\infty}^0 f(s')R_3(z, \tau, \tau_0, s')\, ds'. \tag{3.23}$$

Equations (3.22) and (3.23) lead to

$$\begin{aligned} \frac{\partial \bar{\varphi}}{\partial \tau_0}(z, \tau, \tau_0, s) &= -c(\tau_0)\bar{\varphi}(\tau_0, \tau, \tau_0, s)\int_0^\infty \underline{\varphi}(z, \tau, \tau_0, s')k(s')\, ds', \\ \underline{\varphi}(\tau_0, \tau_0, \tau_0, s) &= 1, \quad \tilde{\tau}_0 \leqq \tau_0 \leqq \tilde{\tau}, \end{aligned} \tag{3.24}$$

where

$$\underline{\varphi}(z, \tau, \tau_0, s) = \tilde{R}(z, \tau, \tau_0, s). \tag{3.25}$$

Similar arguments can be used to find two more equations:

$$\begin{aligned} (\partial \underline{\varphi}/\partial \tau_0)(z, \tau, \tau_0, s) &= a(s)\underline{\varphi}(z, \tau, \tau_0, s) \\ &\quad - c(\tau_0)\underline{\varphi}(\tau_0,\tau,\tau_0, s)\int_0^\infty k(s')\underline{\varphi}(z, \tau, \tau_0, s')\, ds', \end{aligned} \tag{3.26}$$

and

$$\frac{\partial \underline{\varphi}}{\partial \tau}(z, \tau, \tau_0, s) = c(\tau)\underline{\varphi}(\tau, \tau, \tau_0, s)\int_{-\infty}^{0} \bar{\varphi}(z, \tau, \tau_0, s')k(s')\, ds'. \tag{3.27}$$

To obtain the analogues of Chandrasekhar's X and Y equations we set

$$\begin{aligned} \bar{X}(\tau, \tau_0, s) &= \bar{\varphi}(\tau, \tau, \tau_0, s), && s < 0,\\ \bar{Y}(\tau, \tau_0, s) &= \bar{\varphi}(\tau_0, \tau, \tau_0, s), && s < 0,\\ \underline{X}(\tau, \tau_0, s) &= \underline{\varphi}(\tau_0, \tau, \tau_0, s), && s > 0,\\ \underline{Y}(\tau, \tau_0, s) &= \underline{\varphi}(\tau, \tau, \tau_0, s), && s > 0. \end{aligned} \tag{3.28}$$

Choosing z appropriately in equations (3.19), (3.24), (3.26), (3.27) then yields

$$\begin{aligned} &\bar{Y}_1(\tau, \tau_0, s) = -a(s)\bar{Y}(\tau, \tau_0, s) + c(\tau)\bar{X}(\tau, \tau_0, s)\int_{-\infty}^{0} \bar{Y}(\tau, \tau_0, s')k(s')\, ds',\\ &\bar{X}_2(\tau, \tau_0, s) = -c(\tau_0)\bar{Y}(\tau, \tau_0, s)\int_0^{\infty} \underline{Y}(\tau, \tau_0, s')k(s')\, ds',\\ &\underline{Y}_2(\tau, \tau_0, s) = a(s)\underline{Y}(\tau, \tau_0, s) - c(\tau_0)\underline{X}(\tau, \tau_0, s)\int_0^{\infty} \underline{Y}(\tau, \tau_0, s')k(s')\, ds',\\ &\underline{X}_1(\tau, \tau_0, s) = c(\tau)\underline{Y}(\tau, \tau_0, s)\int_{-\infty}^{0} \bar{Y}(\tau, \tau_0, s')k(s')\, ds',\\ &\quad \bar{X}(\xi, \xi, s) = \bar{Y}(\xi, \xi, s) = 1, \qquad s < 0,\\ &\quad \underline{X}(\xi, \xi, s) = \underline{Y}(\xi, \xi, s) = 1, \qquad s > 0,\\ &\qquad \tilde{\tau}_0 \leqq \xi \leqq \tilde{\tau}. \end{aligned} \tag{3.29}$$

THEOREM 3.3. *The functions* $R(\tau, \tau, \tau_0, s) = \bar{X}(\tau, \tau_0, s)$, $R(\tau_0, \tau, \tau_0, s) = \bar{Y}(\tau, \tau_0, s)$, $\tilde{R}(\tau_0, \tau, \tau_0, s) = \underline{X}(\tau, \tau_0, s)$, *and* $\tilde{R}(\tau, \tau, \tau_0, s) = \underline{Y}(\tau, \tau_0, s)$ *satisfy the system* (3.29) *at all points* s *save where* R *and* $\tilde{R}$ *have discontinuities. Similarly the functions* $R(z, \tau, \tau_0, s) = \bar{\varphi}(z, \tau, \tau_0, s)$ *and* $\tilde{R}(z, \tau, \tau_0, s) = \underline{\varphi}(z, \tau, \tau_0, s)$ *satisfy* (3.19) *and* (3.26).

It is clear that knowledge of the functions of Theorem 3.3 completely determines the solution to (3.1). Unfortunately, integration of (3.29) is a formidable task. If, however, c is somewhat restricted, then a tractable system is obtained. (It should be noted that in the cases treated by Chandrasekhar, c is identically constant.)

Let us suppose $\tau_0 = -\tau$ and $c(z) = c(-z)$. Then relatively easy symmetry arguments show that $\bar{X}(\tau, -\tau, -s) = \underline{X}(\tau, -\tau, s)$ and $\bar{Y}(\tau, -\tau, -s) = \underline{Y}(\tau, -\tau, s)$.

THEOREM 3.4. *Suppose $c(z)$ is even. Define $X(\tau, s) = \underline{X}(\tau, -\tau, s)$, $Y(\tau, s) = \underline{Y}(\tau, -\tau, s)$, $s > 0$. Then at all points s of continuity,*

$$\frac{\partial X}{\partial \tau}(\tau, s) = 2c(\tau)Y(\tau, s)\int_0^\infty Y(\tau, s')k(s')\,ds',$$

$$\frac{\partial Y}{\partial \tau}(\tau, s) = -2a(s)Y(\tau, s) + 2c(\tau)X(\tau, s)\int_0^\infty Y(\tau, s')k(s')\,ds', \tag{3.30}$$

$$X(0, s) = Y(0, s) = 1.$$

Equations (3.30) are precisely the X and Y equations of Chandrasekhar in the event $c \equiv$ constant except for the factors of 2 on the right hand sides. These arise because our z interval extends from $-\tau$ to $+\tau$ instead of from zero to τ.

Theorems 3.3 and 3.4 provide not only a means of actually effectively computing solutions of (3.1) (see **[3]**, **[26]**) but they also may be used to calculate the first eigenvalue of the associated homogeneous equation. For convenience let us discuss the case in which $c(z)$ is even and $\tau_0 = -\tau$. Write for fixed $\lambda > 0$

$$\lambda\psi(z) = \int_{-\tau}^{\tau} c(z')K(|z - z'|)\psi(z')\,dz'. \tag{3.31}$$

Define X and Y by equations (3.30) with k replaced by k/λ. Now let equations (3.30) be integrated from $\tau = 0$ to that value $\tilde{\tau}$ at which the solutions cease to exist. Then $\lambda = \lambda_1(\tilde{\tau})$, the highest eigenvalue associated with (3.31) for $\tau = \tilde{\tau}$. By choosing a set of values λ and finding the corresponding numbers $\tilde{\tau}$ one obtains numerically $\lambda_1(\tilde{\tau})$ as a function of $\tilde{\tau}$.

This process has actually been carried out by R. Kalaba and H. Kagiwada at RAND and by R. C. Allen and the author at the University of New Mexico using a variety of kernels. In the cases tested the device seems to provide a satisfactory scheme for numerically evaluating the highest eigenvalue of equations like (3.31). While no effort has been made to thoroughly test the efficacy of the method in comparison with the many standard ways of computing eigenvalues some checks were made against a classical iteration scheme. The present device proved very competitive and superior in some respects. In the table are given eigenvalues computed in this new way for the kernel

$$K(|z'' - z'|) = \int_0^1 e^{-5s}e^{-s|z''-z'|}\,ds. \tag{3.32}$$

The values of $\bar{\lambda}_1$, upper bounds on λ_1, were obtained by the method of §2.

$\tilde{\tau}$	$\lambda_1(\tilde{\tau})$	$\bar{\lambda}_1(\tilde{\tau})$
0.06836	0.0250	0.0269
0.13281	0.0500	0.0519
0.26465	0.1000	0.1017
0.54199	0.2000	0.2017
0.83691	0.3000	0.3015
1.15137	0.4000	0.4016
1.48633	0.5000	0.5020

IV. **Eigenvalues of Schrödinger-like equations.** In this section we shall consider the system

$$(4.1)\qquad \begin{aligned} u''(z) - \lambda^2 u(z) &= f(z)u(z), \quad 0 \leqq z < \infty, \\ u(0) = 0, &\quad u'(0) = 1, \end{aligned}$$

where

$$(4.2)\qquad \int_0^\infty |f(z)|\, dz < \infty,$$

and $f(z)$ is continuous and bounded for $0 \leqq z < \infty$. We seek those positive values of λ such that

$$(4.3)\qquad \lim_{z\to\infty} u(z) = 0.$$

By employing some of the same concepts as used in the preceding section we shall obtain a function of λ whose zeros are precisely the eigenvalues of the system (4.1), (4.3). Moreover, this function is quite amenable to computation.

The basic idea of §3—and of the whole imbedding method—is to carry out a perturbation on the "size of the system." Thus the integral in equation (3.1) was not considered to run between fixed limits, as is classically the case in the study of Fredholm equations, but rather between variable limits τ_0 and τ. In the course of the derivation these limits were *perturbed* by differentiating some of the quantities of interest with respect to τ_0 and τ. It is now desirable to find some parameter or parameters in (4.1) which may similarly be perturbed. We accomplish this in what seems to be a somewhat artificial manner. A bit of reflection reveals, however, that the trick is not really significantly less transparent than that of replacing the classically constant limits in an integral equation by variables.

Specifically, we replace (4.1) by a new system

$$\begin{aligned} &(4.4a)\quad u'' - \lambda^2 u = f(z)u, \\ &(4.4b)\quad u(x) = \cos\theta, \quad u'(x) = \sin\theta, \end{aligned} \qquad 0 \leqq x \leqq z < \infty, \quad 0 \leqq \theta < 2\pi.$$

Here x and θ are the "artificial" parameters. The problem (4.1) has been *imbedded* in a family of related problems. To emphasize the rôles of x and θ we now write

$$(4.5)\qquad u = u(z, x, \theta), \qquad u' = (\partial/\partial z)u(z, x, \theta).$$

The condition (4.2) assures the existence of two fundamental solutions of (4.4a):

$$(4.6)\qquad u_1 = e^{\lambda z}(1 + o(1)), \qquad u_2 = e^{-\lambda z}(1 + o(1)).$$

A more careful analysis of (4.4) along classical lines yields the following theorem (see [1], [27]):

THEOREM 4.1. *The solution* $u(z, x, \theta)$ *of* (4.4) *for fixed* $\lambda > 0$ *can be written as*

$$(4.7)\qquad u(z, x, \theta) = e^{\lambda(z-x)}\{A(x, \theta, \lambda) + \tau(z, x, \theta, \lambda)\}$$

where A has continuous first partial derivatives in x and θ. Moreover

$$\lim_{z\to\infty} \tau(z, x, \theta, \lambda) = 0 \tag{4.8}$$

uniformly in x and θ.

The behavior of A for large x is of considerable interest. To see intuitively what this should be, suppose $f(z) \equiv 0$ for $z > X$. The problem (4.4) reduces to

$$\begin{aligned} &u'' - \lambda^2 u = 0, \\ &u(x, x, \theta) = \cos\theta, \quad u'(x, x, \theta) = \sin\theta, \end{aligned} \qquad X < x \leqq z < \infty. \tag{4.9}$$

The solution to (4.9) is elementary:

$$u(z, x, \theta) = \frac{\lambda\cos\theta + \sin\theta}{2\lambda} e^{\lambda(z-x)} + \frac{\lambda\cos\theta - \sin\theta}{2\lambda} e^{-\lambda(z-x)}. \tag{4.10}$$

Hence in this special case

$$A(x, \theta, \lambda) = (\lambda\cos\theta + \sin\theta)/2\lambda, \qquad x > X. \tag{4.11}$$

It is not hard to prove that this remains "asymptotically true" for f merely "small" at infinity. Thus we have

THEOREM 4.2. *If f satisfies* (4.2) *then*

$$\lim_{x\to\infty} A(x, \theta, \lambda) = \frac{\lambda\cos\theta + \sin\theta}{2\lambda}. \tag{4.12}$$

We are now ready to apply a perturbation technique. Elementary calculation yields

$$\begin{aligned} u(x + \Delta, x, \theta) &= \cos\theta + \Delta\sin\theta + o(\Delta), \\ u'(x + \Delta, x, \theta) &= \sin\theta + \Delta[f(x) + \lambda^2]\cos\theta + o(\Delta). \end{aligned} \tag{4.13}$$

It is desirable to introduce a new initial value problem on $x < x + \Delta \leqq z < \infty$. (It will be convenient to suppose $\Delta > 0$.) In fact, we should like to write

$$\begin{aligned} u(x + \Delta, x, \theta) &= u(x + \Delta, x + \Delta, \psi), \\ u'(x + \Delta, x, \theta) &= u(x + \Delta, x + \Delta, \psi), \end{aligned} \tag{4.14}$$

for some ψ. This, however, is not likely to be possible since u is so normalized that

$$u^2(x + \Delta, x + \Delta, \psi) + u'^2(x + \Delta, x + \Delta, \psi) = 1. \tag{4.15}$$

To overcome this difficulty we put

$$\begin{aligned} u(x + \Delta, x, \theta) &= \eta(x, \theta, \Delta)u(x + \Delta, x + \Delta, \psi(x, \theta, \Delta)) \\ &= \eta(x, \theta, \Delta)\cos\psi(x, \theta, \Delta), \\ u'(x + \Delta, x, \theta) &= \eta(x, \theta, \Delta)u'(x + \Delta, x + \Delta, \psi(x, \theta, \Delta)) \\ &= \eta(x, \theta, \Delta)\sin\psi(x, \theta, \Delta). \end{aligned} \tag{4.16}$$

Trivial calculations using (4.13) and (4.16) now lead to

$$(4.17)\qquad \begin{aligned}\psi(x, \theta, \Delta) &= \theta + \Delta\{(f(x) + \lambda^2)\cos^2\theta - \sin^2\theta\} + o(\Delta),\\ \eta(x, \theta, \Delta) &= 1 + \Delta(f(x) + \lambda^2)\cos\theta\sin\theta + o(\Delta).\end{aligned}$$

From the very definitions of the function $u(z, x, \theta)$, together with the linearity of the system (4.4), we conclude

$$(4.18)\qquad u(z, x, \theta) \equiv \eta(x, \theta, \Delta)u(z, x + \Delta, \psi(x, \theta, \Delta))$$

for $z \geqq x + \Delta$. Equation (4.7) readily leads to

$$(4.19)\qquad A(x, \theta, \lambda) = e^{-\lambda\Delta}\eta(x, \theta, \Delta)A(x + \Delta, \psi(x, \theta, \Delta), \lambda).$$

It is clear that (4.19) together with (4.17) and the known differentiability of A yield a partial differential equation for A. For reasons of convenience we prefer to deal with

$$(4.20)\qquad B(x, \theta, \lambda) = A(x, \theta, \lambda) - (\lambda\cos\theta + \sin\theta)/2\lambda.$$

Then

$$(4.21a)\qquad \begin{aligned}&\partial B/\partial x + [(f(x) + 1 + \lambda^2)\cos^2\theta - 1]\,\partial B/\partial\theta\\ &\qquad = [\lambda - (f(x) + 1 + \lambda^2)\cos\theta\sin\theta]B(x, \theta, \lambda) - f(x)\cos\theta/2\lambda\end{aligned}$$

and (4.12) gives

$$(4.21b)\qquad \lim_{k\to\infty} B(x, \theta, \lambda) = 0.$$

Use of the method of characteristics on (4.21a) leads to

$$(4.22)\qquad \begin{aligned}&dx/ds = 1, \qquad d\theta/ds = (f(x) + 1 + \lambda^2)\cos^2\theta - 1,\\ &dB/ds = [\lambda - (f(x) + 1 + \lambda^2)\cos\theta\sin\theta]B - f(x)\cos\theta/2\lambda\end{aligned}$$

and hence *formally* to

$$(4.23a)\qquad \begin{aligned}B(x, \theta, \lambda) = \frac{1}{2\lambda}\int_x^\infty & f(s)\cos\tilde\theta(s)\\ &\cdot\exp\left\{\int_x^s [(f(p) + 1 + \lambda^2)\cos\tilde\theta(p)\sin\tilde\theta(p) - \lambda]\,dp\right\} ds,\end{aligned}$$

$$(4.23b)\qquad (d\tilde\theta/ds)(s) = (f(s) + 1 + \lambda^2)\cos^2\tilde\theta(s) - 1,$$

$$(4.23c)\qquad \tilde\theta(s) = \theta.$$

This must be considered as a *formal* result since the convergence of the infinite integral defining B is by no means obvious. In fact, unless further restrictions are put on f *direct* proof of the validity of (4.23) seems very difficult. There are several ways to avoid this difficulty (see **[1]**) but it is not appropriate to delve into the details here. We state simply

THEOREM 4.3. *Provided f is such that the integral* (4.23a) *converges the function* $B(x, \theta, \lambda)$ *and hence the function* $A(x, \theta, \lambda)$ *are completely determined by* (4.23) *and* (4.20).

We have not yet added the condition that λ be an eigenvalue of the system—that is, that (4.3) should hold. It is clear from (4.7) that if λ is such that $A(x, \theta, \lambda) \neq 0$ then $u(z, x, \theta)$ becomes infinite as $z \to \infty$. The vital result of this section is the following theorem.

THEOREM 4.4. *A necessary and sufficient condition that* $\lambda > 0$ *be an eigenvalue of the system* (4.9) *is that* $A(x, \theta, \lambda) = 0$.

PROOF. The necessity has just been noted so we need only consider the sufficiency. If $A = 0$ then (4.7) becomes

$$u(z, x, \theta) = e^{\lambda(z-x)}\tau(z, x, \theta, \lambda). \tag{4.24}$$

However, we also know from (4.6) that there exist constants C_1 and C_2 such that

$$u(z, x, \theta) = C_1 e^{\lambda z}(1 + o(1)) + C_2 e^{-\lambda z}(1 + o(1)). \tag{4.25}$$

Now (4.24) implies $e^{-\lambda z}u(z, x, \theta) \to 0$ as $z \to \infty$. Therefore $C_1 = 0$, and the sufficiency is established.

A study of the function A reveals various classical results concerning the behavior of the eigenvalues of the system [7], [**20**]. Of greatest interest here is the fact that equations (4.23) may be handled numerically with relatively little difficulty and provide then a method of actually computing these eigenvalues by simply determining the zeros of A, assuming x and θ given.

R. C. Allen and the author have considered several examples [**1**]. Let us examine one. Suppose

$$f(z) = \alpha(1 - \alpha)/4 \cosh^2 (z/2), \qquad \alpha \text{ a positive integer.} \tag{4.26}$$

This is certainly an admissible f. Choose $x = 0$ and $\theta = \pi/2$. The corresponding system (4.1) can be solved analytically and its eigenvalues are hence known explicitly [**20**]. For $\alpha = 5$ there are exactly two eigenvalues, $\lambda_1 = 0.5$ and $\lambda_2 = 1.5$. For $\alpha = 11$ there are five, $\lambda_j = j - 0.5, j = 1, 2, \ldots, 5$. The system (4.23) was integrated for λ ranging from 0.10 to 1.60 in steps of 0.10 for the first case, and from 0.10 to 5.00 in steps of 0.10 for the second case. The zeros of A were located approximately and then various refinements were employed to obtain better accuracy. A fourth order Runge-Kutta method with single precision arithmetic on an IBM 709 gave excellent results with only a few minutes computing time required. In all cases λ's were calculated correct to four decimal places with little difficulty.

Other problems have been tried quite successfully although in one case double precision arithmetic was needed. No effort has been made to construct a "production type" code since our efforts were strictly experimental.

The requirement (4.2) is a fairly stringent one since it excludes certain "potential" functions of interest to the physicists, in particular the Coulomb potential.

(It should be remarked that only small changes in the arguments already described need be made to allow f to have mild discontinuities.) F. Hagin has studied the generalization

$$u'' - (\lambda^2 + g(z))u = f(z)u \tag{4.27}$$

subject to the same boundary conditions we have been discussing [7]. Here f is as before and g is absolutely continuous with

$$\int_0^\infty |g'(z)|\,dz < \infty, \qquad \int_0^\infty |g(z)|^2\,dz < \infty, \qquad g(z) \to 0 \quad \text{as} \quad z \to 0. \tag{4.28}$$

In particular, one notes that if $g(z)$ behaves like $1/z$ for z large, then g is admissible. Equation (4.22) now has two fundamental solutions of the form

$$\begin{gathered} u_1(z) = e^{\alpha(z,\lambda)}(1 + o(1)), \qquad u_2(z) = e^{-\alpha(z,\lambda)}(1 + o(1)), \\ \alpha(z, \lambda) = \lambda \int_b^z \left\{1 + \frac{g(t)}{2\lambda}\right\} dt, \quad b \text{ large.} \end{gathered} \tag{4.29}$$

The kind of arguments presented can now be carried out for this more general case although the details are more complicated and more delicate. Moreover, a certain amount of singular behavior in f and g in the neighborhood of the initial value can be tolerated. The final formulas are again amenable to computation. Hagin considered the case

$$g(z) = 1/(z + 1), \qquad f(z) = \beta(\beta + 1)/(z + 1)^2, \qquad x = 0. \tag{4.30}$$

When θ is chosen appropriately certain eigenvalues λ can be found analytically. Thus, for example, if $\theta = \tan^{-1}(-\frac{5}{6})$, then $\lambda = 1$ is an eigenvalue of the system. Computational results were again excellent.

One wonders whether the restrictions on f and g may not be considerably relaxed. They are currently imposed largely in order to prove convergence of the integrals defining B and its analogues (see (4.23)). As an experiment the problem

$$\begin{gathered} u'' - (k^2z^2 + k)u = 0, \\ u(x) = \cos\theta, \qquad u'(x) = \sin\theta, \end{gathered} \tag{4.31}$$

was studied [7]. It was *not* possible to establish the convergence or divergence of the integrals arising in the treatment. However, these expressions were evaluated on the IBM 709 and the *numerical* convergence was *extremely* rapid. Values obtained for k for $x = 0$ and various θ's were found to agree very well with analytical results. This suggests that the method may have very wide applicability.

It is difficult to compare the efficacy of this imbedding method for finding eigenvalues with those devices commonly in numerical use. Certainly such a comparative study would be valuable. Hagin has noted, however, that the method seems to become more efficient and more stable as u_1 (see (4.29)) becomes more badly behaved. Thus in the problem discussed in the preceding paragraph $u_1 = e^{(1/2)kz^2}$, yet the k values calculated by the formal procedure were found with ease.

V. **Summary.** In this discussion I have tried to indicate some of the ways in which mathematical ideas originated to handle transport problems—or at least owing their major development to their use in studying such problems—may be of value in attacking other questions in applied mathematics and mathematical physics. As indicated earlier, the treatment has necessarily been confined to a few instances of special personal interest, and I again offer apologies for the many cases slighted or ignored.

Needless to say, the possibilities of future exploration of some of these ideas seem limitless. It is hence my hope that this conference may not only stimulate interest in the subject of Transport Theory but also in the possibility of using the tools of the theory in ever-widening areas of mathematics, numerical analysis, and mathematical physics.

Bibliography

1. R. C. Allen, Jr. and G. M. Wing, *A method for computing eigenvalues of Schrödinger-like equations*, J. Math. Anal. Appl. **15** (1966), 340–354.

2. V. A. Ambarzumian, *Diffuse reflection of light by a foggy medium*, Comptes Rendus (Doklady) de l'Académie de l'URSS, **38** (1943), 229–232.

3. R. E. Bellman, H. H. Kagiwada, and R. E. Kalaba, *Numerical results for the auxiliary equation of radiative transfer*, J. Quant. Spectros. and Radiat. Transfer **6** (1966), 291–310.

4. G. Birkhoff, *Positivity and criticality*, Proc. Sympos. Appl. Math. Vol. 11. Amer. Math. Soc. Providence, R.I., 1961.

5. K. M. Case, *Elementary solutions of the transport equation and their applications*, Ann. Phys. **9** (1960), 1–23.

6. S. Chandrasekhar, *Radiative transfer*, Clarendon Press, Oxford, 1950.

7. F. G. Hagin, *Invariant imbedding and the asymptotic behavior of solutions to initial value problems*, doctoral dissertation, University of Colorado, 1966.

8. T. E. Harris, *The theory of branching processes*, Springer-Prentice-Hall, Berlin, Englewood Cliffs, N.J., 1963.

9. J. Hersch, *Caractérisation variationelle d'une somme de valeurs propres consécutive; généralisation d'inégalités de Pólya-Schiffer et de Weyl*, C. R. Acad. Sci., **252** (1961), 1714–1716.

10. H. H. Kagiwada and R. E. Kalaba, *An initial value method for Fredholm integral equations of convolution type*, RAND Corporation Memorandum RM-5186-PR (1966).

11. A. Leonard and T. W. Mullikin, *A spectral analysis of the anisotropic neutron transport kernel in slab geometry with applications*, J. Math. Phys. **5** (1964), 399–411.

12. ———*An application of singular integral-equation theory to a linearized problem in Couette flow*, Ann. Phys. **30** (1964), 235–248.

13. ———*Integral equations with difference kernels on finite intervals*, Trans. Amer. Math. Soc. **116** (1965), 465–473.

14. T. W. Mullikin, "Nonlinear integral equations of radiative transfer," in *Nonlinear integral equations*, Univ. of Wisconsin Press, Madison, Wis., 1964, pp. 345–374.

15. ———*Estimates of critical dimensions of spherical and slab reactors*, J. Math. Anal. Appl. **5** (1962), 184–199.

16. T. W. Mullikin and I. Selim, *The likelihood-ratio filter for the detection of Gaussian signals in white noise*, RAND Corporation, RM-4207-PR, 1964.

17. A. Roark and G. M. Wing, *A method for computing eigenvalues of certain integral equations*, Numer. Math. **7** (1965), 159–170.

18. J. C. Stewart, *On the X- and Y-functions and Green's function for a finite slab*, Joint Institute for Laboratory Astrophysics, preprint.

19. G. G. Stokes, *Mathematical and physical papers*. Vol. 4, Cambridge University Press, New York, 1880.

20. E. C. Titchmarsh, *Eigenfunction expansions associated with second-order differential equations*. Part I, Oxford, London, 1946.

21. N. G. Van Kampen, *On the theory of stationary waves in plasmas*, Physica **21** (1955), 949–963.

22. N. Wiener and E. Hopf, *Über eine Klasse singulärer Integralgleichungen*, Sitzung Berichtung der Preussische Akademie der Wissenschaften (1931), 696–706.

23. G. M. Wing, *On a method for obtaining bounds on the eigenvalues of certain integral equations*, J. Math. Anal. Appl. **11** (1965), 160–175.

24. ———*Some convexity theorems for eigenvalues of Fredholm integral equations*, J. Math. Appl. **19** (1967), 330–338.

25. ———*An introduction to transport theory*, Wiley, New York, 1962.

26. ———*On certain Fredholm integral equations reducible to initial value problems*, SIAM Review, **9** (1967), 655–670.

27. ———*Invariant imbedding and the asymptotic behavior of solutions to initial value problems*, J. Math. Anal. Appl. **9** (1964), 85–98.

UNIVERSITY OF NEW MEXICO

III. Stochastic Aspects

Stochastic formulations of neutron transport[1]

George I. Bell

Abstract. Neutron transport is considered as a Markov branching process, continuous in space, velocity, and time. The probabilities of elementary events are assumed to be given by known neutron cross sections and we limit our attention to zero-power systems. In the Introduction, various formulations of stochastic neutron transport theory are noted, while in the body of the review, we develop the theory in terms of the probability generating function (§B) and low order moments thereof (§C).

In §B we consider the probability, $p_n(R, t_f; \boldsymbol{x}, \boldsymbol{v}, t)$, that in a multiplying system, a neutron at position $\boldsymbol{x}$ with velocity $\boldsymbol{v}$ at time t will lead to exactly n neutrons in region R of $(\boldsymbol{x}, \boldsymbol{v})$ space at time t_f. An integrodifferential equation for the probability generating function is given, together with final and boundary conditions. The relationship of the generating function to solutions of the adjoint linear transport equation is examined and the qualitative behavior of solutions for subcritical and supercritical systems is indicated. An equation for the extinction probability in supercritical systems is derived. We show that from this generating function the probability distribution of neutrons in a system with a neutron source can be found.

In §C, low order moments of the probability distribution are considered. Equations for them are derived from the equation for the generating function and it is seen that one can solve for the moments successively rather than simultaneously. Moreover, it is shown that the moments can be found from a knowledge of the Green's function of the corresponding linear transport problem. The application of these methods to analysis of experiments is considered briefly. A problem in which counts are detected from two absorption detectors is formulated and other experiments are mentioned.

[1] Work performed under the auspices of the U.S. Atomic Energy Commission.

A. **Introduction.** Neutron transport theory is ordinarily concerned with the mean or expected value of a neutron population. Thus the theory [**1**] is usually formulated in terms of $\bar{n}(\boldsymbol{x}, \boldsymbol{v}, t)\, d\boldsymbol{x}\, d\boldsymbol{v}$ which is to be interpreted as the mean or ensemble expectation value of the number of neutrons having at time t, positions within $d\boldsymbol{x}$ about $\boldsymbol{x}$ and velocities within $d\boldsymbol{v}$ about $\boldsymbol{v}$.

The actual neutron population in a system may or may not resemble the mean value of the population; roughly speaking if there are few neutrons in a system, then the actual population is unlikely to resemble the mean population; whereas if there are many neutrons in a system, the actual population is unlikely to be very different from the mean population. We shall try to give a little more precise meaning to these ideas later on, but at the outset some extreme cases are intuitively clear. (1) Consider, as a first example, a supercritical system which contains only a weak neutron source. For such a system the mean neutron population will be proportional to the source and it will increase exponentially in time. The actual neutron population, however, will be strictly zero until the first source neutron appears. Thereafter, the population may increase very rapidly or it may die out to zero again. In neither case is the actual neutron population likely to resemble the mean population. (2) A power reactor may contain of the order of 10^{15} neutrons. For such a large number of neutrons, one would probably expect that the fluctuations in the population would be small compared to the mean value. Although relatively small, the fluctuations may nevertheless be quite interesting and their study may be fruitful.

These two extreme examples typify the two most important types of practical questions which have motivated the development of a stochastic theory of neutron transport. In the first case one seeks to compute probabilities of extreme departures from the mean value of a population, for example, to compute the probability that the population becomes extinct. Such problems are evidently related to the classical Galton-Watson problem of the extinction of surnames in the theory of Markov processes [**2**]. In problems of the second type, one studies the fluctuation in the neutron population in a steady state system with the aim of deducing properties of the system from the fluctuations. Here, by a steady state system, I mean one in which the equations for the mean neutron population have a time independent solution. A subcritical system with a steady source or a critical system are examples of steady systems. In such systems the time average of the neutron population is assumed to be identical to the mean population and one studies fluctuations in the population (or actually fluctuations in detector counting rates) in order to deduce such characteristics of the system as the degree of subcriticality and the lifetimes of neutrons and delayed neutron precursors. The number of counts recorded by a detector in an interval of time is typically analysed, together with the correlations between counts in the same detector at different times or between different detectors at the same or different times.

That neutron transport is a stochastic process is at once apparent when one recognizes that the macroscopic cross sections which always appear in transport theory are simply *probabilities* of events happening to a neutron per unit distance of neutron travel. The statistics of the development of a neutron chain will be governed by these elementary processes, that is, by the neutron cross sections. In power reactors, there are, in addition, other sources of fluctuations which arise from such phenomena as vibrations of the fuel rods or turbulence and boiling of coolant. These effects give rise to more or less random variations of the cross sections and they will not be considered in this review. They have generally been treated only in a very approximate manner [3] as a source of noise in a space independent noise theory. By ignoring such effects, we confine our attention to low power or zero power reactors, and we consider the neutron cross sections to be given and known so that the event probabilities governing the stochastic process are known.

Neutron transport is then a Markov process, i.e. a stochastic process in which all information about the population is given by its present state and in which the future evolution is independent of the state at earlier times. If delayed neutrons are to be included in the treatment, then the number and location of the delayed neutron precursors must be included in the description of the present state. The state of the neutron population must be described as a function of two continuous (three vector) variables, space, $\boldsymbol{x}$, and velocity, $\boldsymbol{v}$, and it will then evolve as a function of time.

The space and velocity variables can be made discrete by introducing energy groups and a space mesh in a manner similar to that used in ordinary transport theory, but we will not consider such treatments in this review. More interesting and fruitful have been studies in which time was not retained as a continuous variable but where the neutrons were treated a generation at a time, and we will briefly review such treatments before taking up the formulation with continuous time. When neutrons are classified by generations, an elementary event of some particular kind such as a fission or a collision is taken to divide successive generations of neutrons. In a fundamental study Everett and Ulam [**4**] considered neutrons a generation at a time and in addition they treated space and velocity as discrete rather than continuous variables. Introducing generating functions they showed that the probability distributions for higher generations can be obtained by iteration of a nonlinear generating transformation. By examining this transformation and its fixed points they were able to characterize supercritical, critical, and subcritical systems. More recently, Mullikin [**5**] analyzed the generation problem in (continuous) plane or spherical geometry and showed that generating functions for neutrons in successive generations are related by a nonlinear integral transformation. By applying the Perron-Frobenius-Jentzsch theory of positive operators to this transformation, he was able to distinguish the criticality of systems and to establish the extinction probability for supercritical systems. Somewhat related also is the work of Abu-Shumays [**6**] in applying generating functions to albedo problems in nonmultiplying media. In summary, it appears that there are mathematical advantages in treating neutrons a generation at a time. The neutrons in successive generations are simply related and their probability generating functions are connected by a transformation which can be studied. However, for comparison with experiments these theories are no substitute for treatment with continuous time.

We turn now to consider neutron transport as a Markov process, continuous in space, velocity and time. There is one other similar Markov process in nature that has been extensively studied, namely the cascade processes wherein cosmic ray showers are formed, and the theories which were developed, especially by Jánossay and Ramakrishnan [**2**], for treating the cosmic ray cascades have influenced the development of a stochastic theory of neutron transport. Thus Pál [**7**] patterned his pioneering studies after the work of Jánossay using probability generating functions and considering the point of first collision as a regeneration point. Similarly, several authors have been influenced by the product density functions introduced by Bhabha and Ramakrishnan. We will see shortly how these concepts arise in the stochastic theory of neutron transport.

To my knowledge, the first development of a stochastic theory of neutron transport (with continuous time) was made by Feynman, in unpublished Los Alamos reports. Characteristically, his work was highly original and unencumbered by any knowledge of Galton-Watson processes, cascade theory or the like. The first published stochastic transport theory was that of Pál [**7**], who framed his theory in terms of the probability that a single neutron would lead to some number (n) of neutrons at a later time. By considering the various possible first collisions, and introducing the generating function, Pál derived the nonlinear integrodifferential equation satisfied by the generating function, which is a backwards Kolmogorov equation for the stochastic process and he discussed some of its properties. This general approach will be followed in this review. When making application of his theory to the interpretation of experiments, Pál ignored the space and energy dependence of his theory, thereby ending with a point reactor model. Unfortunately, Pál's fundamental work remained largely unknown to reactor physicists for several (nearly five) years.

A rather similar and general development was undertaken by Matthes [**8**] who was primarily concerned with the correlations between detector outputs or, as will be seen later, with the first two moments of the probability distribution. Moreover, his work emphasized a discrete space-energy mesh.

More recently a number of workers have expanded, clarified, and applied the generating function approach. We may mention the work of Borgwald [**9**] in formulating the theory of detector correlations in a unified way, that of Bell [**10**] in clarifying the relationship of the extinction probability to the generating function, and that of Otsuka and Saito [**11**] in elucidating the relationhsip between generating functions and product densities of all orders. We call attention also to the related work of Raievsky [**12**], Harris [**13**], and Govorkov [**14**]. It should be observed that little of this development has been on a formal mathematical basis.

Two related but different approaches to stochastic theories of neutron transport should be mentioned. In all of the work which has been referred to so far it was assumed that a classical description of neutron transport was adequate and that a neutron could be treated as a point particle, fully specified by its position and velocity. The cross sections were assumed to be known interaction probabilities. Now in fact, neutrons have a spin and a finite wavelength and one might wonder whether a quantum mechanical analysis of neutron transport wouldn't be preferable to a classical treatment. Such a quantum mechanical transport theory has been developed by Osborn **[15]** and colleagues and they have applied it to stochastic theory—phrased in terms of singlet and doublet densities. To date, so far as I am aware, this quantum mechanical stochastic theory has not been used to obtain any results which are not more simply found from classical stochastic theory. However, there would seem to be no doubt as to its wider validity in principle. In this review we will consider in detail only classical theory.

A different approach has been through the use of a Langevin equation. It will be recalled that in Langevin's original theory of Brownian motion **[16]** a random force was postulated to account for the erratic motion of Brownian particles. In the Langevin form of neutron transport it is postulated that the difference between the actual neutron population and the mean one is maintained by a random source which arises from the fluctuations in the neutron population itself **[17]**, **[18]**. The physical basis of the Langevin approach seems a good deal less rigorous than the method which we will employ, but an interesting equivalence between that approach and the use of the first two moments or product densities has recently been established **[18]**.

B. **Formulation of the problem with generating functions.** In this section we will derive an equation satisfied by a probability generating function, note its relation to the ordinary Boltzmann equation, and see how the extinction probability can be found.

Let us consider neutron transport within a finite region of space bounded by a convex surface, B. Within this region we assume that the neutron interaction cross sections are known, and for simplicity we assume that the cross sections are independent of time and we neglect delayed neutrons. As usual, we neglect neutron-neutron collisions. Let $\sigma(\boldsymbol{x}, \boldsymbol{v})$ be the total macroscopic cross section; this is the probability that a neutron at position $\boldsymbol{x}$, velocity $\boldsymbol{v}$ will have a collision per unit distance of neutron travel. Let $c(\boldsymbol{x}, \boldsymbol{v} \to \boldsymbol{v}')\, d\boldsymbol{v}'$ be the mean number of neutrons which emerge from the collision at x having velocities within the range $d\boldsymbol{v}'$ about $\boldsymbol{v}'$.

We must now characterize collisions involving fission, which we assume are the only kind of collision which can lead to an increase in the number of neutrons in the system. Let $\sigma_f(\boldsymbol{x}, \boldsymbol{v})$ be the fission cross section for a neutron having $(\boldsymbol{x}, \boldsymbol{v})$ and let $P_\nu(\boldsymbol{x}, \boldsymbol{v})$ be the probability that fission at $\boldsymbol{x}$, induced by a neutron with velocity $\boldsymbol{v}$, will lead to the instantaneous (prompt) emission of ν neutrons. For simplicity we assume that each fission neutron is emitted isotropically with a probability $\chi(v)\, dv$ of having speeds in the interval dv about v.

With this notation, the ordinary transport equation **[1]** for the mean or ensemble average neutron density $\bar{n}(\boldsymbol{x}, \boldsymbol{v}, t)$ is

$$(1) \qquad \left[\frac{\partial}{\partial t} + \boldsymbol{v} \cdot \nabla + v\sigma(\boldsymbol{x}, \boldsymbol{v})\right]\bar{n}(\boldsymbol{x}, \boldsymbol{v}, t) - \int d\boldsymbol{v}'\sigma(\boldsymbol{x}, \boldsymbol{v}')c(\boldsymbol{x}, \boldsymbol{v}' \to \boldsymbol{v})v'\bar{n}(\boldsymbol{x}, \boldsymbol{v}', t) = 0$$

which we abbreviate as

$$L(\boldsymbol{x}, \boldsymbol{v}, t)\bar{n}(\boldsymbol{x}, \boldsymbol{v}, t) = 0. \tag{2}$$

Here we have assumed for the moment that there are no neutron sources in the systems, and we have in mind an initial value problem in which $\bar{n}(\boldsymbol{x}, \boldsymbol{v}, 0)$ is given, boundary conditions are given (typically, no incoming neutrons or $\bar{n} = 0$ for $\boldsymbol{x} \in B$ and $\boldsymbol{v} \cdot \hat{n} \leq 0$ with $\hat{n}$ an outward normal) and then (1) determines the evolution of the mean population.

For future reference, it is worth defining an operator, $L^\dagger$, adjoint to L, where

$$\begin{aligned} L^\dagger(\boldsymbol{x}, \boldsymbol{v}, t)m^\dagger(\boldsymbol{x}, \boldsymbol{v}, t) &= \left[-\frac{\partial}{\partial t} - \boldsymbol{v} \cdot \nabla + v\sigma(\boldsymbol{x}, \boldsymbol{v})\right] m^\dagger(\boldsymbol{x}, \boldsymbol{v}, t) \\ &\quad - v\sigma(\boldsymbol{x}, \boldsymbol{v}) \int d\boldsymbol{v}' c(\boldsymbol{x}, \boldsymbol{v} \to \boldsymbol{v}') m^\dagger(\boldsymbol{x}, \boldsymbol{v}', t) \end{aligned} \tag{3}$$

and the functions $m^\dagger$ satisfy appropriate final conditions and adjoint boundary conditions (typically, no "outgoing" $m^\dagger$, or $m^\dagger = 0$ for $\boldsymbol{x} \in B$ and $\boldsymbol{v} \cdot \hat{n} > 0$). If we then define an inner product $(m^\dagger, \bar{n})$ as an integral over space, velocity, and time, we find that

$$(L^\dagger m^\dagger, \bar{n}) = (m^\dagger, L\bar{n}) \tag{4}$$

or, more generally, boundary values of $\bar{n}$ and $m^\dagger$ may be involved **[10]**. As is well known, $m^\dagger$ can be interpreted in terms of the importance of neutrons (having $(\boldsymbol{x}, \boldsymbol{v})$ at time t) in contributing to some final state.

If there is a nonfission source of neutrons in the system such that $S(\boldsymbol{x}, \boldsymbol{v}, t)\, d\boldsymbol{x}\, d\boldsymbol{v}$ is the probability that a neutron appears per unit time, then the mean neutron population will satisfy

$$L\bar{n} = S. \tag{5}$$

For the particular case where S is a product of delta functions representing a single neutron born at position $\boldsymbol{x}_0$, and time t_0, with velocity $\boldsymbol{v}_0$, then the mean population, $\bar{n}$, which is obtained from (5) will be a Green's function which we denote

$$\mathrm{Gr}\,(\boldsymbol{x}_0, \boldsymbol{v}_0, t_0 \to \boldsymbol{x}, \boldsymbol{v}, t),$$

and

$$L(\boldsymbol{x}, \boldsymbol{v}, t)\,\mathrm{Gr}\,(\boldsymbol{x}_0, \boldsymbol{v}_0, t_0 \to \boldsymbol{x}, \boldsymbol{v}, t) = \delta(\boldsymbol{x} - \boldsymbol{x}_0)\delta(\boldsymbol{v} - \boldsymbol{v}_0)\delta(t - t_0) \tag{6}$$

where we consider $t \geq t_0$. Since L is a linear operator, we will have for an arbitrary source

$$\bar{n}(\boldsymbol{x}, \boldsymbol{v}, t) = \iint d\boldsymbol{x}_0\, d\boldsymbol{v}_0 \int_{-\infty}^{t} dt_0 S(\boldsymbol{x}_0, \boldsymbol{v}_0, t_0)\,\mathrm{Gr}\,(\boldsymbol{x}_0, \boldsymbol{v}_0, t_0 \to \boldsymbol{x}, \boldsymbol{v}, t). \tag{7}$$

Moreover, the adjoint operator will have a Green's function which is the same function, except that the meaning of the variables is interchanged,

$$L^\dagger(\boldsymbol{x}_0, \boldsymbol{v}_0, t_0)\,\mathrm{Gr}\,(\boldsymbol{x}_0, \boldsymbol{v}_0, t_0 \to \boldsymbol{x}, \boldsymbol{v}, t) = \delta(\boldsymbol{x} - \boldsymbol{x}_0)\delta(\boldsymbol{v} - \boldsymbol{v}_0)\delta(t - t_0) \tag{6a}$$

where still $t \geq t_0$. The adjoint problem with a source $L^\dagger m^\dagger = S^\dagger$ will then have the solution

(7a) $$m^\dagger(\boldsymbol{x}_0, \boldsymbol{v}_0, t_0) = \iint d\boldsymbol{x}\, d\boldsymbol{v} \int_{t_0}^{\infty} dt S^\dagger(\boldsymbol{x}, \boldsymbol{v}, t) \operatorname{Gr}(\boldsymbol{x}_0, \boldsymbol{v}_0, t_0 \to \boldsymbol{x}, \boldsymbol{v}, t).$$

We will encounter these Green's functions again when we consider moments of the probability distribution.

Let us now seek the probability $p_n(R, t_f; \boldsymbol{x}, \boldsymbol{v}, t)$ that a neutron at position $\boldsymbol{x}$ with velocity $\boldsymbol{v}$ at time t will lead to exactly n neutrons in region R of $\boldsymbol{x}, \boldsymbol{v}$ space at time t_f. We may formulate this in terms of the probability of various outcomes of the *first* collision. A typical term will include, for example, the probability that the collision occurs at a position $\boldsymbol{x}'$ (along the vector $\boldsymbol{x}' = \boldsymbol{x} + \boldsymbol{v}s$ $(s > 0)$) multiplied by the probability that the collision leads to fission with the emission of ν neutrons $(\sigma_f/\sigma)P_\nu$ multiplied by the probability that these ν neutrons jointly lead to n neutrons in R at t_f. Summing over all possibilities and introducing the probability generating function

(8) $$G(z, R, t_f; \boldsymbol{x}, \boldsymbol{v}, t) = \sum_{n=0}^{\infty} z^n p_n(R, t_f; \boldsymbol{x}, \boldsymbol{v}, t),$$

one can derive [7], **[10]** an equation which is satisfied by the generating function. A somewhat simpler equation is found for the function

(9) $$\mathscr{G}(z, R, t_f; \boldsymbol{x}, \boldsymbol{v}, t) = 1 - G$$

namely,

(10) $$\begin{aligned} L^\dagger(\boldsymbol{x}, \boldsymbol{v}, t)\mathscr{G}(z, R, t_f; \boldsymbol{x}, \boldsymbol{v}, t) &= -\sum_{j=2}^{\infty} \frac{(-1)^j}{j!} \overline{\nu(\nu-1)\cdots(\nu-j+1)}\, \sigma_f(\boldsymbol{x}, \boldsymbol{v})v \\ &\quad \times \left[\int \mathscr{G}(z, R, t_f; \boldsymbol{x}, \boldsymbol{v}', t)\chi(v')\, d\boldsymbol{v}'\right]^j \end{aligned}$$

where

(11) $$\overline{\nu(\nu-1)\cdots(\nu-j+1)} = \sum_{\nu=j}^{\nu_{\max}} \nu(\nu-1)\cdots(\nu-j+1)P_\nu(\boldsymbol{x}, \boldsymbol{v}).$$

The nonlinear terms in (10) arise only from fission, with pairs of neutrons giving rise to the quadratic term and so forth; that is, the coefficient of the quadratic term $\overline{\nu(\nu-1)}/2!$, is the expected number of pairs emitted in a fission,

$$\overline{\nu(\nu-1)(\nu-2)}/3!$$

is the mean number of triplets, and so forth.

The function $\mathscr{G}$ also satisfies certain final conditions and boundary conditions. At the final time, when $t = t_f$, $p_n = \delta_{n1}$ *if* the neutron is in R, $\boldsymbol{x}, \boldsymbol{v} \in R$, and $p_n = \delta_{n0}$ if $\boldsymbol{x}, \boldsymbol{v} \notin R$. Therefore by definition, the final condition on $\mathscr{G}$ is

(12) $$\begin{aligned} \mathscr{G}(z, R, t_f; \boldsymbol{x}, \boldsymbol{v}, t_f) &= 1 - z \quad \text{if} \quad \boldsymbol{x}, \boldsymbol{v} \in R, \\ &= 0 \qquad\ \ \text{if} \quad \boldsymbol{x}, \boldsymbol{v} \notin R. \end{aligned}$$

The boundary condition is easily found by noting that if the initial neutron is on B and moving outwards ($\boldsymbol{v} \cdot \hat{n} > 0$) then the number of neutrons within B at later times is sure to be zero. Hence, for such initial neutrons $p_0 = 1$ and

$$\mathscr{G}(z, R, t_f; \boldsymbol{x}, \boldsymbol{v}, t) = 0 \quad \text{if} \quad \boldsymbol{x} \in B, \boldsymbol{v} \cdot \hat{n} > 0. \tag{13}$$

We thus see that one minus the probability generating function satisfies a nonlinear integrodifferential equation (10), together with final conditions (12), and boundary conditions. In principle, at least, one can solve (10) by integrating backwards in time starting from the final conditions. Equation (10) can be regarded as a backwards Kolmogorov equation for the stochastic process. The forward Kolmogorov equation has also been considered by Matthes [**19**], who concluded that it was much less useful.

A formal mathematical analysis has not been made of the nonlinear transport equation for $\mathscr{G}$. However, I believe that from knowledge of solutions of the linearized equation, together with physical arguments, one can deduce many properties of the solution $\mathscr{G}$ to (10). First of all let us consider the solution, $\mathscr{G}_l$, to the linearized version of (10), i.e. with the right-hand side set equal to zero but with the same final and boundary conditions. Evidently $\mathscr{G}_l$ satisfies the source free adjoint transport equation, $L^\dagger \mathscr{G}_l = 0$, together with the usual adjoint boundary conditions. If one thinks of solving this equation for $\mathscr{G}_l$ backwards in time, its asymptotic behavior in time will be exponential [**10**], [**20**], $\sim e^{\alpha t}$ where $\alpha < 0$ if the system is supercritical, $\alpha > 0$ if the system is subcritical, and $\alpha = 0$ if the system is critical.

Secondly, it can be shown [**10**] that the nonlinear terms which form the right-hand side of (10) are nonpositive for z real and $0 \leq z \leq 1$. We can now see that these nonlinear terms will occur in (10) with opposite sign to the term involving c. They will thus effectively reduce the number of neutrons emerging from a collision and their effect on the solution can be seen by regarding them as a nonlinear absorption process.

We can now describe the general behavior of the solution to (10), which would be obtained by integrating backwards in time from the value $1 - z$ in R. Let us first consider a subcritical system. For $1 - z$ sufficiently small, $\mathscr{G}$ will satisfy the linearized equation, initially at least. After transients, $\mathscr{G}$ will decrease $\sim e^{\alpha t}$ as t decreases and hence the linear equation will continue to hold and $\mathscr{G}$ will decrease exponentially to zero as the solution is carried backwards in time. If we start from a value of $1 - z$ which is not small, then the nonlinear terms are no longer negligible, but since they act as absorptions they will make the system even more subcritical. Hence $\mathscr{G}$ should decrease more rapidly with the nonlinear terms than without them. Eventually $\mathscr{G}$ will become small enough that the nonlinear terms can be neglected, and for any fixed z ($0 \leq z \leq 1$), $\mathscr{G}$ will decrease asymptotically at early times in just the same manner as $\mathscr{G}_l$.

This behavior of $\mathscr{G}$ for a subcritical system is just what we expect on physical grounds. When $t_f - t$ is large and increasing we would expect that all of the probabilities, p_n, will be small and decreasing, except for p_0 which will be approaching unity.

The situation is altogether different for a supercritical system. As one integrates backwards in time from a small value of $1 - z$, $\mathcal{G}$ will increase and the nonlinear terms will become increasingly important. However, since the nonlinear terms act as absorption, the increase in $\mathcal{G}$ will be bounded and in fact $\mathcal{G}$ will approach a steady state, $\mathcal{G}_0$, which will evidently be a solution of (10) with the time derivative deleted from $L^\dagger$. Thus for a supercritical system

$$\mathcal{G}(z, R, t_f; \boldsymbol{x}, \boldsymbol{v}, t) \xrightarrow[t_f - t \to \infty]{} \mathcal{G}_0(\boldsymbol{x}, \boldsymbol{v}) \tag{14}$$

independent of z $(0 \leq z < 1)$ and independent of R.

The physical interpretation of the steady state solution, G_0, is quite clear. If a neutron is introduced into a supercritical system it will either lead to a persisting (divergent) chain reaction or else the population will become extinct. G_0 is simply the probability that a neutron at $\boldsymbol{x}$ with velocity $\boldsymbol{v}$ will lead to a persisting chain reaction while $1 - \mathcal{G}_0$ is the extinction probability, i.e. the probability that no neutrons will be in the system long after the initial neutron was introduced. This is just the implication of (14). Since $\mathcal{G} \to \mathcal{G}_0$ independent of z $(0 \leq z < 1)$, the probabilities, p_n, must be zero for finite n. Thus n must either be zero or approaching infinity. (Actually for any finite $t_f - t$ there will be a range of z values, $0 \leq 1 - z \leq \epsilon$, for which $\mathcal{G}$ is appreciably different from $\mathcal{G}_0$, but as $t_f - t \to \infty$, $\epsilon \to 0$.)

It can also be shown by direct derivation **[10]**, **[21]** that the probability of obtaining a persisting chain reaction satisfies the time independent version of (10) and is identically $\mathcal{G}_0$. Moreover it has been found **[10]** that $\mathcal{G}_0$ can be computed numerically with about the same amount of effort as is required to find an eigenvalue of the ordinary neutron transport equation. Thus it appears that the nonlinear terms of (10) do not substantially increase the difficulties of numerical solution, at least for the time independent problem. Time dependent transport problems without nonlinear terms are rather formidable computationally, and it would be premature to conclude that the nonlinear are as innocuous in general.

Thus we have seen that by regarding the nonlinear terms in (10) as absorptions we have been able to gain considerable insight into the solutions. While our conclusions haven't been formally proved, they are so clearly in accord with intuitive feelings regarding multiplicative processes and with the rigorous results obtained when neutrons were treated a generation at a time **[4]**, **[5]** that we may be confident that they would be confirmed by a detailed analysis.

Rather little has been done towards obtaining the time dependent generating function from (10). It is natural to think of reducing the nonlinear transport equation to multigroup diffusion form and seeking a solution in terms of eigenfunctions of the linearized equation. Some steps have been taken in these directions **[7]**, **[8]**, **[10]** but without conclusive results.

So far we have been considering probabilities that a single neutron will lead to various final conditions. Although this is a fundamental concern, it is evidently highly idealized and more generally we should consider that the system contains a source of neutrons. Letting $S(\boldsymbol{x}, \boldsymbol{v}, t)\, d\boldsymbol{x}\, d\boldsymbol{v}\, dt$ be the probability that a neutron appears (from events other than neutron collisions) with position in $d\boldsymbol{x}$ about $\boldsymbol{x}$, velocity $d\boldsymbol{v}$ about $\boldsymbol{v}$ and time dt about t, and ignoring the possibility of several neutrons appearing at once, we can seek the probability that such a source leads to n neutrons in R at t_f. Denoting the corresponding generating function (defined as in (8)) by $G(z, S)$ where we have suppressed the R and t_f dependence of G, it can be shown that **[7]**, **[9]**, **[10]**

$$G(z, S) = \exp\left\{\int_{-\infty}^{t_f} dt \int d\boldsymbol{x} \int d\boldsymbol{v} S(\boldsymbol{x}, \boldsymbol{v}, t)(G(z; \boldsymbol{x}, \boldsymbol{v}, t) - 1)\right\}. \tag{15}$$

This shows that once one knows the single neutron generating function, the generating function for a source can be found by quadrature.

C. **Moments of the probability distribution.** For many applications one does not need to know the probability generating function, but only the first two moments of the probability distribution. These can be found from the generating function by differentiating G one or two times with respect to z and evaluating the result at $z = 1$, and the equations satisfied by the moments can be found by differentiating (10) with respect to z and evaluating the result at $z = 1$. The equations satisfied by the moments, of course, can be derived directly **[9]**, **[11]**, **[13]** but we prefer to derive them from (10), in the interests of a unified presentation. It will be seen that the moments are easier to calculate than the generating function. This is because, first of all, the equation for a particular moment involves only that moment and those of lower order **[11]** so that we may solve for the moments successively instead of simultaneously, and secondly, because the equations can be solved if one knows the Green's function of the linear problem.

Following Otsuka and Saito, let us introduce the moments γ_n by means of the factorial moment expansion

$$G(z, R, t_f; \boldsymbol{x}, \boldsymbol{v}, t) = \sum_{n=0}^{\infty} \frac{(z-1)^n}{n!} \gamma_n(R, t_f; \boldsymbol{x}, \boldsymbol{v}, t), \tag{16}$$

where by differentiating (16) and using the definition of G in (8), we find

$$\gamma_m = \sum_{n=m}^{\infty} n(n-1)\cdots(n-m+1)p_n. \tag{17}$$

From (12) and (13) we see that the final and boundary conditions on the moments are:

$$\begin{aligned} \gamma_0 &\equiv 1 \\ \gamma_1(R, t_f; \boldsymbol{x}, \boldsymbol{v}, t_f) &= 1 \quad \text{if } \boldsymbol{x}, \boldsymbol{v} \in R, \\ &= 0 \quad \text{otherwise}, \\ \gamma_n(R, t_f; \boldsymbol{x}, \boldsymbol{v}, t_f) &= 0 \quad \text{for } n \geq 2, \end{aligned} \tag{18}$$

and

$$\gamma_n(R, t_f; \boldsymbol{x}, \boldsymbol{v}, t) = 0 \quad \text{if } \boldsymbol{x} \in B \quad \text{and} \quad \boldsymbol{v} \cdot \hat{n} > 0.$$

The γ's are moments for a population arising from a single neutron. For a population arising from a source, $S(\boldsymbol{x}, \boldsymbol{v}, t)$ we may insert the expansion (16) into (15), thereby obtaining a (factorial cumulant) expansion for $G(z, S)$

$$G(z, S) = \exp\left\{\sum_{n=1}^{\infty} \frac{(z-1)^n}{n!} \tilde{\sigma}_n(R, t_f; S)\right\} \tag{19}$$

where

$$\tilde{\sigma}_n(R, t_f; S) = \int_{-\infty}^{t_f} dt \int d\boldsymbol{x} \int d\boldsymbol{v} \gamma_n(R, t_f; \boldsymbol{x}, \boldsymbol{v}, t) S(\boldsymbol{x}, \boldsymbol{v}, t). \tag{20}$$

Once more, by differentiating (19) one or two times and setting $z = 1$ we find

$$\tilde{\sigma}_1(R, t_f; S) = \sum_{n=0}^{\infty} n p_n(R, t_f; S) = \bar{n}(R, t_f; S) = \langle n \rangle \tag{21}$$

so that $\tilde{\sigma}_1$ is just the expectation value of the number of neutrons in R at t_f due to the source, S. Furthermore

$$\begin{aligned} \tilde{\sigma}_2(R, t_f; S) &= \sum_{n=0}^{\infty} n(n-1) p_n - \left(\sum_{n=0}^{\infty} n p_n\right)^2 \\ &= \sigma(R, t_f; S) - \bar{n}(R, t_f; S) \end{aligned} \tag{22}$$

where σ is the variance, $\langle n^2 \rangle - \langle n \rangle^2$.

We may now find the equations which are satisfied by the moments, γ_n, by differentiating (10) with respect to z and setting $z = 1$. We find

$$L^\dagger(\boldsymbol{x}, \boldsymbol{v}, t) \gamma_1(R, t_f; \boldsymbol{x}, \boldsymbol{v}, t) = 0 \tag{23}$$

$$\begin{aligned} L^\dagger(\boldsymbol{x}, \boldsymbol{v}, t) \gamma_2(R, t_f; \boldsymbol{x}, \boldsymbol{v}, t) &= \overline{\nu(\nu - 1)}\, \sigma_f(\boldsymbol{x}, \boldsymbol{v}, t) v \\ &\quad \times \left[\int \gamma_1(R, t_f; \boldsymbol{x}, \boldsymbol{v}', t) \chi(v')\, d\boldsymbol{v}'\right]^2. \end{aligned} \tag{24}$$

Similar but more complicated equations can be formed for the higher moments **[11]**. In each equation, the right-hand side contains only lower order moments. These equations are to be solved subject to the final conditions and boundary conditions given in (18). Equation (23) tells us nothing new; it is simply the adjoint transport equation for the mean number of neutrons. In the equation (24) for γ_2 we see that there is a source term (nonzero right-hand side) which arises from the emission of pairs of fission neutrons $(\overline{\nu(\nu - 1)})$ and can be readily computed once γ_1 is known. Without fission, clearly γ_2 would equal zero. In principle, the set of equations for any finite number of γ's could be solved recursively. Formally we may write down the solutions for γ_1 and γ_2 in terms of the Green's functions of equations (6) and (7). Thus

$$\gamma_1(R, t_f; \boldsymbol{x}, \boldsymbol{v}, t) = \iint_{\boldsymbol{x}', \boldsymbol{v}' \in R} d\boldsymbol{x}'\, d\boldsymbol{v}'\, \mathrm{Gr}\,(\boldsymbol{x}, \boldsymbol{v}, t \to \boldsymbol{x}'\boldsymbol{v}', t_f) \tag{25}$$

is the solution of (23), plus the final condition on γ_1, while

$$\gamma_2(R, t_f; \boldsymbol{x}, \boldsymbol{v}, t) = \int_{-\infty}^{t_f} dt' \int d\boldsymbol{x}' \left[\int d\boldsymbol{v}'' \gamma_1(R, t_f; \boldsymbol{x}', \boldsymbol{v}'', t') \chi(\boldsymbol{v}'') \right]^2 \times \int d\boldsymbol{v}' \, \overline{\nu(\nu - 1)\, \sigma_f(\boldsymbol{x}', \boldsymbol{v}', t')} v' \,\mathrm{Gr}\, (\boldsymbol{x}, \boldsymbol{v}, t \rightarrow \boldsymbol{x}', \boldsymbol{v}', t') \tag{26}$$

is the solution of (24). Higher order equations have also been studied by Saito [**11**]. From these equations we see that if the Green's function of the linear transport problem is known, then the moments of the probability distribution can be found recursively.

Moments for a system with a source can now be found by inserting equations (25) or (26) in (20). The mean neutron population, in R at t_f, is then

$$\bar{n}(R, t_f; S) = \iint_{\boldsymbol{x}, \boldsymbol{v} \in R} d\boldsymbol{x}\, d\boldsymbol{v} f_1(\boldsymbol{x}, \boldsymbol{v}, t_f), \tag{27}$$

where we have defined

$$f_1(\boldsymbol{x}, \boldsymbol{v}, t_f) = \int_{-\infty}^{t_f} dt \int d\boldsymbol{x}' \int d\boldsymbol{v}' \,\mathrm{Gr}\, (\boldsymbol{x}', \boldsymbol{v}', t \rightarrow \boldsymbol{x}, \boldsymbol{v}, t_f) S(\boldsymbol{x}', \boldsymbol{v}', t), \tag{28}$$

and f_1 is evidently just the expected value of the neutron density at t_f resulting from the source S. It has also been called the singlet density [**15**].

The variance, σ, can also be expressed in terms of the Green's function by inserting equations (25) and (26) in (20) and (22). By freely interchanging orders of integration we may express the result as

$$\sigma(R, t_f; S) = \iint_{\boldsymbol{x}_1, \boldsymbol{v}_1 \in R} d\boldsymbol{x}_1\, d\boldsymbol{v}_1 \iint_{\boldsymbol{x}_2, \boldsymbol{v}_2 \in R} d\boldsymbol{x}_2\, d\boldsymbol{v}_2 [\tilde{f}_2(\boldsymbol{x}_1, \boldsymbol{v}_1, \boldsymbol{x}_2, \boldsymbol{v}_2, t_f) + f_1(\boldsymbol{x}_1, \boldsymbol{v}_1, t_f) \delta(\boldsymbol{x}_1 - \boldsymbol{x}_2) \delta(\boldsymbol{v}_1 - \boldsymbol{v}_2)] \tag{29}$$

where

$$\tilde{f}_2(\boldsymbol{x}_1, \boldsymbol{v}_1, \boldsymbol{x}_2, \boldsymbol{v}_2, t_f) = \int_{-\infty}^{t} dt' \int d\boldsymbol{x}' \int d\boldsymbol{v}_1'' \int d\boldsymbol{v}_2'' \chi(\boldsymbol{v}_1'') \chi(\boldsymbol{v}_2'') \times \mathrm{Gr}\, (\boldsymbol{x}', \boldsymbol{v}_1'', t' \rightarrow \boldsymbol{x}_1, \boldsymbol{v}_1, t_f) \,\mathrm{Gr}\, (\boldsymbol{x}', \boldsymbol{v}_2'', t' \rightarrow \boldsymbol{x}_2, \boldsymbol{v}_2, t_f) \times \int d\boldsymbol{v}' f_1(\boldsymbol{x}', \boldsymbol{v}', t') \overline{\nu(\nu - 1)\, \sigma_f(\boldsymbol{x}', \boldsymbol{v}', t')} v'. \tag{30}$$

Such quantities as $\tilde{f}_2$ play a central role in the theory of moments and can be given a simple physical interpretation. It can be shown [**11**] that if we define the function f_2 by

$$f_2(\boldsymbol{x}_1, \boldsymbol{v}_1, t; \boldsymbol{x}_2, \boldsymbol{v}_2, t) = \tilde{f}_2(\boldsymbol{x}_1, \boldsymbol{v}_1, \boldsymbol{x}_2, \boldsymbol{v}_2, t) + f_1(\boldsymbol{x}_1, \boldsymbol{v}_1, t) f_1(\boldsymbol{x}_2, \boldsymbol{v}_2, t) + \delta(\boldsymbol{x}_1 - \boldsymbol{x}_2) \delta(\boldsymbol{v}_1 - \boldsymbol{v}_2) f_1(\boldsymbol{x}_1, \boldsymbol{v}_1, t) \tag{31}$$

then the quantity $f_2(\boldsymbol{x}_1, \boldsymbol{v}_1, t; \boldsymbol{x}_2, \boldsymbol{v}_2, t)\, d\boldsymbol{x}_1\, d\boldsymbol{v}_1\, d\boldsymbol{x}_2\, d\boldsymbol{v}_2$ is the ensembled average of the product of the number of neutrons in $d\boldsymbol{x}_1$ about $\boldsymbol{x}_1$ and $d\boldsymbol{v}_1$ about $\boldsymbol{v}_1$ times the number of neutrons in $d\boldsymbol{x}_2$ about $\boldsymbol{x}_2$ and $d\boldsymbol{v}_2$ about $\boldsymbol{v}_2$, both at time t. It has also been called the doublet density **[12]** and the quantity $\tilde{f}_2 + f_1 f_2$ is the product density of degree two **[11]** which was introduced in cosmic ray theory by Bhabha and Ramakrishnan.

It is apparent that the difficulties in determining moments depend to a large extent on how complicated a Green's function is taken for the system. For purposes of orientation it is sometimes useful to use a trivial Green's function in which there is no space and velocity dependence and only the asymptotic time dependence is taken into account. Of course, there is no point in going through a stochastic transport theory merely to use such a Green's function but it is a useful check that one's equations make sense. Thus if we let

$$\int G(\boldsymbol{x}', \boldsymbol{v}', t \to \boldsymbol{x}, \boldsymbol{v}, t_f)\, d\boldsymbol{x}\, d\boldsymbol{v} = \exp\{\alpha(t_f - t)\},$$

$$\int S(\boldsymbol{x}', \boldsymbol{v}', t)\, d\boldsymbol{x}'\, d\boldsymbol{v}' = S,$$

and let R include all velocities and space inside B, we find for a subcritical system (for which $\alpha < 0$)

$$\bar{n} = S/-\alpha,$$

(32)

$$\sigma = \left(\frac{\overline{\nu(\nu - 1)}}{2} \frac{\overline{\sigma_f v}}{-\alpha} + 1\right) \frac{S}{-\alpha},$$

where $\overline{\sigma_f v}$ is a reciprocal neutron lifetime against fission. Except for very subcritical systems, $\overline{\sigma_f v}/-\alpha \gg 1$ and the first term dominates the expression for σ. Note that $\sigma/\bar{n}^2 \sim 1/S$ so that the relative fluctuations in a neutron population are most extreme for weak sources and decrease for strong sources. The physical phenomena responsible for this are clear. For a weak source the neutron population will consist of a series of individual multiplicative chains, each triggered by a different source neutron, and the chains will seldom overlap in time. For stronger sources the chains will begin to overlap so that the neutron population at a particular time will be likely to include neutrons from several independent chains, and the population will thereby be a less violently fluctuating function of time.

It is clear that from a general expression such as (29) one could obtain more accurate and complicated expressions for σ by using more realistic Green's functions.

In order to apply this formalism to the interpretation of experiments, it is necessary to generalize the methods. Thus, experimentally, one does not measure the probability distribution of neutrons in a region R but rather one measures the output from one or more detectors. To a considerable extent, detectors can be accommodated by changing the final conditions **[8]**–**[11]** and by using a vector z in the generating function, with one component per detector. However, the generating function still satisfies an equation very similar to (10) and various moments, or correlations between detector outputs can be obtained much as above. Sometimes delayed neutrons must explicitly be considered **[8]**–**[11]**.

In order to clarify these remarks, let us consider a possible experiment. Suppose that our system contains two detectors, D_1 and D_2, either one of which may upon absorbing a neutron give rise to a signal which is registered as a count. The detectors must be included in the specification of cross sections for the system. In addition, let $c_i(\boldsymbol{x}, \boldsymbol{v}, t)$ be the probability that in a collision a neutron of velocity $\boldsymbol{v}$ at position $\boldsymbol{x}$ time t will be absorbed and will be registered as a count in detector i $(i = 1, 2)$. Let us then seek the joint probability that in the time interval $(t_{f_1} \leq t \leq t_{f_1} + \delta t_{f_1})$ exactly n_1 counts will be registered by detector D_1 *and* exactly n_2 counts will be registered by the detector D_2 in the time interval $(t_{f_2} \leq t \leq t_{f_2} + \delta t_{f_2})$. This probability we denote by $p_{n_1 n_2}(D_1(t_{f_1}, \delta t_{f_1}), D_2(t_{f_2}, \delta t_{f_2}); \boldsymbol{x}, \boldsymbol{v}, t)$ and we introduce the corresponding generating function

$$G(z_1, z_2, D_1, D_2; \boldsymbol{x}, \boldsymbol{v}, t) = \sum_{n_1=0}^{\infty} \sum_{n_2=0}^{\infty} z_1^{n_1} z_2^{n_2} p_{n_1 n_2}. \tag{33}$$

By considering all possible first collision events we may again derive an equation satisfied by the generating function. For times preceding the earlier of t_{f_1} and t_{f_2} the equation for $1 - G = \mathscr{G}$ will be identical to (10), but for times during which counts can be recorded by either detector we must take into account the possibility that a count will be registered at the first collision. The result is to add two terms to the right-hand side of (10) so that we obtain

$$\begin{aligned} L^\dagger \mathscr{G} = &-\sum_{j=2}^{\infty} \frac{(-1)^j}{j!} \overline{\nu(\nu - 1) \cdots (\nu - j + 1)}\, \sigma_f v \left[\int \mathscr{G} \chi(v')\, dv' \right]^j \\ &+ \sigma(\boldsymbol{x}, \boldsymbol{v})[(1 - z_1)c_1(\boldsymbol{x}, \boldsymbol{v}, t) + (1 - z_2)c_2(\boldsymbol{x}, \boldsymbol{v}, z)] \end{aligned} \tag{34}$$

where it should be noted that c_i is defined to be nonzero only if $t_{f_i} \leq t \leq t_{f_i} + \delta t_{f_i}$. Boundary conditions on $\mathscr{G}$ are as before, see (13), and the final condition is that $\mathscr{G} = 0$ when t equals the greater of $t_{f_i} + \delta t_{f_i}$ $(i = 1, 2)$. Solutions of (34) can be sought by integrating backwards in time and the solution will evidently pick up its dependence on z_1 and z_2 from the new terms.

Moments of the joint probability distribution can be obtained much as before. Thus we may introduce the joint factorial moment expansion

$$G(z_1, z_2) = \sum_{n_1=0}^{\infty} \sum_{n_2=0}^{\infty} \frac{(z_1 - 1)^{n_1}}{n_1!} \frac{(z_2 - 1)^{n_2}}{n_2!} \gamma_{n_1 n_2}(D_1, D_2; \boldsymbol{x}, \boldsymbol{v}, t), \tag{35}$$

and equations for the low order moments can be obtained by differentiating (34) one or more times with respect to z_1 or z_2 and setting $z_1 = z_2 = 1$.

For example, we see that γ_{10}, which is the mean number of counts registered by detector D_1, satisfies

$$L^\dagger \gamma_{10} = \sigma(\boldsymbol{x}, \boldsymbol{v}) c_1(\boldsymbol{x}, \boldsymbol{v}, t) \tag{36}$$

which has the solution (assuming c_1 to be constant while the detector is counting)

$$(37)\quad \gamma_{10}(\boldsymbol{x}, \boldsymbol{v}, t) = \int_{t_{f1}}^{t_{f1}+\delta t_{f2}} dt' \int d\boldsymbol{x}' \int d\boldsymbol{v}' \sigma(\boldsymbol{x}', \boldsymbol{v}') c_1(\boldsymbol{x}', \boldsymbol{v}') \mathrm{Gr}\,(\boldsymbol{x}, \boldsymbol{v}, t \to \boldsymbol{x}', \boldsymbol{v}', t).$$

The marked similarity of this solution to that in (25) is to be noted.

Similarly, the second order moments γ_{02}, γ_{11}, and γ_{20} may be expressed in terms of the Green's function. γ_{20} will be found in terms of γ_{10}, as in (26), while γ_{11}, which represents a cross-correlation between the outputs of the two detectors, will involve both γ_{10} and γ_{01}. The moments, $\tilde{\sigma}$, for a system with a source may be found as before.

These results, for a detector in which counts are associated with absorptions, are somewhat simpler than would be found for a fission detector, but the same methods may be employed for fission detectors.

In conclusion, it may be worthwhile to note a few types of experiments which may be analyzed using the methods which we have discussed in this section. For more details, one can consult, for example, [7], [9], [19]. In most of these experiments, the delayed neutrons are not important and one is measuring the statistical properties of individual prompt neutron chains in subcritical reactors.

In the "Rossi-α" experiment [22] one or more detectors may be used to measure the time distribution of pairs of counts due to neutrons originating from a common ancestor in a neutron chain. The probability $p(\tau)$ of obtaining a time interval, τ, between the successive counts is measured as a function of τ. If the source is weak so that individual chains seldom overlap, then successive counts are likely to be related by a common ancestor and, except for transients, $p(\tau) \sim e^{\alpha\tau}$. Thus, from the measurement, α and the degree of subcriticality of a system can be found.

In the Feynman [23], or variance to mean methods, one measures the mean and the variance of the number of counts recorded in a time, T. As is plausible from (32), from the variance and mean, as functions of T, one can deduce α and other characteristics of the system.

In the p_0 method, proposed by Mogil'ner and Zolotukhin [24], the probability, p_0, that no counts are recorded by a detector in some period, T, is measured. By comparing this with the mean counting rate one can deduce properties of the system.

Recently a number of experiments have been performed of the auto-correlation and cross-correlation functions of detector outputs, and of their Fourier transforms, the power spectral densities. The discussion of such experiments is beyond the scope of this review and we refer the reader elsewhere for descriptions of the experiments and analysis [8], [9], [25], [26]. However, the analysis may be made within the framework which we have outlined.

Finally, we mention a quite different type of experiment [27] in which a system with a weak source is brought to a supercritical state and the distribution of wait times, before appearance of the first divergent chain, is observed. From such experiments and an analysis along the lines of §B, one can determine source requirements for pulsed reactors and one can specify safe startup programs for reactors with weak sources. Further references are given in [10] and [13].

REFERENCES

1. B. Davison and J. B. Sykes, *Neutron transport theory*, Oxford Univ. Press, New York, 1957.

2. T. E. Harris, *The theory of branching processes*, Springer-Verlag, Berlin, and Prentice-Hall, Englewood Cliffs, N.J., 1963.[2]

3. J. A. Thie, *Reactor noise*, Rowman and Littlefield, New York, 1963.

4. C. J. Everett and S. Ulan, *Multiplicative systems in several variables*, Los Alamos Reports LA-683, 690, and 707 (1948).

5. T. W. Mullikin, *Neutron branching processes*, Rand Corp. Report RM-2693 (1961).

6. I. K. Abu-Shumays, *Generating functions and transport theory*, Ph.D. Thesis, Harvard University, 1966.

7a. L. Pál, *On the theory of stochastic processes in nuclear reactors*, Nuovo Cimento (10) **7** (1958), supplement 25.

7b. ———, *Statistical theory of chain reactions in nuclear reactors.* I, II, III, Acta Physica Hungar. **14** (1962), 345. English transl. by V. Shibayev, Harwell, NP-TR-951, 1962.

7c. ———, *Statistical theory of neutron chain reactors*, Proc. Third Internat. Conf. Peaceful Uses of Atomic Energy, Geneva **2** (1964), 218.

8. W. Matthes, *Statistical fluctuations and their correlation in reactor neutron distributions*, Nukleonik **4** (1962), 213.

9a. H. Borgwaldt and D. Sanitz, *Die Impulskorrelation zwier Neutronendetcktoren im stationären Reaktor*, Nukleonik **5** (1963), 239.

9b. H. Borgwaldt and D. Stegemann, *A common theory for neutronic noise analysis experiments in nuclear reactors*, Nukleonik **7** (1965), 313.

10. G. Bell, *On the stochastic theory of neutron transport*, Nuclear Sci. Eng. **21** (1965), 390.

11. M. Otsuka and K. Saito, *Neutron fluctuations in a multiplying medium*, Nuclear Sci. Eng. **24** (1966), 410.[3]

12a. V. Raievski, *Fluctuations statistiques du nombre de neutrons dans une pile*, CEA report 917 (1958).

12b. ———, *Fluctuations statistiques dans les piles*, CEA 1673 (1960).

13. D. R. Harris, *Neutron fluctuations ina reactor of finite size*, Nuclear Sci. Eng. **21** (1965), 369.

14. A. B. Govorkov, *Statistical reactor kinetics equations*, Atomnaya Énergiya **17** (1964), 474; **13** (1962), 152.

15a. R. K. Osborn and S. Yip, *Foundations of neutron transport theory* (1963), ANS and USAEC Monograph, Gordon and Breach, New York, 1966.

15b. R. K. Osborn and M. Natelson, *Kinetic equations for neutron distributions*, J. Nuclear Energy **19** (1965), 619.

16. P. Langevin, *Sur la théorie du mouvement brownian*, Comptus Rendus **146** (1908), 530.[4]

17. J. R. Sheff and R. W. Albrecht, *The space dependence of reactor noise I—theory*, Nuclear Sci. Eng. **24** (1966), 246.

18. A. Z. Akcasu and R. K. Osborn, *Applications of Langevin's technique to space- and energy-dependent noise analysis*, Nuclear Sci. Eng. **26** (1966), 13.

19. W. Matthes, *Theory of fluctuations in neutron fields*, Nukleonik **8** (1966), 87.

20. K. Jörgens, *An asymptotic expansion in the theory of neutron transport*, Comm. Pure Appl. Math. **11** (1958), 219.

21. G. E. Hansen, *Assembly of fissionable material in the presence of a weak neutron source*, Nuclear Sci. Eng. **8** (1960), 709.

22. J. D. Orndoff, *Prompt neutron periods of metal critical assemblies*, Nuclear Sci. Eng. **2** (1957), 450.

23a. R. P. Feynman, F. de Hoffmann, and R. Serber, *Dispersion of the neutron emission in U-235 fission*, J. Nuclear Energy **3** (1956), 64.

---

[2] Much of this material is given in simpler form in, for example, A. T. Bharucha-Reid, *Elements of the theory of Markov processes and their applications*, McGraw-Hill, New York, 1960.

[3] Other references therein, and K. Saito and Y. Taji, *Theory of branching processes of neutrons in a multiplying medium* 1967, Nuclear Sci. Eng. **30** (1967), 54.

[4] The theory is nicely presented in S. Chandrasekhar, *Stochastic problems in physics and astronomy*, Rev. Mod. Phys. **15** (1943), 1.

23b. F. de Hoffmann, *The science and engineering of nuclear power.* Vol. II, Addison-Wesley, Reading, Mass., 1949, chapter 9.

24. A. I. Mogil'ner and V. G. Zolotukhin, *Measuring the characteristics of kinetics of a reactor by the statistical p-method,* Atomnaya Énergiya, **10** (1961), 377, 379, and **15** (1963), 11.

25. M. Natelson, R. K. Osborn, and F. Shure, *Space and energy effects in reactor fluctuation experiments,* J. Nuclear Energy **20** (1966), 557.

26. R. E. Uhrig, *Reactivity measurements using noise techniques,* Trans. Amer. Nuclear Soc. **9** (1966), 522.

27. T. F. Wimett, R. H. White, W. R. Stratton, and D. P. Wood, *Godiva* II—*An unmoderated pulse-irradiation reactor,* Nuclear Sci. Eng. **8** (1960), 691.

UNIVERSITY OF CALIFORNIA,
LOS ALAMOS SCIENTIFIC LABORATORY,
LOS ALAMOS, NEW MEXICO

Multiplicative first-passage processes and transport theory[1]

J. E. Moyal

1. **Introduction.** We shall be dealing in this article with mathematical models for a certain class of transport processes. We are concerned with a population of particles, to be known as the *radiation*, interacting with a fixed medium; our first assumption is that the disturbance of the medium by the radiation is negligible. The interaction consists of a discrete set of *atomic events*, or less accurately *collisions*, leading to scattering, absorption, annihilation of radiation particles, or the creation of new particles; we make the further assumptions that the duration of such atomic events is negligible, so that they can be deemed to occur instantaneously, and that the cross-sections for these events depend only on the instantaneous states and positions of the radiation particles involved, and possibly on the time, which implies the *Markovian* (nonhereditary) character of the process. We shall also allow for the possibility of continuous changes of state of the radiation between atomic events due to external forces or Markovian processes such as diffusion or loss of energy by ionization. Finally, we assume that the interaction between radiation particles can be neglected. Thus, starting at any given time, each radiation particle gives rise to an independent population; this implies the *multiplicative* character of the process (*branching process*, see [3]). Physical processes for which these models appear adequate are: the diffusion and multiplication of neutrons in fissionable material; the development of electron-photon and nucleon cascades in matter.

We shall concentrate here on problems connected with the changes of state of the whole radiation as the process develops in space and time, assuming that the

[1] Work performed under the auspices of the U.S. Atomic Energy Commission.

cross-sections for atomic events are given, and also, when relevant, that the transition probabilities for continuous changes of state between atomic events are known. We adopt a resolutely stochastic point of view and set up a (somewhat complicated) formalism involving equations for the transition and first-passage probability distributions of the whole population of radiation particles. We reap the benefit of this wholesale approach when we are able to prove by elementary methods that these equations always have an admissible solution, which is unique if it is normalized to unity (*stable process*). The position is not so favorable as regards the usual transport equations for the distribution of mean numbers of particles, which may fail to have an admissible solution in the unstable case (the *supercritical* case for neutron multiplication). The connection between equations for the mean (and higher-order moment) distributions and the population distribution equations is studied in §4. The relation between stability and criticality is discussed in §6 in connection with the concept of an associated Markov chain. The example of a simple "rod" model of neutron multiplication is considered in §7 as an illustration of the general theory.

2. **Multiplicative first-passage processes.** We begin with an account of the notation (see [9]):

x denotes the position of a single radiation particle.

X is the set of all such positions, i.e. the space in which the radiation develops (in the applications we have in mind X is either 3-space R^3 or space-time R^4).

u denotes the *state* (energy, momentum, spin, etc.) of a single radiation particle (if there is more than one kind of particle in the radiation, then the *type* of particle is included in the specification of u).

U is the *individual particle state space.*

$\omega = (u, x)$ is the *phase* of a single particle.

$\Omega = U \times X$ is the individual particle phase-space.

$\omega^n = (\omega_1, \ldots, \omega_n)$, the ordered set of phases of the n particles in the radiation, is the *population phase.*

$\Omega^n = \Omega \times \Omega \times \cdots \times \Omega$, n times, is the set of all population phases ω^n for a fixed number n, with Ω^0 denoting conventionally an empty population.

$\mathbf{\Omega} = \bigcup_{n=0}^{\infty} \Omega^n$ is the *population phase-space;* $\mathbf{U} = \bigcup_{n=0}^{\infty} U^n$ is the *population state space;* $\mathbf{B}(\mathbf{\Omega})$ is a suitably defined σ-field of *measurable* sets in $\mathbf{\Omega}$.

Λ, $\mathbf{\Lambda}$ denote measurable subsets of Ω, $\mathbf{\Omega}$ respectively.

$X[\tau]$ is a closed and measurable set in X bounded by a measurable *surface* τ.

$x[\tau]$ denotes a point on τ and $\omega[\tau] = (u, x[\tau])$.

$\Omega[\tau] = U \times X[\tau]$ and $\mathbf{\Omega}[\tau] = \bigcup_{n=0}^{\infty} \Omega^n[\tau]$.

$\Sigma[\tau] = U \times \tau$ and $\mathbf{\Sigma}[\tau] = \bigcup_{n=0}^{\infty} \sum^n[\tau]$.

$\Gamma[\tau]$, $\mathbf{\Gamma}[\tau]$ denote measurable subsets of $\Sigma[\tau]$, $\mathbf{\Sigma}[\tau]$, respectively.

A *population probability distribution* is a probability distribution P on $\mathbf{B}(\mathbf{\Omega})$; its nth component $P^{(n)}$ is its restriction to $\mathbf{B}(\Omega^n)$. We take care of the fact that radiation particles are indistinguishable by restricting ourselves to *symmetric* distributions (see [7] for the general theory of population processes).

We are interested in the following problem (cf. [**9**], [**10**]): given the cross-sections for atomic events and (when required) the transition probabilities for "continuous" changes of phase between such events, to calculate the *first-passage distributions* P, where P is a conditional probability distribution on $\mathbf{B}(\mathbf{\Sigma}[\tau]) \times \mathbf{\Omega}[\tau]$ (i.e. $P(\cdot \mid \omega^k[\tau_0])$ is a probability distribution on $\mathbf{B}(\mathbf{\Sigma}[\tau])$ for each fixed $\omega^k[\tau_0] \in \mathbf{\Omega}[\tau]$, and $P(\mathbf{\Gamma}[\tau] \mid \cdot)$ is a measurable function on $\mathbf{\Omega}[\tau]$ for each fixed $\mathbf{\Gamma}[\tau] \in \mathbf{B}(\mathbf{\Sigma}[\tau])$), $P(d\omega^n[\tau] \mid \omega^k[\tau_0])$ is the probability that eventually n radiation particles will effect a first passage through τ in the phase elements $d\omega_1[\tau] \cdots d\omega_n[\tau] = d\omega^n[\tau]$, conditional on an initial population of k particles in the phases $(\omega_1[\tau_0], \ldots, \omega_k[\tau_0]) = \omega^k[\tau_0]$, where $X[\tau_0] \subseteq X[\tau]$.[2] This is essentially a generalization of the classical Milne problem to multiplicative processes. The distributional aspects of Markov process theory can be generalized to deal with this type of problem by choosing as the independent variable of the process the boundaries τ of a collection T of sets $X[\tau]$, partially ordered by inclusion (thus $\tau \geq \tau_0$ if $X[\tau] \supseteq X[\tau_0]$). A *first-passage process* is a family of first-passage distributions P, defined for all pairs $\tau \geq \tau_0$, which satisfy for every ordered triple $\tau \geq \tau_1 \geq \tau_0$ a generalized Chapman-Kolmogorov relation (briefly, a C.K. relation).

$$(2.1) \qquad P(\mathbf{\Gamma}[\tau] \mid \omega^k[\tau_0]) = \sum_{n=0}^{\infty} \int_{\Sigma^n[\tau_1]} P(\mathbf{\Gamma}[\tau] \mid \omega^n[\tau_1]) P^{(n)}(d\omega^n[\tau_1] \mid \omega^k[\tau_0]).$$

The theory of first-passage processes includes that of the familiar *temporal* Markov processes as a special case: if we take $X = \mathbf{R}^4$, the Minkowski space-time, and let T be a collection of suitably chosen space-like surfaces, we obtain in fact a Lorentz-invariant formulation of the theory of temporal Markov processes. If we let T be the set of all planes normal to the time-axis in a particular reference frame, then we can identify τ with the time and the first-passage distribution with the usual temporal transition probability. Furthermore, by making a suitable choice of the surfaces τ in $\mathbf{R}^4$, we can also include in the theory first-passage problems involving the time, e.g. the first-passage distributions through a surface in 3-space prior to a given time (§7 contains an example with just one space dimension and time).

In dealing with multiplicative processes, it is convenient to work with *probability generating functionals* (p.g.f. for short). A complex-valued function $\boldsymbol{f}$ on $\mathbf{\Omega}$ assigns a unique complex number $\boldsymbol{f}(\omega^n)$ to each population phase ω^n ($n = 0, 1, 2, \ldots$). Let ζ be a measurable function on Ω, and define $\boldsymbol{\zeta}$ on $\mathbf{\Omega}$ such that $\boldsymbol{\zeta}(\omega^0) = 1$, $\boldsymbol{\zeta}(\omega^n) = \prod_{i=1}^n \zeta(\omega_i)$ ($n = 1, 2, \ldots$). The p.g.f. G of a given first-passage distribution P is then the expectation value of $\boldsymbol{\zeta}$ relative to P:

$$(2.2) \quad G[\zeta, \tau \mid \omega^k[\tau_0]] = \mathsf{E}[\boldsymbol{\zeta} \mid \omega^k[\tau_0]] = \sum_{n=0}^{\infty} \int_{\Sigma^n[\tau]} \prod_{i=1}^{n} \zeta(\omega_i[\tau]) P^{(n)}(d\omega^n[\tau] \mid \omega^k[\tau_0]).$$

Obviously, G exists for all ζ such that $\|\zeta\| = \sup_{\omega \in \Omega} |\zeta(\omega)| \leq 1$; one shows

[2] Reentrant particles and their descendants are excluded.

(cf. **[7]**) that if P is symmetric, then it is uniquely determined by G. The multiplicative character of the process is then neatly expressed by the condition that

$$G[\zeta, \tau \mid \omega^k[\tau_0]] = \prod_{i=1}^{k} G[\zeta, \tau \mid \omega_i[\tau_0]]. \tag{2.3}$$

It follows that the process is uniquely characterized by the family of p.g.f.'s $G[\zeta, \tau \mid \omega[\tau_0]]$ conditional on the phases $\omega[\tau_0]$ of a single "ancestor." In terms of the p.g.f., the C.K. relation (2.1) becomes the functional relation

$$\begin{aligned} G[\zeta, \tau \mid \omega[\tau_0]] &= \sum_{n=0}^{\infty} \int_{\Sigma^n[\tau_1]} \prod_{i=1}^{k} G[\zeta, \tau \mid \omega_i[\tau_1]] P^{(n)}(d\omega^n[\tau_1] \mid \omega[\tau_0]) \\ &= G\{G[\zeta, \tau \mid \cdot], \tau_1 \mid \omega[\tau_0]\}, \qquad (\tau \geq \tau_1 \geq \tau_0); \end{aligned} \tag{2.4}$$

(the last expression in (2.4) means that we substitute $G[\zeta, \tau \mid \cdot]$ for ζ in $G[\zeta, \tau_1 \mid \omega[\tau_0]]$).

3. **The first-collision integral equation.** From the cross-sections for atomic events and the transition probabilities for changes of phase between such events we obtain more or less directly two functions P_0, Q with the following properties (see **[5]**, **[8]**, **[9]**, **[10]**):

(1) The first-passage distribution P_0 with *no* atomic event, which is an incomplete conditional probability distribution on $\mathbf{B}(\Sigma[\tau]) \times \Omega[\tau]$ satisfying the C.K. relation (2.1). P_0 is incomplete in the sense that $K_0(\tau \mid \omega[\tau_0]) = P_0(\Sigma[\tau] \mid \omega[\tau_0]) \leq 1$, i.e. it is normalized to ≤ 1; K_0 is the total probability that the particle initially in phase $\omega[\tau_0]$ will effect a first passage through τ prior to any atomic events; since there are no atomic events, and hence no multiplication of particles, the support of P_0 is restricted to $\mathbf{B}(\Sigma[\tau])$.

(2) The first collision and consequent phase distribution Q, which is an incomplete conditional probability distribution on $\mathbf{B}[\mathbf{\Omega}] \times \Omega[\tau]$, where $Q(d\omega^n, \tau \mid \omega[\tau_0])$ is the probability that the particle with initial phase $\omega[\tau_0] \in \Omega[\tau]$ will suffer its first atomic event before effecting a passage through τ and give rise as an immediate result of this event to n particles in the phase element $d\omega_1 \cdots d\omega_n = d\omega^n$, is assumed to have its support in $\mathbf{\Omega}[\tau]$: i.e. $Q(\mathbf{\Omega}, \tau \mid \omega[\tau_0]) = Q(\mathbf{\Omega}[\tau], \tau \mid \omega[\tau_0]) = \theta(\tau \mid \omega[\tau_0])$. θ is the probability that at least one atomic event will occur before the whole radiation effects a first passage through τ.

(3) P_0 and Q are related by the consistency conditions

$$K_0(\tau \mid \omega[\tau_0]) + \theta(\tau \mid \omega[\tau_0]) \equiv 1, \tag{3.1}$$

$$Q(\mathbf{\Lambda}, \tau \mid \omega[\tau_0]) = Q(\mathbf{\Lambda}, \tau_1 \mid \omega[\tau_0]) + \int_{\Sigma[\tau_1]} Q(\mathbf{\Lambda}, \tau \mid \omega[\tau_1]) P_0(d\omega[\tau_1] \mid \omega[\tau_0]), \tag{3.2}$$
$$(\tau \geq \tau_1 \geq \tau_0).$$

The reason why P_0 and Q must satisfy these conditions is obvious: the probabilities K_0, θ that the initial particle suffers respectively no atomic event or at least one such before it effects a first passage through τ must always sum up to 1. Similarly, if $\tau \geq \tau_1 \geq \tau_0$, then given an initial particle in phase $\omega[\tau_0]$, the probability that the first atomic event will occur before a first passage through τ and leave the radiation in some phase belonging to $\mathbf{\Lambda}$ (the left-hand side of (3.2)) is the sum of the probabilities of the same thing prior to a first passage through τ_1 (first term in the right-hand side) and after a first passage through τ_1 but prior to a first passage through τ (second term in the right-hand side).

Given P_0 and Q satisfying these conditions, the first-passage distribution of the process must satisfy the *first-collision integral equation*

$$(3.3)\quad P(\mathbf{\Gamma}[\tau] \mid \omega[\tau_0]) = P_0(\mathbf{\Gamma}[\tau] \mid \omega[\tau_0]) + \sum_{n=0}^{\infty} \int_{\Omega^n[\tau]} P(\mathbf{\Gamma}[\tau] \mid \omega^n) Q^{(n)}(d\omega^n, \tau \mid \omega[\tau_0]).$$

The reason why P must satisfy (3.3) is clear: given an initial particle in phase $\omega[\tau_0]$, the probability that the radiation will effect a first passage through τ in some phase belonging to $\mathbf{\Gamma}[\tau]$ after an arbitrary number of atomic events (the left-hand side of (3.3)) is the sum of the probabilities of the same thing with, respectively, no atomic events (first term in the right-hand side) and at least one atomic event (second term in the right-hand side).

Our first problem is therefore to seek for a solution of (3.3) which is a (possibly incomplete) first-passage probability distribution satisfying the C.K. relation (2.1). It is convenient to do this in terms of the p.g.f. In these terms, (3.3) becomes

$$(3.4)\qquad G[\zeta, \tau \mid \omega[\tau_0]] = G_0[\zeta, \tau \mid \omega[\tau_0]] + H\{G[\zeta, \tau \mid \cdot], \tau \mid \omega[\tau_0]\},$$

where G_0 and H are the p.g.f.'s of respectively P_0 and Q, i.e.

$$(3.5)\qquad G_0[\zeta, \tau \mid \omega[\tau_0]] = \int_{\Sigma[\tau]} \zeta(\omega[\tau]) P_0(d\omega[\tau] \mid \omega[\tau_0]),$$

$$(3.6)\qquad H[\zeta, \tau \mid \omega[\tau_0]] = \sum_{n=0}^{\infty} \int_{\Omega^n[\tau]} \prod_{i=1}^{n} \zeta(\omega_i) Q^{(n)}(d\omega^n, \tau \mid \omega[\tau_0]).$$

We simplify the notation by writing (3.4) in the abbreviated form $G = G_0 + H[G]$. If $0 \leq \zeta \leq 1$, then it can be shown (cf. **[8]**, **[10]**) that the sequence of nonnegative functionals $\{G_n\}$ defined by the iteration relation

$$(3.7)\qquad G_{n+1} = G_0 + H[G_n]$$

is nondecreasing and bounded by 1 and hence converges to a functional $G_R \leq 1$, which is the smallest nonnegative solution of (3.4) and also satisfies the C.K. relation (2.4). Furthermore, each G_n is the p.g.f. of an incomplete conditional probability distribution P_n, the sequence $\{P_n\}$ is nondecreasing and bounded by 1, and hence converges to a conditional probability distribution $P_R \leq 1$. Finally, one shows that G_R is in fact the p.g.f. of P_R and hence that P_R satisfies (3.3) and (2.1). We call P_R and G_R the *regular solutions* of (3.3) and (3.4), respectively.

The question of the uniqueness of the regular solution is important, particularly in connection with problems of criticality. We shall call a solution G of (3.4) *admissible* if $0 \leqq G \leqq 1$ whenever $0 \leqq \zeta \leqq 1$. Let

$$K(\tau \mid \omega[\tau_0]) = P(\mathbf{\Sigma}[\tau] \mid \omega[\tau_0]) = G[1, \tau \mid \omega[\tau_0]] \leq 1,$$

then K satisfies the integral equation

$$K = K_0 + H[K]. \tag{3.8}$$

By the same token, we shall call a solution K of (3.8) *admissible* if $0 \leq K \leq 1$. It is shown in **[10]** that $G_R[\zeta \cdot \mid \cdot\,]$ is the unique admissible solution of (3.4) for each fixed ζ such that $0 \leqq \zeta \leqq 1$ if and only if $K_R(\tau \mid \omega[\tau_0]) = G_R[1, \tau \mid \omega[\tau_0]] \equiv 1$ or, equivalently, if and only if $K \equiv 1$ is the only admissible solution of (3.8). The deficiency $1 - K_R(\tau \mid \omega[\tau_0])$ is interpreted as the probability that given an initial particle in the phase $\omega[\tau_0]$, the radiation will suffer an infinite number of atomic events before effecting a first passage through τ. We shall call the process *stable* when $K_R \equiv 1$ and *unstable* otherwise.

4. **First-collision equation for mean distributions.** Transport theory is usually concerned with *mean distributions* of particles (see [7] for the general theory of mean and higher-order moment distributions). The mean distribution connected with a first-passage probability distribution P is

$$M(\Gamma[\tau] \mid \omega^k[\tau_0]) = \sum_{n=1}^{\infty} nP^{(n)}(\Gamma[\tau] \times \Sigma^{n-1}[\tau] \mid \omega^k[\tau_0]). \tag{4.1}$$

The left-hand side of (4.1) is the expectation value relative to P of the number $N(\Gamma[\tau])$ of particles which effect a first passage through τ with phases in the set $\Gamma[\tau]$, conditional on an initial population in phase $\omega^k[\tau_0]$. M can be obtained also by functional differentiation from the p.g.f. G of P (cf. [7]):

$$\begin{aligned} \int_{\Sigma[\tau]} \zeta(\omega[\tau]) M(d\omega[\tau] \mid \omega^k[\tau_0]) &= \lim_{\eta \to 1} \delta_\zeta G[\eta, \tau \mid \omega^k[\tau_0]] \\ &= \lim_{\eta \to 1} \left\{ \frac{\partial}{\partial \lambda} G[\eta + \lambda\zeta, \tau \mid \omega^k[\tau_0]] \right\}_{\lambda=0}; \end{aligned} \tag{4.2}$$

set $\zeta(\omega) = \delta(\Gamma[\tau] \mid \omega)$, the characteristic function of the set $\Gamma[\tau]$, and we obtain $M(\Gamma[\tau] \mid \omega^k[\tau_0])$.

If the process is multiplicative and P is normalized to 1, then we can see from (2.4) and (4.2) that

$$M(\Gamma[\tau] \mid \omega^k[\tau_0]) = \sum_{i=1}^{k} M(\Gamma[\tau] \mid \omega_i[\tau_0]), \tag{4.3}$$

and it can be shown (cf. **[8]**, **[10]**) that in consequence M satisfies the generalized C.K. relation

$$M(\Gamma[\tau] \mid \omega[\tau_0]) = \int_{\Sigma[\tau_1]} M(\Gamma[\tau] \mid \omega[\tau_1]) M(d\omega[\tau_1] \mid \omega[\tau_0]), \qquad (\tau \geq \tau_1 \geq \tau_0). \tag{4.4}$$

We see thus that M has the character and properties of the first-passage probability distribution of a single-particle process, except that it need not be bounded by 1. We shall say that M *exists* as a *finite* respectively *σ-finite conditional measure* on $\mathbf{B}(\Sigma[\tau]) \times \Omega[\tau]$ if $M(\cdot \mid \omega[\tau_0])$ is a finite respectively σ-finite measure on $\mathbf{B}(\Sigma[\tau])$

for each fixed $\omega[\tau_0] \in \Omega[\tau]$, and $M(\Gamma[\tau] \mid \cdot)$ is a measurable function on $\Omega[\tau]$ for each fixed $\Gamma[\tau] \in \mathbf{B}(\Sigma[\tau])$.

Let L be the mean distribution of the first collision and consequent phase distribution Q, i.e.

$$L(\Lambda, \tau \mid \omega[\tau_0]) = \sum_{n=1}^{\infty} nQ^{(n)}(\Lambda \times \Omega^{n-1}, \tau \mid \omega[\tau_0]), \tag{4.5}$$

and let $M_0 = P_0$, the first-passage distribution with no atomic events. It can be shown (cf. [**10**]) that the consistency relation (3.2) implies the following relation between L and M_0:

$$L(\Lambda, \tau \mid \omega[\tau_0]) = L(\Lambda, \tau_1 \mid \omega[\tau_0]) + \int_{\Sigma[\tau]} L(\Lambda, \tau \mid \omega[\tau_1]) M_0(d\omega[\tau_1] \mid \omega[\tau_0]) \qquad (\tau \geq \tau_1 \geq \tau_0). \tag{4.6}$$

If P is normalized to 1 and satisfies the integral equation (3.3), and if M and L exist as σ-finite conditional measures, then it can also be shown that M satisfies a first-collision integral equation

$$M(\Gamma[\tau] \mid \omega[\tau_0]) = M_0(\Gamma[\tau] \mid \omega[\tau_0]) + \int_{\Omega[\tau]} M(\Gamma[\tau] \mid \omega) L(d\omega, \tau \mid \omega[\tau_0]), \tag{4.7}$$

which is the analogue of the first-collision integral equation for the first-passage probability distribution of a single-particle process, except that L and M are no longer necessarily bounded by 1; hence the situation is no longer so favorable as regards the existence and uniqueness of solutions of (4.7). Let us call such a solution *admissible* if it is a σ-finite conditional measure. We simplify the notation by writing (4.7) in the abbreviated form

$$M = M_0 + M * L$$

and define a sequence $\{M_n\}$ by the iteration relation

$$M_{n+1} = M_0 + M_n * L. \tag{4.8}$$

Then (under the assumption that L is a σ-finite conditional measure satisfying (4.6)) we can show that $\{M_n\}$ is a nondecreasing sequence of σ-finite conditional measures, and that if (4.7) has an admissible solution at all, then $\{M_n\}$ converges to a limit M_R which is the smallest admissible solution of (4.7) and which satisfies the C.K. relation (4.4). We call M_R the *regular solution* of (4.7). Furthermore, it can be shown that if the process is stable ($K_R \equiv 1$), then the regular solution P_R of (3.3) has a σ-finite mean distribution M if and only if $M = M_R$. It is shown in [**10**] that under certain not too restricted conditions imposed on L and M_0 the converse is also true: i.e. if the process is unstable, then $\{M_n\}$ diverges.

These results can be extended to higher order moment distributions and hence used to calculate expected fluctuations from the mean. Thus, let $M^{(2)}$ and $L^{(2)}$ be the *second-order factorial moment distributions* of, respectively, P and Q, i.e.

$$M^{(2)}(\Gamma^{(2)}[\tau] \mid \omega[\tau_0]) = \sum_{n=2}^{\infty} n(n-1)P^{(n)}(\Gamma^{(2)}[\tau] \times \Sigma^{n-2}[\tau] \mid \omega[\tau_0]), \tag{4.9}$$

$$L^{(2)}(\Lambda^{(2)}, \tau \mid \omega[\tau_0]) = \sum_{n=2}^{\infty} n(n-1)Q^{(n)}(\Lambda^{(2)} \times \Omega^{n-2}, \tau \mid \omega[\tau_0]), \tag{4.10}$$

where $\Gamma^{(2)}[\tau]$, $\Lambda^{(2)}$ are measurable subsets of respectively $\Sigma^2[\tau]$ and Ω^2; let $M^{[2]} = M \times M$ and let

$$M_0^{(2)}(\Gamma^{(2)}[\tau] \mid \omega[\tau_0]) = \int_{\Omega^2[\tau]} M^{[2]}(\Gamma^{(2)}[\tau] \mid \omega^2)L^{(2)}(d\omega^2, \tau \mid \omega[\tau_0]). \tag{4.11}$$

If P is normalized to 1 and satisfies (2.3), and if $M^{(2)}$ and $L^{(2)}$ exist as σ-finite conditional measures, then $M^{(2)}$ satisfies the first-collision equation

$$M^{(2)} = M_0^{(2)} + M^{(2)} * L, \tag{4.12}$$

which is like (4.7) with $M_0^{(2)}$ substituted for M_0. Because of this similar results hold; the sequence $\{M_n^{(2)}\}$ defined by the iteration relation $M_{n+1}^{(2)} = M_0^{(2)} + M_n^{(2)} * L$ is nondecreasing and converges to the smallest admissible solution $M_R^{(2)}$ of (4.12) if such a solution exists at all. If the process is stable, then P_R has a σ-finite second-order factorial moment distribution $M^{(2)}$ if and only if $M^{(2)} = M_R^{(2)}$. The results obtained above for $M^{(2)}$ generalize in an obvious way to higher order factorial normed distributions.

5. **Processes involving only atomic events.** In the case where the states of all radiation particles remain constant between atomic events and only their positions change, the functions P_0 and Q of §3 have simple expressions in terms of cross-sections. If a particle initially at x suffers no atomic event while travelling a path of length s, its "position" vector becomes $x + \mu s$, where: (1) if X is ordinary 3-space, then μ is the unit vector in the direction of motion; (2) if X is space-time, then μ is the vector whose spatial components are as in (1), while its time component is $1/v$, where v is the absolute velocity.

It is convenient to parametrize u in such a way that we can write $u = (\gamma, \mu)$, where γ stands for the remaining state parameters. From the cross-sections, we can easily obtain the total atomic event rate $\lambda(u, x) = \lambda(\gamma, \mu, x)$ and the transition probability ϕ conditional on an atomic event: thus $\phi(du^k \mid u_0, x_0)$ is the probability that if a particle in state u_0 suffers an atomic event at x_0, then k particles will be produced in the element of population state space $du_1 \cdots du_2 = du^k$. Let

$$R(\mu, x, \tau) = \inf \{s \mid x + \mu s \in \tau, x \in X[\tau]\}, \tag{5.1}$$

where we set $R = \infty$ if $x + \mu s$ does not lie on τ for any finite s. Then

$$\begin{aligned} &P_0(A \times S \mid \mu_0, x_0, \tau) \\ &= \exp\left(-\int_0^{R(\mu_0,x_0,\tau)} \lambda(x_0 + \mu_0 s)\, ds\right)\delta(A \mid \gamma_0, \mu_0)\delta(S \mid x_0 + \mu_0 R(\mu_0, x_0, \tau)), \end{aligned} \tag{5.2}$$

(5.3)
$$Q(\mathbf{A}\times E\mid\gamma_0,\mu_0,x_0)=\int_0^{R(\mu_0,x_0,\tau)}\phi(\mathbf{A}\mid\gamma_0,\mu_0,x_0+\mu_0s)\delta(E\mid x_0+\mu_0s)\times\exp\left(-\int_0^s\lambda(x_0+\mu_0s')\,ds'\right)\lambda(x_0+\mu_0s)\,ds,$$

where A, $\mathbf{A}$, S and E are measurable subsets of respectively U, U, τ and $X[\tau]$, $\delta(A\mid\gamma,\mu)$, $\delta(S\mid x)$ and $\delta(E\mid x)$ are the characteristic functions of respectively, A, S and E and where we have for brevity written x_0 for $x[\tau_0]$ and $\lambda(x_0+\mu_0s)$ for $\lambda(\gamma_0,\mu_0,x_0+\mu_0s)$. It can be shown (cf. **[10]**) that P_0 and Q, thus defined, satisfy the consistency relations (3.1) and (3.2). The first-collision equation (3.4) becomes (suppressing for brevity the variables ζ, τ in the notation for G)

(5.4)
$$G(\gamma_0,\mu_0,x_0)=G_0(\gamma_0,\mu_0,x_0)+\int_0^{R(\mu_0,x_0,\tau)}g[G(\cdot,\cdot,x_0+\mu_0s)\mid\gamma_0,\mu_0,x_0+\mu_0s]\times\exp\left(-\int_0^s\lambda(x_0+\mu_0s')\,ds'\right)\lambda(x_0+\mu_0s)\,ds,$$

where g is the p.g.f. associated with ϕ:

(5.5)
$$g[\eta\mid\gamma_0,\mu_0,x_0]\sum_{n=0}^{\infty}\int_{U^n}\prod_{i=1}^{n}\eta(\gamma_i,\mu_i)\phi^{(n)}(d\gamma^n\,d\mu^n\mid\gamma_0,\mu_0,x_0).$$

The iteration relation (3.7) takes the form

(5.6)
$$G_{n+1}(\gamma_0,\mu_0,x_0)=G_0(\gamma_0,\mu_0,x_0)+\int_0^{R(\mu_0,x_0,\tau)}g[G_n(\cdot,\cdot,x_0+\mu_0s)\mid\gamma_0,\mu_0,x_0+\mu_0s]\times\exp\left(-\int_0^s\lambda(x_0+\mu_0s')\,ds'\right)\lambda(x_0+\mu_0s)\,ds.$$

From (5.3) we derive the "backward" integro-differential equation

(5.7) $$(\mu_0\cdot(\partial/\partial x_0))G(\gamma_0,\mu_0,x_0)=\lambda(x_0)\{G(\gamma_0,\mu_0,x_0)-g[G(\cdot,\cdot,x_0)\mid\gamma_0,\mu_0,x_0]\}.$$

Note then, while (5.3) implies (5.7), the converse is only true when suitable boundary conditions are imposed on (5.7).

Suppose that the p.g.f. G satisfying (5.4) is normalized to 1, and that the mean distributions M, ψ of, respectively, G, ϕ exist as σ-finite conditional measures; here

(5.8)
$$\psi(A\mid\gamma_0,\mu_0,x_0)=\sum_{n=1}^{\infty}n\phi^{(n)}(A\times U^{n-1}\mid\gamma_0,\mu_0,x_0);$$

then M satisfies the first-collision equation (4.7), which now takes the form (suppressing the variable $\Gamma(\tau)$ in the notation for M)

(5.9)
$$M(\gamma_0,\mu_0,x_0)=M_0(\gamma_0,\mu_0,x_0)+\int_U\int_0^{R(\mu_0,x_0,\tau)}M(\gamma,\mu,x_0+\mu_0s)\psi(d\gamma\,d\mu\mid\gamma_0,\mu_0,x_0+\mu_0s)\times\exp\left(-\int_0^s\lambda(x_0+\mu_0s')\,ds'\right)\lambda(x_0+\mu_0s)\,ds;$$

and (5.9) implies the "backward" equation

$$(5.10)\qquad \begin{aligned}&(\mu_0 \cdot (\partial/\partial x_0))M(\gamma_0, \mu_0, x_0)\\ &\quad = \lambda(x_0)\left\{M(\gamma_0, \mu_0, x_0) - \int_U M(\gamma, \mu, x_0)\psi(d\gamma\, d\mu \mid \gamma_0, \mu_0, x_0)\right\}.\end{aligned}$$

If in the time-dependent case we rewrite (5.10) in the form

$$(5.11)\qquad \begin{aligned}&\left(\frac{\partial}{\partial t_0} + \mathrm{v}_0 \,.\, \nabla x_0\right) M(\alpha_0, \mathrm{v}_0, \mathrm{x}_0, t_0)\\ &\quad = v_0\lambda(\mathrm{x}_0, t_0)\left\{M(\alpha_0, \mathrm{v}_0, \mathrm{x}_0, t_0) - \int_u M(\alpha, \mathrm{v}, \mathrm{x}_0, t_0)\psi(d\alpha\, d\mathrm{v} \mid \alpha_0, \mathrm{v}_0, \mathrm{x}_0, t_0)\right\}\end{aligned}$$

where $x = (\mathrm{x}, t)$, $u = (\alpha, \mathrm{x})$, x is the 3-space position vector, t is the time, v is the velocity and $v = |\mathrm{v}|$, we see that (5.11) is the formal adjoint of the usual "forward" transport equation. The dependence on t drops out in the case of first-passage processes in 3-space, and (5.10) becomes the so-called steady-state transport equation.

6. **The associated multiplicative chain.** Consider a multiplicative process characterized by the functions P_0, Q as in §3. We introduce a probability distribution W for the "first generation" of particles in $X[\tau]$ produced by an initial particle in phase $\omega[\tau_0] \in X[\tau]$, whose p.g.f. is

$$(6.1)\qquad \begin{aligned}Z[\zeta, \tau \mid \omega[\tau_0]] &= K_0(\tau \mid \omega[\tau_0]) + H[\zeta, \tau \mid \omega[\tau_0]]\\ &= K_0(\tau \mid \omega[\tau_0]) + Q^{(0)}(\tau \mid \omega[\tau_0])\\ &\quad + \sum_{n=1}^{\infty} \int_{\Omega^n[\tau]} \prod_{i=1}^{n} \zeta(\omega_i) Q^{(n)}(d\omega^n, \tau \mid \omega[\tau_0]),\end{aligned}$$

where H was defined in (3.6) and $K_0(\tau \mid \omega[\tau_0]) = P_0(\Sigma[\tau] \mid \omega[\tau_0])$. The "extinction probability" in this distribution is $W^{(0)} = K_0 + Q^{(0)}$, the sum of the probabilities K_0 of a first passage through τ of the initial particle prior to any atomic event and $Q^{(0)}$ of the extinction or absorption of the initial particle at the first atomic event before a first passage through τ. It follows from the consistency relation (3.1) that W is normalized to 1, since

$$(6.2)\qquad \begin{aligned}Z[1, \tau \mid \omega[\tau_0]] &= K_0(\tau \mid \omega[\tau_0]) + H[1, \tau \mid \omega[\tau_0]]\\ &= K_0(\tau \mid \omega[\tau_0]) + \theta(\tau \mid \omega[\tau_0]) \equiv 1.\end{aligned}$$

The iteration relation

$$(6.3)\qquad Z_{n+1}[\zeta, \tau \mid \omega[\tau_0]] = Z\{Z_n[\zeta, \tau \mid \cdot], \tau \mid \omega[\tau_0]\}$$

defines the sequence $\{Z_n\}$ of p.g.f.'s of a *multiplicative chain* corresponding to successive generations of particles in $X[\tau]$, with $Z_1 = Z$; we call it the chain *associated* with the original process characterized by P_0 and Q. It is known from the general theory of multiplicative chains (cf. **[3]**, **[6]**) that the corresponding sequence of extinction probabilities $W\{_n^{(0)}\}$ is nondecreasing and bounded by 1 and that it converges to an *asymptotic extinction probability* $W_{\infty}^{(0)}$, which is the smallest nonnegative solution of the functional equation

$$\zeta(\omega[\tau_0]) = Z[\zeta, \tau \mid \omega[\tau_0]]. \tag{6.4}$$

But (6.4) is clearly identical with equation (3.8), and we saw in §3 that the smallest nonnegative solution of (3.8) is $K_R(\tau \mid \omega[\tau_0]) = G_R[1, \tau \mid \omega[\tau_0]] = P_R(\Sigma[\tau] \mid \omega[\tau_0])$; hence $K_R \equiv W_{\infty}^{(0)}$, and we conclude that the process is stable if and only if the asymptotic extinction probability $W_{\infty}^{(0)}$ of the associated chain $\{Z_n\}$ is identically unity.

The mean distribution L of Q defined in (4.6) is obviously also the mean distribution of W. If $L(\Omega, \tau \mid \omega_0)$ is bounded, then L can be considered as the kernel of an integral positive linear operator on the Banach space of all bounded measurable complex functions on $\Omega[\tau]$ with the usual sup norm; the nth power L^n of L is then clearly the mean distribution associated with Z_n. It can be shown (cf. **[4]**) that the spectral radius $\alpha_M = \lim_{n\to\infty} \|L^n\|^{1/n}$ of L is in the point spectrum of the adjoint operator L^*; under additional restrictions, α_M is also the largest eigenvalue of

$$\alpha\psi(\omega_0) = \int_{\Omega[\tau]} \psi(\omega)L(d\omega, \tau \mid \omega_0). \tag{6.5}$$

We assume that W has finite moment distributions up to the third order. If $\alpha_M < 1$, then $Q_{\infty}^{(0)} \equiv 1$ (Mullikin, unpublished). With some additional conditions on L, one shows that this result is still true when $\alpha_M = 1$ (cf. **[12]**) and that $Q_{\infty}^{(0)} \not\equiv 1$ when $\alpha_M > 1$ (cf. **[10]**).

The problem of criticality in models for neutron multiplication is often discussed in terms of the chain associated with the time-independent first-passage process (see e.g. **[3]**, **[11]**, **[12]**, **[13]**, **[14]** and **[15]**). In such models, the mean distribution L defined in (4.6) is bounded, and for simple geometries one can associate a parameter, say a, with the surfaces τ (the thickness for slabs, radius for spheres), which has a critical value a_c such that if α_M is the largest positive eigenvalue of equation (6.5), then $\alpha_M < 1$ for $a < a_c$ (the *subcritical* case), $\alpha_M = 1$ for $a = a_c$ (the *critical* case), and $\alpha_M > 1$ for $a > a_c$ (the *supercritical* case). This implies under certain conditions that the asymptotic extinction probability $Q_{\infty}^{(0)} \equiv 1$ or, equivalently, that the process is stable ($K_R \equiv 1$) if and only if $a \leq a_c$. Another criterion of criticality has been proposed (see e.g. **[14]**) in terms of equation (4.7) or some equivalent equation such as (5.10) for the time-independent mean distribution: one can prove in certain cases that there exists a critical value a_c such that this equation has a finite nonnegative solution if and only if $a < a_c$.

One obtains perhaps a better insight into the physical meaning of criticality by examining the time-dependent version of these models (see **[14]** and **[15]**) which are always stable (i.e. $K_R \equiv 1$ for all finite times t) and such that the mean number M of particles effecting a first passage before t is always finite. One can show at least for certain simplified models that as $t \to \infty$, M converges to a finite number when $a < a_c$ but diverges exponentially if $a > a_c$; this exponential increase in the mean corresponds to our usual idea of a supercritical reactor. There is no contradiction in the fact that the time-dependent process is always stable while the time-independent one is unstable in the supercritical case, because the latter refers to eventual first-passages and hence corresponds to an infinite lapse of time. In fact, one can show in the case of "rod" models (see §7 and **[15]**) that all these criteria of criticality are strictly equivalent and that for $a < a_c$ the time-dependent mean converges as $t \to \infty$ to the regular solution of the equation for the time-independent mean. It is further shown in **[10]** that these results are generally valid for reasonable models of neutron multiplication.

7. **A "rod" model.** A simple nontrivial example illustrating the foregoing theory is a one-dimensional "rod" model apparently first considered in **[1]**; see also **[2]**, **[3]**, and **[14]** (where further references will be found) and **[15]**. In this model all the radiation particles move on a line with constant absolute velocity, so that the only individual state variable is the direction of motion μ, which can take only two values: $\mu = 1$ for motion to the right and $\mu = -1$ for motion to the left. The "surfaces" τ are simply the end points of intervals and we can, without undue loss of generality, restrict ourselves to the set of all finite intervals $(-a, a)$, where $a \geq 0$; thus $\tau_a = \{-a, a\}$ and the phase $\omega[\tau_a]$ has only two possible values: $\omega[\tau_a] = (1, a)$ or $\omega[\tau_a] = (-1, -a)$. We assume further a constant mean free path (which we can set equal to 1 by a suitable choice of unit) and a probability q_{ij} that at each collision the particle splits into $i + j$ particles, with i moving in the same and j in the reverse direction as the "parent," all with the same absolute velocity, with $\sum_i \sum_j q_{ij} = 1$. Let $P_{k,n-k}(\tau_a, \mu, x)$ be the probability conditional on a single initial particle at x moving in direction μ that eventually n particles will cross the boundaries a first time with k moving to the right through a and $n - k$ moving to the left through $-a$. If we write $\zeta(1, a) = z_1$ and $\zeta(-1, -a) = z_{-1}$, where z_1 and z_{-1} are complex variables of modulus ≤ 1, then expression (2.2) for the p.g.f. becomes

$$G[\zeta, \tau_a \mid \mu, x] = \sum_{n=0}^{\infty} \sum_{k=0}^{n} P_{k,n-k}(\tau_a, \mu, x) z_1^k z_{-1}^{n-k}. \tag{7.1}$$

Let

$$g(z_1, z_{-1}) = \sum_{i=0}^{\infty} \sum_{j=0}^{\infty} q_{ij} z_1^i z_{-1}^j; \tag{7.2}$$

the first-collision equation (5.6) becomes (suppressing the variables ζ, τ_a in the notation for G)

$$G[\mu, x] = z_\mu e^{-(a-\mu x)} + \int_0^{a-\mu x} g(G[\mu, x + \mu s], G[-\mu, x + \mu s]) e^{-s} ds; \tag{7.3}$$

the "backward" equation (5.7) becomes

$$\mu(\partial/\partial x) G[\mu, x] = G[\mu, x] - g(G[\mu, x], G[-\mu, x]) \tag{7.4}$$

with the boundary conditions $G[\mu, \mu a] = z_\mu$.

An explicit expression for the regular solution G_R of (7.3) or (7.4) is available in the special case where

$$q_{ij} = q_j \delta_{i,j+1}, \qquad j = 1, 2, 3 \ldots \tag{7.5}$$

(see [2]). We simplify the problem slightly by setting $z_1 = z_{-1} = z$, so that G_R is the p.g.f. for the total number of particles issuing from the rod. Let

$$\phi_a(\theta, \mu x) = \exp\left\{(\mu x - a)\sum_{n=1}^{\infty} q_n(1 - \theta^n)\right\}; \tag{7.6}$$

then for $|z| < 1$

$$G_R[z, \tau_a \mid \mu, x] = z\phi_a[K(z), x], \tag{7.7}$$

where $K(z)$ is the (unique) solution of the functional equation

$$\theta = z^2\phi_a(\theta, -a). \tag{7.8}$$

The expansion of G_R in powers of z can be written in the form

$$G_R[z, \tau_a \mid \mu, x] = z\sum_{n=0}^{\infty} \frac{z^n}{n!}\left\{\left(\frac{d}{d\theta}\right)^{n-1}\left[\phi_a^n(\theta, -a)\frac{d}{d\theta}\phi_a(\theta, \mu x)\right]\right\}_{\theta=0}, \tag{7.9}$$

where conventionally $(d/d\theta)^{-1}(d/d\theta) = 1$.

Turning to the question of criticality, we note that the expressions above define G_R only for $|z| < 1$; it can be shown, however, that $G_R \to K_R$ as $z \to 1$, and that $K_R(\tau_a \mid \mu, x) \equiv 1$ if $a \leq \frac{1}{2}m$, while $K_R(\tau_a \mid \mu, x) < 1$ if $a > \frac{1}{2}m$ and $|x| < a$, where $m = \sum_{n=1}^{\infty} nq_n$. We have thus exhibited in this particular case the existence of a critical length $a_c = \frac{1}{2}m$.

We can investigate criticality for the more general "rod" model via the eigenvalue equation (6.5) (see [3]), which becomes in this case

$$\alpha\psi(\mu, x) = \int_0^{a-\mu x} \{m_1\psi(\mu, x + \mu s) + m_{-1}\psi(-\mu, x + \mu s)\}e^{-s}\, ds, \tag{7.10}$$

where $m_1 = \sum_i \sum_j iq_{ij}$ and $m_{-1} = \sum_i \sum_j jq_{ij}$; the equivalent "backward" equation is

$$\alpha(\mu(\partial/\partial x) - 1)\psi(\mu, x) = -m_1\psi(\mu, x) - m_{-1}\psi(-\mu, x) \tag{7.11}$$

with the boundary conditions $\psi(\mu, \mu a) = 0$. Let α_M be its largest positive eigenvalue: clearly α_M depends on a. One finds then that if $m_1 + m_{-1} \leq 1$, then $\alpha_M \leq 1$ for all a, while if $m_1 + m_{-1} > 1$, then there exists a critical length

$$\begin{aligned} a_c &= (1/2\,|\gamma|)\operatorname{arctanh}|\gamma/(m_1 - 1)| && \text{if } m_1 - 1 > m_{-1} \\ &= \tfrac{1}{2}m && \text{if } m_1 - 1 = m_{-1} = m, \\ &= (1/2\,|\gamma|)\arctan|\gamma/(m_1 - 1)| && \text{if } m_1 - 1 < m_{-1} \end{aligned} \tag{7.12}$$

where

$$\gamma = \sqrt{(m_1 - 1)^2 - m_{-1}^2}, \tag{7.13}$$

such that $\alpha_M \leq 1$ if and only if $a \leq a_c$. In the particular case where q_{ij} is given by (7.5), we have $m_1 - 1 = m_{-1} = \sum_n nq_n = m$, and hence $K_R \equiv 1$ if and only if $a \leq a_c$; it can be shown that this last is in fact true for general q_{ij}.

We can investigate the criticality criteria in terms of the equation for the total mean number of particles $M(\tau_a \mid \mu, x)$ which effect a first passage through τ_a (see [**14**]). The first-collision equation (5.9) becomes (suppressing the variable τ_a in the notation for M)

(7.14)
$$M(\mu, x) = e^{-(a-\mu x)} + \int_0^{a-\mu x} \{m_1 M(\mu, x + \mu s) + m_{-1} M(-\mu, x + \mu s)\} e^{-s}\, ds,$$

which is equivalent to the "backward" equation

(7.15) $$\mu \partial M(\mu, x)/\partial x = -(m_1 - 1)M(\mu, x) - m_{-1} M(-\mu, x)$$

with boundary conditions $M(\mu, \mu a) = 1$. The solution

(7.16) $M(\mu, x)$
$$= \frac{\gamma \cosh (a + \mu x)\gamma - (m_1 - 1) \sinh (a + \mu x)\gamma + m_{-1} \sinh (a - \mu x)}{\cosh 2a - (m_1 - 1) \sinh 2a} \quad \text{if, } \gamma \neq 0,$$
$$= \frac{1 - 2m\mu x}{1 - ma} \quad \text{if } m_1 - 1 = m_{-1} = m \text{ and hence } \gamma = 0,$$

where γ is given in (7.13), is bounded and nonnegative for all μ, x and for all a if $m_1 + m_{-1} \leq 1$, and only for $a < a_c$ if $m_1 + m_{-1} > 1$, where a_c is given by (7.12); it can become negative or infinite for some x if $a \geq a_c$.

We can also formulate time-dependent first-passage problems in the general rod model by taking $X = R \times T$, where $R = (-\infty, \infty)$ is the set of all positions of a particle, $T = [0, \infty]$ is the time axis, and choosing the one-parameter family of surfaces $\tau_a = T_a \cup T_{-a}$, where $T_{\pm a} = \{a\} \times T$; hence the phase $\omega[\tau_a] = (\mu, t)$, where t is a point on $T_{\mu a}$ and $\zeta(\omega[\tau_a]) = \zeta(\mu, t)$. The first-collision expression (5.6) becomes

(7.17)
$$G[\mu, x, t] = \zeta(\mu, t + a - \mu x)e^{-(a-\mu x)} + \int_0^{a-\mu x} g(G[\mu, x + \mu s, t + s], G[-\mu, x + \mu s, t + s])e^{-s}\, ds,$$

where g was defined in (7.2). If we set $\zeta(\mu, t) = z$ when $t \leq t_1$, $\zeta(\mu, t) = 1$ when $t > t_1$, then

(7.18) $$M(t_1 \mid \mu, x, t_0) = \{\partial G[\mu, x, t_0]/\partial z\}_{z=1}$$

is the mean number of particles effecting a first passage before time t_1 conditional on an initial particle at t_0 in position x moving in direction μ. It is seen that M depends only on $t_1 - t_0 = t$, and that the "backward" equation for M takes the form

(7.19) $$(\partial/\partial t - \mu(\partial/\partial x))M(\mu, x, t) = (m_1 - 1)M(\mu, x, t) + m_{-1}M(-\mu, x, t),$$

with the boundary conditions $M(\mu, \mu a, t) = 1$ and $M(\mu, x, 0) = 0$ for $|x| < a$. The process is stable and one can prove that the regular solution of (7.19) is finite. Taking Laplace transforms of both sides of (7.19) relative to t and using the boundary conditions, one finds that

$$(7.20) \quad m(\mu, x, s) = \int_0^\infty M(\mu, x, t)e^{-st}\,dt = \frac{(s - m_1 + 1)\sinh(a + \mu x)\theta + \theta\cosh(a + \mu x)\theta + m_{-1}\sinh(a - \mu x)\theta}{s[(s - m_1 + 1)\sinh 2a\theta + \theta\cosh 2a\theta]}$$

where $\theta = ((s - m_1 + 1)^2 - m_{-1}^2)^{1/2}$. It can be shown (see **[14]** for the special case where $m_1 = n_1 = 1$ and **[15]** for the general case) that as $t \to \infty$, M converges to the (bounded and nonnegative) solution (7.16) of equation (7.14) when $a < a_c$, and diverges to $+\infty$, linearly when $a = a_c$ and exponentially when $a > a_c$.

Acknowledgements. I am much indebted to Dr. P. J. Brockwell for his help in drafting §7 of this article, which is partly based on unpublished results of ours, and to Professor T. W. Mullikin for many stimulating discussions and for the communication of his unpublished results quoted in §6.

References

1. R. Bellman, R. Kalaba and G. M. Wing, *On the principle of invariant embedding and one-dimensional neutron multiplication*, Proc. Nat. Acad. Sci. USA **43** (1957), 517–520.

2. P. J. Brockwell, *Multiplicative transport processes in one dimension*, J. Math. Anal. Appl. **16** (1966), 173–187.

3. T. E. Harris, *The theory of branching processes*, Springer, New York, 1963.

4. M. A. Krasnosel'skii, *Positive solutions of operator equations*, Noordhoff, Groningen, 1964.

5. J. E. Moyal, *Discontinuous Markoff processes*, Acta Math. **98** (1957), 221–264.

6. ———, *Multiplicative population chains*, Proc. Roy. Soc. A**266** (1962), 518–526.

7. ———, *The general theory of stochastic population processes*, Acta Math. **108** (1962), 1–31.

8. ———, *Multiplicative population processes*, J. Appl. Prob. **1** (1964), 267–283.

9. ———, *A general theory of first-passage distributions in transport and multiplicative processes*, J. Math. Phys. **7** (1966), 464–473.

10. ———, *Multiplicative first-passage processes* (to appear).

11. T. W. Mullikin, *Neutron branching processes*, J. Math. Anal. Appl. **3** (1961), 507–525.

12. ———, *Limiting distributions for critical multitype branching processes with discrete time*, Trans. Amer. Math. Soc. **106** (1963), 469–494.

13. A. M. Weinberg and E. P. Wigner, *The physical theory of neutron chain reactors*, Univ. of Chicago Press, Chicago, Ill., 1958.

14. G. M. Wing, *An introduction to transport theory*, Wiley, New York, 1962.

15. P. J. Brockwell and J. E. Moyal, *The characterization of criticality for one-dimensional transport processes*, J. Math. Anal. Appl. **22** (1968), 25–44.

Kinetic theory of transport and fluctuation phenomena

R. K. Osborn

Introduction. Transport theory, and its generalizations to fluctuation theory, comprises mathematical descriptions of complex systems which correspond to a visualization of the system as a collection of particle-like entities. "Particle-like" merely implies a high relative degree of localization. This is in contrast to field, or wave-like, mathematical descriptions of precisely the same systems. More explicitly, transport theory (including its generalizations) is a theory of the expected values of monomials of number operators—hermitian, quadratic functionals of creation and destruction operators; whereas, field or wave theories are theories of expected values of hermitian functionals of monomials of odd degree in creation and destruction operators. The classic example of this distinction is provided by the wave and photon descriptions of electromagnetic radiation. In the former case, the electric and magnetic vectors satisfying Maxwell's field equations are the expected values of hermitian, vector, functionals of first degree monomials of creation and destruction operators, while in the second case the photon density which satisfies a transport equation is the expected value of a number operator—a hermitian, quadratic function of creation and destruction operators. Strict analogies exist also in the wave and transport descriptions of collections of quanta with finite rest mass, e.g. neutrons and electrons.

In principle, neither of these ways of describing complex systems is more exact than the other. Furthermore, in a number of practical cases, neither can be regarded as exclusive or inclusive of the other. For example, the interpretation of the diffraction of x-rays or slow neutrons by crystals is equally satisfactory in terms of either wave or transport theory. Conversely, the photo-electric effect is difficult to understand in terms of the wave theory of electromagnetic radiation; whereas the refraction of light by a prism is hard to understand from the point of view of transport theory.

Just as the wave theory of electromagnetic radiation has appeared to be identified with Maxwell's field equations, transport theory has seemingly been identified with theories of singlet densities. Higher order wave and transport theories are generally lumped under the heading of theories of noise, or fluctuations, or correlations. Our concern here will be with various orders of transport theory—particularly the first and second, i.e. singlet densities and their variances—and their applications to a variety of systems. Particular attention will be paid to the underlying unity of the variety of transport and fluctuation problems that have been examined theoretically, and, in several instances, experimentally as well.

I. **General theory.** A primary function of theory is to inform and guide observation. The point of contact between theory and observation is the calculation, or measurement, of an expectation value. The relevant expectation values to transport theory or measurement are of density operators, or multiples thereof. Thus, the first theoretical problem to be faced is the one of choosing an appropriate operator representative for a density.

Classically, such a choice presents no difficulty. It is obvious that, for example, singlet density operators are approximately represented by

$$\rho(\boldsymbol{x}, t) = e^{-tL} \sum_{\sigma} \delta(\boldsymbol{x} - \boldsymbol{x}^{\sigma}) \tag{1}$$

in configuration space, and by

$$\rho(\boldsymbol{x}, \boldsymbol{v}, t) = e^{-tL} \sum_{\sigma} \delta(\boldsymbol{x} - \boldsymbol{x}^{\sigma})\delta(\boldsymbol{v} - \boldsymbol{v}^{\sigma}) \tag{2}$$

in phase space. In these formulae, $(\boldsymbol{x}, \boldsymbol{v}, t)$ are laboratory parameters and $(\boldsymbol{x}^{\sigma}, \boldsymbol{v}^{\sigma})$ are position and velocity coordinates assigned to particles. The operator, L, is the Poisson-Bracket-operator defined by (A any function of dynamical variables)

$$LA \equiv \sum_{j=1}^{3} \sum_{\sigma} \left(\frac{\partial H}{\partial x_j^{\sigma}} \frac{\partial}{\partial p_j^{\sigma}} - \frac{\partial H}{\partial p_j^{\sigma}} \frac{\partial}{\partial x_j^{\sigma}} \right) A, \tag{3}$$

where H is the hamiltonian for the system.

Quantum mechanically, the choice does not seem quite so obvious, though we will argue here for one that seems both natural and usable. Since many transport phenomena do not appear to depend explicitly on quantum effects, one can ask why use the quantum mechanics at all except when absolutely necessary. Without exploring the subtleties of the situation, our defense for the use of quantum mechanics here rests on three propositions: (1) for practical purposes the analysis is just as simple one way as the other; (2) the results include those obtained classically where relevant; and (3) there is no way known to me to adapt classical arguments to the description of systems in which particles are not conserved.

We accept Ono's choice [**1**] for an operator representative for a density, i.e.

$$\begin{aligned} \rho(\boldsymbol{X}, \boldsymbol{V}, \alpha, t) &= U^{\dagger}(t) a^{\dagger}(\boldsymbol{X}, \boldsymbol{V}, \alpha) a(\boldsymbol{X}, \boldsymbol{V}, \alpha) U(t) \\ &= U^{\dagger}(t) \rho(\boldsymbol{X}, \boldsymbol{V}, \alpha) U(t). \end{aligned} \tag{4}$$

In this formula $(\boldsymbol{X}, \boldsymbol{V})$ are points in a discretized, laboratory phase space corresponding to the centers of hypercells of volume $(h/m)^3$, where h is Planck's constant and m is the mass of the particles of interest. For massless particles, the natural phase coordinates are in configuration and "wave-vector" space, i.e. $(\boldsymbol{X}, \boldsymbol{K})$, for which the elementary hypercells have volume $= (2\pi)^3$. Since the details surrounding this matter have been discussed extensively elsewhere, they will be eschewed here. The parameters, α, are the remaining labels needed to complete the specification of the particles, e.g. spins for electrons, photons, neutrons, etc., and internal states for atoms and molecules. The operators $a^\dagger$ and a create and destroy particles of kind α in the cell centered at $(\boldsymbol{X}, \boldsymbol{V})$. Thus $\rho = a^\dagger a$ is a number operator which, in a diagonalizing representation, takes on positive, integral values or zero. The operator $U(t)$ is the time evolution operator defined by

$$(5) \qquad U(t) = \exp(-iHt/\hbar); \qquad \hbar = h/2\pi.$$

Accordingly,

$$(6) \qquad \Psi(\boldsymbol{X}, \boldsymbol{V}, \alpha, t) = \operatorname{Tr} \rho(\boldsymbol{X}, \boldsymbol{V}, \alpha, t)D(0) = \operatorname{Tr} \rho(\boldsymbol{X}, \boldsymbol{V}, \alpha)D(t),$$

is a positive definite function representing the expected number of particles of kind α to be found in the phase cell centered at $(\boldsymbol{X}, \boldsymbol{V})$.

Of course, the equations describing the dependence of Ψ or ρ on laboratory parameters will be sum-difference equations. However, in most practical situations, the cell dimensions will be very small compared to the observable, phase-space scale lengths. In such instances, the equations may be approximated as integro-differential equations.

To simplify the present discussion, no velocity dependent potentials will be included in H, e.g. no magnetic fields. Their inclusion occasions no difficulty in principle [**2**], merely adding considerable bulk to the analysis.

In general, the hamiltonian breaks up into two classes of terms when expressed in terms of the cell operators—one class describing particle streaming, including self-consistent and external field effects, and the other class describing the interactions of particles in the same cell. Thus we display

$$(7) \qquad H = H^Q + H^C,$$

where H^Q contains the former and H^C the latter class of terms. Approximate equations descriptive of $\rho(\boldsymbol{X}, \boldsymbol{V}, \alpha, t)$ will be derived by making opposite kinds of approximations with respect to the influence of H^Q and H^C.

The coarse-graining of phase space has to be supplemented by a coarse-graining in time as well. We display

$$(8) \qquad \rho(t + \tau) = U^\dagger(\tau)\rho(t)U(\tau) \cong \rho(t) + \tau\, \partial\rho/\partial t,$$

i.e.

$$\partial\rho/\partial t \cong \tau^{-1}[U^\dagger(\tau)\rho(t)U(\tau) - \rho(t)].$$

We then further approximate

$$
\begin{aligned}
U(\tau) &= \exp(-i(H^Q + H^C)\tau/\hbar) \\
&\cong \exp(-iH^Q\tau/\hbar)\exp(-iH^C\tau/\hbar) \qquad (9)\\
&= U^Q(\tau)U^C(\tau).
\end{aligned}
$$

The evolution of ρ over the time interval τ is now estimated according to

$$
\begin{aligned}
U^\dagger(\tau)\rho(t)U(\tau) &= U^\dagger(t)\rho(\tau)U(t) \\
&= U^\dagger(t)U^\dagger(\tau)\rho(0)U(\tau)U(t) \\
&\cong U^\dagger(t)U^{C\dagger}(\tau)U^{Q\dagger}(\tau)\rho(0)U^Q(\tau)U^C(\tau)U(t) \qquad (10)\\
&\cong U^\dagger(t)U^{C\dagger}(\tau)\{\rho(0) + (i\tau/\hbar)[H^Q, \rho(0)]\}U^C(\tau)U(t) \\
&= U^\dagger(t)U^{C\dagger}(\tau)\rho(0)U^C(\tau)U(t) + (i\tau/\hbar)U^\dagger(t)[H^Q, \rho(0)]U(t) + O(\tau^2).
\end{aligned}
$$

Here we have again made use of the smallness of τ relative to the macroscopic time scale, and have neglected the effects of streaming during collision and vice versa. The commutator in (10) is readily worked out to first order in displacements in phase space, after which equation (8) becomes

$$
(11)\quad \partial\rho/\partial t + \partial V_j\rho/\partial X_j + \partial a_j\rho/\partial V_j \cong U^\dagger(t)\tau^{-1}[U^{C\dagger}(\tau)\rho(0)U^C(\tau) - \rho(0)]U(t).
$$

The approximations required to obtain equation (11) are many, and few of them have been examined quantitatively. However, one or two of them are easily discussed qualitatively, and rather nicely illustrate the sense in which various approximations are squeezed against each other. The validity of the approximations involved in treating streaming to the level discussed here depends on, among other things, the smallness of the parameter $\tau/(L/V)$, i.e. the ratio of the coarse-graining time to the time required for a particle to cross a configuration-space cell. Obviously, for a given cell size and particle speed, this quantity can be made as small as desired by simply reducing τ. However, as will be evident later, if not so already, the validity of useful treatments of collisions depends on the smallness of (τ_C/τ), where τ_C is a characteristic collision time. In these instances, large values of τ are desired. Clearly, the accuracy of the description of streaming competes with the accuracy in the description of collisions.

In the light of the remarks about the hamiltonian, the accelerations are expressible as

$$
(12)\quad a_j = -\frac{1}{m_\alpha}\frac{\partial\Phi}{\partial X_j} - \frac{1}{m_\alpha}\sum_{X',V',\alpha'(X'\neq X)} \rho(X', V', \alpha', t)\,\frac{\partial V^{\alpha\alpha'}(|X - X'|)}{\partial X_j},
$$

where $V^{\alpha\alpha'}$ is a potential describing the interaction of a particle of the αth kind with another of the α'th kind, and Φ is an external potential. If the external potentials are time dependent, they must vary negligibly over time intervals of order τ for the above arguments to be at all meaningful.

Introducing the notation, $\boldsymbol{P} \sim X, V$, and noting that $\rho(\boldsymbol{P}, \alpha, t)\rho(\boldsymbol{P}', \alpha', t)$ is a doublet density operator in the sense that

$$\Psi(\boldsymbol{P}, \alpha; \boldsymbol{P}', \alpha'; t) = \mathrm{Tr}\, \rho(\boldsymbol{P}, \alpha, t)\rho(\boldsymbol{P}', \alpha', t)D(0) \tag{13}$$

is the one-time doublet density for particles of kind α in cell $\boldsymbol{P}$ and particles of kind α' in cell $\boldsymbol{P}'$; we find

$$\begin{aligned} &\left(\frac{\partial}{\partial t} + \frac{\partial}{\partial X_j} V_j + \frac{\partial}{\partial X'_j} V'_j + \frac{\partial}{\partial V_j} a_j + \frac{\partial}{\partial V'_j} a'_j\right) \rho(\boldsymbol{P}, \alpha, t)\rho(\boldsymbol{P}', \alpha', t) \\ &\quad = U^\dagger(t)\tau^{-1}[U^{C\dagger}(\tau)\rho(\boldsymbol{P}, \alpha, 0)\rho(\boldsymbol{P}', \alpha', 0)U^C(\tau) - \rho(\boldsymbol{P}, \alpha, 0)\rho(\boldsymbol{P}', \alpha', 0)]U(t). \end{aligned} \tag{14}$$

The structure of the equations for still higher order, one-time products of density operators is obvious from (11) and (14). However, there does not seem to be any forceful demand for densities of higher than second order.

In concluding this section, we note an important bit of unfinished business. To complete the theory at the present level, it would be desirable to find useful equations descriptive of two-time, two-point densities or density operators, i.e. for $\rho(\boldsymbol{P}, \alpha, t)\rho(\boldsymbol{P}', \alpha', t')$ or

$$\begin{aligned} \Psi_2(\boldsymbol{P}, \alpha, t; \boldsymbol{P}', \alpha', t') &= \mathrm{Tr}\, [\rho(\boldsymbol{P}, \alpha, t), \rho(\boldsymbol{P}', \alpha', t')]_+ D(0) \\ &= \mathrm{Tr}\, [\rho(\boldsymbol{P}, \alpha, 0), \rho(\boldsymbol{P}', \alpha', t' - t)]_+ D(t). \end{aligned} \tag{15}$$

(The symmetrization specified by the anticommutator, $[\quad]_+$, is necessary if Ψ_2 is to be real.) However, if collisional effects (such as are encompassed by the right-hand sides of (11) and (14), for example) are important, tractable equations for such quantities apparently have yet to be derived in the context of kinetic theory. They have been obtained—with and without collisions—by a variety of semi-phenomenological arguments (particularly for stationary systems for which $D(t) = D(0)$, and furthermore have been shown to be useful in several instances [**3**]–[**6**]. Also, classical kinetic theory has been employed by Klimonovitch [**7**], Rosenbluth and Rostoker [**8**], and Dupree [**9**] for the derivation of multiple time densities—mainly for application to plasma studies. Of course, it is easy to find a relation between the operator introduced in (15) above and various other density operators bearing some structural similarity to equation (14). But, so far, I have been unable to show that such relations are of any computational utility.

II. **Applications.** In this section we explore some of the implications of equations (11) and (14). In particular, we examine their application to studies of transport and fluctuations in neutron distributions, plasmas, and photon distributions.

In the neutron case, it is convenient to develop equations for the densities—in particular the singlet and doublet, one-time neutron densities—defined as

$$\Psi_1(\boldsymbol{P}, s, t) = L^{-3}\, \mathrm{Tr}\, \rho(\boldsymbol{P}, s)D(t), \tag{16a}$$

and

$$\Psi_2(\boldsymbol{P}, s; \boldsymbol{P}', s'; t) = L^{-6}\, \mathrm{Tr}\, \rho(\boldsymbol{P}, s)\rho(\boldsymbol{P}', s')D(t). \tag{16b}$$

The label 's' designates the spin state of the neutrons, and the factors of $(1/L^3)$ and $(1/L^6)$ have been introduced to conform to convention, e.g. so that $\Psi_1(\boldsymbol{P}, s, t) = \Psi_1(\boldsymbol{X}, \boldsymbol{V}, s, t)$ represents the expected number per unit volume of neutrons of spin state 's' having velocity $\boldsymbol{V}$ in the space cell centered at $\boldsymbol{X}$ at time t. Usually the neutron spin states are not accounted for in transport and fluctuation measurements; hence, we will regard the densities defined in (16a) and (16b) as summed over spin labels from here on. Using equations (11) and (14) in conjunction with equations (16a) and (16b) respectively, we find that

$$\frac{\partial \Psi_1}{\partial t} + V_j \frac{\partial \Psi_1}{\partial X_j} = \sum_{nn'} [\rho_{n'n'} - \rho_{nn}] T_{n'n} D_{nn}(t) \tag{17a}$$

and

$$\begin{aligned} &\left(\frac{\partial}{\partial t} + V_j \frac{\partial}{\partial X_j} + V'_j \frac{\partial}{\partial X'_j}\right) \Psi_2(\boldsymbol{P}, \boldsymbol{P}', t) \\ &\qquad = \sum_{nn'} [\rho_{n'n'}(\boldsymbol{P})\rho_{n'n'}(\boldsymbol{P}') - \rho_{nn}(\boldsymbol{P})\rho_{nn}(\boldsymbol{P}')] T_{n'n} D_{nn}(t). \end{aligned} \tag{17b}$$

In these equations, $T_{n'n} = \tau^{-1} |U^C_{n'n}(\tau)|^2$, and ρ_{nn} is a diagonal matrix element in the number representation, i.e. $\rho_{nn}(\boldsymbol{P}) = n_{\boldsymbol{P}}$. One further approximation has been employed here, namely,

$$\begin{aligned} \operatorname{Tr} U^\dagger(t) U^{C\dagger}(\tau) \rho(0) U^C(\tau) U(t) D(0) &= \operatorname{Tr} \rho(0) U^C(\tau) D(t) U^{C\dagger}(\tau) \\ &= \sum_{nn'n''} \rho_{n'n'} U^C_{n'n} D_{nn''} U^{C\dagger}_{n''n'} \\ &\cong \sum_{nn'} \rho_{n'n'} U^C_{n'n} D_{nn} U^{C\dagger}_{nn'} \\ &= \sum_{nn'} \rho_{n'n'} |U^C_{n'n}|^2 D_{nn}. \end{aligned} \tag{18}$$

This neglect of the off-diagonal elements of D is a form of the random phase approximation. No accelerations appear in (17a) and (17b), since it is rare for long range forces to influence neutron distributions appreciably.

To proceed further, it is necessary to evaluate the transition probabilities per unit time, $T_{n'n}$. This is a straightforward, but laborious, task which has been discussed in detail elsewhere **[10]**, so we go directly to the results. Furthermore, since it is not expected that the essential coarseness of phase space will ever be manifest in neutron transport and fluctuation measurements, we will present these results in the continuum limit. And still further, since also it is not likely that degenerate neutron distributions will be encountered in practice, we will neglect the dependence of the transition probabilities upon final state neutron occupation numbers. We then find (in the binary collision limit) that equations (17a) and (17b) become,

$$(\partial/\partial t + B)\Psi_1(\boldsymbol{x}, \boldsymbol{v}, t) = Q(\boldsymbol{x}, \boldsymbol{v}, t), \tag{19a}$$

and

$$(\partial/\partial t + B + B')\Psi_2(\boldsymbol{x}, \boldsymbol{v}, \boldsymbol{x}', \boldsymbol{v}', t)$$

(19b)
$$= Q(\boldsymbol{x}, \boldsymbol{v}, t)\Psi_1(\boldsymbol{x}', \boldsymbol{v}', t) + Q(\boldsymbol{x}', \boldsymbol{v}', t)\Psi_1(\boldsymbol{x}, \boldsymbol{v}, t)$$
$$+ \delta(\boldsymbol{x} - \boldsymbol{x}')\,\delta(\boldsymbol{v} - \boldsymbol{v}')Q(\boldsymbol{x}, \boldsymbol{v}, t) + \delta(\boldsymbol{x} - \boldsymbol{x}')\Gamma\Psi_1(\boldsymbol{x}, \boldsymbol{v}, t),$$

where the operators B and Γ are defined by

(20a)
$$B\Psi_1(\boldsymbol{x}, \boldsymbol{v}, t) = (\boldsymbol{v} \cdot \nabla + r_t)\Psi_1(\boldsymbol{x}, \boldsymbol{v}, t) - \int d^3v' r_s(\boldsymbol{v}' \to \boldsymbol{v})\Psi_1(\boldsymbol{x}, \boldsymbol{v}', t)$$
$$- \int d^3v' r_f' \sum_{j\alpha} \alpha B_\alpha^j(\boldsymbol{v}', \boldsymbol{v})\Psi_1(\boldsymbol{x}, \boldsymbol{v}', t),$$

and

(20b)
$$\Gamma\Psi_1(\boldsymbol{x}, \boldsymbol{v}, t) = \left[r_f \sum_{j\alpha} \alpha B_\alpha^j(\boldsymbol{v}, \boldsymbol{v}') - r_s(\boldsymbol{v} \to \boldsymbol{v}') \right] \Psi_1(\boldsymbol{x}, \boldsymbol{v}, t)$$
$$+ \left[r_f' \sum_{j\alpha} B_\alpha^j(\boldsymbol{v}', \boldsymbol{v}) - r_s(\boldsymbol{v}' \to \boldsymbol{v}) \right] \Psi_1(\boldsymbol{x}, \boldsymbol{v}', t)$$
$$+ \int d^3v'' r_f'' \sum_{j\alpha\beta} B_{\alpha\beta}^j(\boldsymbol{v}'' \mid \boldsymbol{v}, \boldsymbol{v}')\Psi_1(\boldsymbol{x}, \boldsymbol{v}'', t) + \delta(\boldsymbol{v} - \boldsymbol{v}')$$
$$\cdot \left[r_t \Psi_1(\boldsymbol{x}, \boldsymbol{v}, t) + \int d^3v'' r_s(\boldsymbol{v}'' \to \boldsymbol{v})\Psi_1(\boldsymbol{x}, \boldsymbol{v}'', t) \right].$$

Here we have introduced the following notations: r to represent the appropriate, macroscopic probabilities per unit time per neutron for an event; $B_\alpha^j(\boldsymbol{v}', \boldsymbol{v})\, d^3v$ to represent the probability that a fission induced by a neutron with velocity $\boldsymbol{v}'$ will produce exactly j neutrons, α of which have velocities in d^3v about $\boldsymbol{v}$; and $B_{\alpha\beta}^j(\boldsymbol{v}'' \mid \boldsymbol{v}, \boldsymbol{v}')\, d^3v\, d^3v'$ to represent the probability that a fission induced by a neutron at v'' will produce exactly j neutrons, α of which have velocities in d^3v about $\boldsymbol{v}$ and β of which have velocities in d^3v' about $\boldsymbol{v}'$. For simplicity we have neglected delayed neutrons.

Of course, equation (19a) is simply the familiar neutron transport equation. Equations (19a) and (19b) have been employed with substantial success in the interpretation of neutron fluctuation measurements **[11]**. Their application to real systems usually requires numerical treatment, but a few interesting results can be obtained analytically for fictitious cases. We mention two of them here.

A steady state solution to (19a) for monoenergetic neutrons in an infinite homogeneous medium with a uniform source is trivially found, i.e.

(21)
$$\Psi_1 = Q/(r_a - vr_f).$$

For the same conditions, the solution to (19b) is not so trivial. However, if the conventional approximations of diffusion theory are made, we obtain

(22)
$$\Psi_2(\boldsymbol{x}, \boldsymbol{x}') = \Psi_1^2 + \delta(\boldsymbol{x} - \boldsymbol{x}')\Psi_1 + \frac{\langle j(j-1)\rangle r_f \Psi_1}{2vD} \frac{\exp(-|\boldsymbol{x} - \boldsymbol{x}'|/(L/(1-k)))}{4\pi\, |\boldsymbol{x} - \boldsymbol{x}'|},$$

where D is the diffusion coefficient, v is the speed of the neutrons, $k = \nu r_f/r_a$ is the infinite medium multiplication factor, $L = (vD/r_a)^{1/2}$ is the diffusion length, and $\langle j \rangle = \nu$ is the average number of neutrons (prompt) produced per fission. It is seen that the neutron distribution is non-Poisson-like, i.e.

$$(\Psi_2(x, x') - \Psi_1^2)/\Psi_1 \neq \delta(x - x'),$$

only because of the fission process. The correlation length is $L/(1 - k)$, where L is the r.m.s. distance neutrons travel from the point of birth by fission to death by absorption. For multi-energy distributions, it may be shown **[11]** that the correlation length for thermal neutrons is approximately $(L^2 + \tau)^{1/2}/(1 - k)$, where τ is the m.s. distance neutrons travel from point of birth at high energy by fission to point of arrival at thermal energy. Thus, practically all of the thermal neutrons in small core, water moderated, thermal reactors are correlated. (Caution: no attempt should be made to interpret formula (22) in the critical limit, i.e. for $k = 1$. All indications are **[12]** that equations (19a) and (19b) do not both have steady state solutions for this case.)

Another case for which (19a) and (19b) are easily, simultaneously solvable is the steady state in a purely scattering medium with ambient nuclei in thermal equilibrium. In such an instance, the neutron singlet density is uniform in space, constant in time, and Maxwellian in velocity (more strictly, Fermi-Dirac, but for the low density, relatively high mean energy distributions encountered in practice, the distinction is meaningless). As seen above, in the absence of fission no long range correlation is to be expected, so the operator B may be represented by

$$B \to r_Q(v) - \int d^3v' r_s(v' \to v) \tag{23}$$

and Ψ_1 satisfies $B\Psi_1 = 0$. Equation (19b) becomes

$$(B + B')\Psi_2(x, v, x', v') = \delta(x - x')\Gamma\Psi_1. \tag{24}$$

By direct substitution and use of the above condition on the singlet density, we find that

$$\Psi_2 = \Psi_1(v)\Psi_1(v') + \delta(x - x')\,\delta(v - v')\Psi_1(v) \tag{25}$$

is a solution. Interpreting $\delta(x - x')$ as a Kronecker delta function in the discretized configuration space, and the cell dimensions as equal to system dimensions, this is just the usual result for density fluctuations in an ideal Maxwellian gas.

For a second example of an application of the basic equations (11) and (14), we consider the two-time density-density correlations in a fully ionized, thermal plasma. This application is of considerable practical importance because the interpretation of the measurements of the scattering of laser light by plasmas **[13]**–**[15]** depends so crucially upon it. The cross-section for the scattering of a photon with initial angular frequency ω, traveling in direction $\boldsymbol{\Omega}$, into $d\omega'$ about

ω' and $d\Omega'$ about Ω' is **[16]**

$$\begin{aligned}(26)\quad &\sigma(\omega,\mathbf{\Omega};\omega',\mathbf{\Omega}')\,d\omega'\,d\Omega' \\ &= \frac{\omega'}{\omega}\,\sigma_T(\mathbf{\Omega}\cdot\mathbf{\Omega}')\,d\omega'\,d\Omega'\exp(\hbar\,\Delta\omega/2\theta)\,\mathrm{Sech}\,\frac{\hbar\,\Delta\omega}{2\theta}\,G(\boldsymbol{\varkappa},\Delta\omega)\end{aligned}$$

where σ_T is the Thomson cross section, $\Delta\omega = \omega - \omega'$ and $\boldsymbol{\varkappa} = \boldsymbol{k} - \boldsymbol{k}'$ ($\boldsymbol{k}$ the photon wave vector), θ is the temperature in energy units, and

$$\begin{aligned}G(\boldsymbol{\varkappa},\Delta\omega) &= \frac{1}{4\pi N}\int_{-\infty}^{\infty} dt\,\exp(-i\,\Delta\omega t)\iint d^3v\,d^3v' \\ (27)\quad &\times\iint d^3x\,d^3x'\exp(i\boldsymbol{\varkappa}\cdot(\boldsymbol{x}-\boldsymbol{x}'))\langle[\rho^e(\boldsymbol{x}',\boldsymbol{v}',0),\rho^e(\boldsymbol{x},\boldsymbol{v},t)]_+\rangle_T.\end{aligned}$$

The symbol $\langle[\ \ ]_+\rangle_T$ implies thermal average of an anticommutator, N is the number of scattering particles (in this case electrons) in the scattering volume, and ρ^e is the electron density operator defined in equation (4). In this example, substantial simplification is realized by virtue of the facts that particles are conserved and that classical approximations are expected to be quite sufficient. In such an event, the right-hand-side of equation (11) can be neglected, all operators may be presumed to commute, and consequently, equation (14) is not needed since equations for any order, multiple-time density operators are now obtainable from (11) and (12) alone. For a single kind of singly ionized ion, the equations needed are (we go immediately to the phase space continuum)

$$(28)\quad \frac{\partial\rho^e}{\partial t} + v_j\frac{\partial\rho^e}{\partial x_j} - \frac{1}{m_e}\frac{\partial\rho^e}{\partial v_j}\int d^3x'\,d^3v'\frac{\partial V^{ee}}{\partial x_j}\rho^{e'} - \frac{1}{m_e}\frac{\partial\rho^e}{\partial v_j}\int d^3x'\,d^3v'\frac{\partial V^{ei}}{\partial x_j}\rho^{i'} = 0,$$

and another obtained from (28) by interchanging $e \rightleftarrows i$, designating electrons and ions respectively. We also have for this case

$$(29)\quad \langle[\rho^e(\boldsymbol{x}',\boldsymbol{v}',0),\rho^e(\boldsymbol{x},\boldsymbol{v},t)]_+\rangle_T = 2\langle\rho^e(\boldsymbol{x}',\boldsymbol{v}',0)\rho^e(\boldsymbol{x},\boldsymbol{v},t)\rangle_T.$$

Introducing fluctuation operators a la Dupree **[9]**, i.e.

$$(30)\quad \delta\rho = \rho - \langle\rho\rangle_T,$$

and assuming that $\langle\rho\rangle_T$ is independent of $\boldsymbol{x}$ and t, we obtain (e.g. for the electrons)

(31)

$$\begin{aligned}&\frac{\partial\,\delta\rho^e}{\partial t} + v_j\frac{\partial\,\delta\rho^e}{\partial x_j} - \frac{1}{m_e}\frac{\partial\langle\rho^e\rangle_T}{\partial v_j}\int d^3x'\,d^3v'\frac{\partial V^{ee}}{\partial x_j}\delta\rho^{e'} - \frac{1}{m_e}\frac{\partial\langle\rho\rangle\,e}{\partial v_j}\int d^3x'\,d^3v'\frac{\partial V^{ei}}{\partial x_j}\delta\rho^{i'} \\ &= \frac{1}{m_e}\frac{\partial\,\delta\rho^e}{\partial v_j}\int d^3x'\,d^3v'\frac{\partial V^{ee}}{\partial x_j}\delta\rho^{e'} + \frac{1}{m_e}\frac{\partial\,\delta\rho^e}{\partial v_j}\int d^3x'\,d^3v'\frac{\partial V^{ei}}{\partial x_j}\delta\rho^{i'}.\end{aligned}$$

Again, an equation for the ion fluctuations is obtained from (31) by interchanging $e \rightleftarrows i$. Since equations (28) and hence also (31) are taken from (11) in the continuum limit, they must be regarded as approximate in the sense that close encounters (collisions) have been neglected, and restricted in the sense that the spatial integrations must not include the singularities of the potentials. However, they may be considered exact, as shown by Dupree [9], without any restriction on the integrations (provided they exist) if the hamiltonian is presumed to include the coulomb self-energies.

We rewrite these equations more compactly as

$$\partial\psi/\partial t + M\psi = Q, \tag{32}$$

where

$$\psi(\boldsymbol{x}, \boldsymbol{v}, t) = \begin{pmatrix} \delta\rho^e(\boldsymbol{x}, \boldsymbol{v}, t) \\ \delta\rho^i(\boldsymbol{x}, \boldsymbol{v}, t) \end{pmatrix}, \tag{33a}$$

$$Q(\boldsymbol{x}, \boldsymbol{v}, t) = \begin{pmatrix} Q^e(\boldsymbol{x}, \boldsymbol{v}, t) \\ Q^i(\boldsymbol{x}, \boldsymbol{v}, t) \end{pmatrix}, \tag{33b}$$

and

$$M = \begin{pmatrix} B^e & \Gamma^{ei} \\ \Gamma^{ie} & B^i \end{pmatrix}. \tag{33c}$$

The elements of the vector, Q, are identified as the right-hand sides of equation (31), and the one obtained from it by the $e \rightleftarrows i$ interchange. Similarly, the integro-differential operators B and Γ are explicitly obtained from (31) and its companion equation for the ions. The transpose of ψ satisfies

$$\partial\tilde{\psi}/\partial t + \tilde{\psi}\tilde{M} = \tilde{Q}, \tag{34}$$

where

$$\tilde{\psi}(\boldsymbol{x}, \boldsymbol{v}, t) = (\delta\rho^e(\boldsymbol{x}, \boldsymbol{v}, t)\ \delta\rho^i(\boldsymbol{x}, \boldsymbol{v}, t)) \tag{35a}$$

and

$$\tilde{M} = \begin{pmatrix} \overleftarrow{B}^e & \overleftarrow{\Gamma}^{ie} \\ \overleftarrow{\Gamma}^{ei} & \overleftarrow{B}^i \end{pmatrix}, \tag{35b}$$

where $\overleftarrow{B}$ and $\overleftarrow{\Gamma}$ are integro-differential operators acting on quantities to the left. Introducing $\boldsymbol{P}$ to represent a phase point, we define a correlation matrix by

$$\Lambda(\boldsymbol{P}', \boldsymbol{P}, t) = \langle \psi(\boldsymbol{P}', 0)\tilde{\psi}(\boldsymbol{P}, t)\rangle_T, \tag{36}$$

where the subscript "T" means thermal average. Using (32) and (34), we then find that

$$\Lambda(\boldsymbol{P}', \boldsymbol{P}, t) = \int_0^\infty dx \int_0^\infty dy e^{-xM'} \langle Q(\boldsymbol{P}, -x)\tilde{Q}(\boldsymbol{P}, t - y)\rangle_T e^{-y\tilde{M}}. \tag{37}$$

The static correlation function is correspondingly given by equation (37) for $t = 0$. Now assume that

$$\langle Q(\boldsymbol{P}', t_1)\tilde{Q}(\boldsymbol{P}, t_2)\rangle_T = g(t_1 - t_2)D(\boldsymbol{P}', \boldsymbol{P}), \tag{38}$$

so that

$$\Lambda(\boldsymbol{P}', \boldsymbol{P}, t) = \int_0^\infty dx \int_0^\infty dy g(t - y + x)e^{-xM'}D(\boldsymbol{P}', \boldsymbol{P})e^{-y\tilde{M}}. \tag{39}$$

Equation (38) represents the first substantial approximation in the analysis of this problem. Since $\langle Q\tilde{Q}\rangle_T$ is a matrix of two-time, fourth order correlation functions, there is no reason to believe that it can be factored as a scalar function of time displacement and a matrix function of phase points. We now make another approximation (which is equivalent to regarding our description of the system as Markoffian [5]), i.e. we set

$$g(t_1 - t_2) = \delta(t_1 - t_2). \tag{40}$$

We then find that

$$\Lambda(\boldsymbol{P}', \boldsymbol{P}, t) = \int_0^\infty dx \int_0^\infty dy\, \delta(t - y + x)e^{-xM'}D(\boldsymbol{P}', \boldsymbol{P})e^{-y\tilde{M}}, \tag{41a}$$

and

$$\Lambda(\boldsymbol{P}', \boldsymbol{P}, 0) \equiv \Lambda(\boldsymbol{P}', \boldsymbol{P}) = \int_0^\infty dx e^{-xM'}D(\boldsymbol{P}', \boldsymbol{P})e^{-x\tilde{M}}. \tag{41b}$$

The latter of these equations can be immediately solved to obtain

$$D(\boldsymbol{P}', \boldsymbol{P}) = M'\Lambda(\boldsymbol{P}', \boldsymbol{P}) + \Lambda(\boldsymbol{P}', \boldsymbol{P})\tilde{M}. \tag{42}$$

Recalling equation (27), we see that it is the space and time fourier transform of $\Lambda_{11}(\boldsymbol{P}', \boldsymbol{P}, t)$ that is desired. Defining

$$\Lambda(\boldsymbol{\varkappa}, \boldsymbol{v}', \boldsymbol{v}, \omega) = \int_{-\infty}^\infty dt e^{-i\omega t} \iint d^3x'\, d^3x e^{i\boldsymbol{\varkappa}\cdot(\boldsymbol{x}-\boldsymbol{x}')}\Lambda(\boldsymbol{x}', \boldsymbol{v}', \boldsymbol{x}, \boldsymbol{v}, t), \tag{43}$$

and using (41a) and (42), we find that

$$\begin{aligned}\Lambda(\boldsymbol{\varkappa}, \boldsymbol{v}', \boldsymbol{v}, \omega) = \{[M(i\boldsymbol{\varkappa}, \boldsymbol{v}) + i\omega I]^{-1}\Lambda(\boldsymbol{\varkappa}, \boldsymbol{v}, \boldsymbol{v}')\}^{\mathrm{T}} \\ + [M(-i\boldsymbol{\varkappa}, \boldsymbol{v}') - i\omega I]^{-1}\Lambda(\boldsymbol{\varkappa}, \boldsymbol{v}', \boldsymbol{v}),\end{aligned} \tag{44}$$

where T denotes transpose, and further

$$\begin{aligned}\Lambda_{11}(\boldsymbol{\varkappa}, \omega) &= \iint d^3v\, d^3v' \Lambda_{11}(\boldsymbol{\varkappa}, \boldsymbol{v}', \boldsymbol{v}, \omega) \\ &= 2\,\mathrm{Re} \iint d^3v\, d^3v' \{[M(-i\boldsymbol{\varkappa}, \boldsymbol{v}) - i\omega I]^{-1}\Lambda(\boldsymbol{\varkappa}, \boldsymbol{v}', \boldsymbol{v})\}_{11}.\end{aligned} \tag{45}$$

Neglecting terms which contribute only to forward scattering (i.e. are proportional to $\delta(\boldsymbol{\varkappa})$), we note that

$$\Lambda_{11}(\boldsymbol{\varkappa}, \Delta\omega) = 2\pi NG(\boldsymbol{\varkappa}, \Delta\omega), \tag{46}$$

where G is the scattering function defined in equation (27).

The manipulations required to reduce (45) to something more recognizable **[8]**,**[17]**,**[18]** are elementary but tedious, so we merely quote the final result, i.e.

$$(47)\quad G(\boldsymbol{\varkappa}, \Delta\omega) = \frac{1}{\kappa\,|\epsilon|^2}\Bigg[\Bigg\{|1+\alpha_i|^2 M^e\left(\frac{\Delta\omega}{\kappa}\right) + |\alpha_e|^2 M^i\left(\frac{\Delta\omega}{\kappa}\right)\Bigg\} - \frac{i\kappa\gamma}{2\Delta\omega}\Bigg\{\frac{2\gamma+1+\beta}{2\gamma}\left(nG^e + \frac{1}{2\gamma+1+\beta}\right)(\alpha_e\epsilon^* - \alpha_e^*\epsilon) + (1-\beta)(nG^e)(\epsilon^*\alpha_e\alpha_i - \epsilon\alpha_e^*\alpha_i^*)\Bigg\}\Bigg].$$

Here we have introduced considerable new notation, i.e.

$$(48a)\quad \alpha(\kappa, \Delta\omega) = \frac{1}{2\gamma}\left[1 + \frac{\Delta\omega}{\kappa}\left(P\int_{-\infty}^{\infty}\frac{du M(u)}{u - \Delta\omega/\kappa} + i\pi M\left(\frac{\Delta\omega}{\kappa}\right)\right)\right],$$

$$(48b)\quad \gamma = (\kappa\lambda_D)^2 = \theta\kappa^2/8\pi n e^2,$$

$$(48c)\quad M(u) = (m/2\pi\theta)^{1/2}\exp(-mu^2/2\theta),$$

$$(48d)\quad \epsilon(\kappa, \Delta\omega) = 1 + \alpha_e + \alpha_i,$$

$$(48e)\quad \langle\rho^e(-\boldsymbol{\varkappa}, \boldsymbol{v}', 0)\rho^e(\boldsymbol{\varkappa}, \boldsymbol{v}, 0)\rangle_T = N\,\delta(\boldsymbol{v} - \boldsymbol{v}')M^e(\boldsymbol{v}) + nNM^e(\boldsymbol{v})M^e(\boldsymbol{v}')G^e(\kappa),$$

$$(48f)\quad N = nV, \qquad V = \text{scattering volume},$$

$$(48g)\quad \langle\rho^e(-\boldsymbol{\varkappa}, \boldsymbol{v}', 0)\rho^i(\boldsymbol{\varkappa}, \boldsymbol{v}, 0)\rangle_T = nNM^e(\boldsymbol{v}')M^i(\boldsymbol{v})G^{ei}(\kappa),$$

$$(48h)\quad [G^{ei}(\kappa) - (2\pi)^3\,\delta(\boldsymbol{\varkappa})] = -\beta[G^e(\kappa) - (2\pi)^3\,\delta(\boldsymbol{\varkappa})],$$

and have assumed that $n^e = n^i = n$. Terms contributing only to forward scattering have been neglected. Of course, the assumption of thermal equilibrium underlies everything, including equations (26) and (27). If now we calculate $G^e(\kappa)$ and $G^{ei}(\kappa)$ in the conventional way **[25]** for an equilibrium plasma, we find that $\beta = 1$ and

$$(49)\quad nG^e(\kappa) = -[2(\gamma+1)]^{-1}.$$

Then the cross section, equation (26), becomes **[8]**,**[17]**,**[18]**

$$(50)\quad \sigma(\omega, \boldsymbol{\Omega}; \omega', \boldsymbol{\Omega}') = (\omega'/\omega)\sigma_T(\boldsymbol{\Omega}\cdot\boldsymbol{\Omega}')\exp(\hbar\Delta\omega/2\theta) \cdot \operatorname{Sech}\frac{\hbar\Delta\omega}{2\theta}\,\frac{1}{\pi\kappa\,|\epsilon|^2}\left[|(1+\alpha_i)|^2 M^e\left(\frac{\Delta\omega}{\kappa}\right) + |\alpha_e|^2 M^i\left(\frac{\Delta\omega}{\kappa}\right)\right].$$

This formula has been compared with a number of experiments **[13]**–**[15]**, and the agreement appears to be quite good. It should be noted, however, that, in a non-equilibrium plasma, neither $\beta = 1$ nor $nG^e = -[2(\gamma+1)]^{-1}$ is necessarily expected to hold. Hence, in such cases, it would seem that the more complicated version of the scattering function, equation (45), would require examination rather than the simplified form leading to equation (50).

The last application that we will consider here is to photon distributions. For high energy photons, e.g. x-rays and gamma-rays, the discussion (excepting details of interaction mechanisms) parallels the one for neutron distributions very closely [**2**], [**19**]. In fact, with due regard for differences in interactions, equations (19a) and (19b) can be used as they stand for treatments of gamma ray distributions in nuclear reactors and reactor shields [**20**]. However, such is not the case when dealing with relatively low mean energy electromagnetic radiation distributions. In this instance, we must return to the more basic relations, (11) and (14).

Recalling an earlier remark to the effect that the natural phase coordinates for photons are points in a wave vector-configuration space, we define a photon singlet density by

$$\Psi_1(\boldsymbol{X}, \boldsymbol{K}, t) = \operatorname{Tr} \sum_{\lambda} \rho(\boldsymbol{X}, \boldsymbol{K}, \lambda, t) D(0), \tag{51}$$

where the sum is over polarization (spin) states of the photons. Using equation (11) we obtain (noting that here, as in the neutron case, long range forces do not come into play appreciably),

$$\frac{\partial \Psi_1}{\partial t} + \frac{\partial V_j \Psi_1}{\partial X_j} = \sum_{nn'} [\rho_{n'n'} - \rho_{nn}] T_{n'n} D_{nn}(t). \tag{52}$$

In free space the velocity of the photons of wave-vector $\boldsymbol{K}$ is $\boldsymbol{V} = c\boldsymbol{\Omega}$ where $\boldsymbol{\Omega} = \boldsymbol{K}/K$. In a medium containing charged particles, the velocity becomes instead $\boldsymbol{V} = \mu c\boldsymbol{\Omega}$, where $\mu = cK/\omega(K)$ is the index of refraction of the medium. This modification of the velocity is not introduced phenomenologically, but emerges naturally in the course of the derivation of equation (11) [**21**]. However, the specific form of the modification is a direct consequence of carrying through the derivation of (11) in a certain way, and the result can hardly be argued to be unique. The explicit form of μ as a function of medium parameters is also determined within the derivation in a self-consistent way by adopting a photon dressing scheme discussed by Mead [**22**]. Of course, in an inhomogeneous medium, μ may vary from point to point.

The reduction of the right-hand side of equation (52) proceeds exactly as the corresponding reduction in the neutron case, except that here we cannot ignore the dependence of the transition probabilities per unit time ($T_{n'n}$) upon final state, photon occupation numbers. Illustratively, the equation that we ultimately obtain, if only simple emission and absorption significantly influences the photon density, is

$$\partial \Psi_1/\partial t + c\Omega_j \, \partial \mu \Psi_1/\partial X_j = -(\alpha - \epsilon)\Psi_1 + \epsilon K^2/4\pi^3. \tag{53}$$

We have introduced α and ϵ to represent the macroscopic probabilities per unit time per photon for absorption and emission respectively. In media for which μ can be approximated by unity, equation (53) is just the conventional equation of radiant energy transfer [**23**], [**24**]. Furthermore, in media in which the particles can be regarded as in local, thermal equilibrium, it is easily shown that

$$\alpha = \exp(\hbar\omega/\theta)\epsilon, \tag{54}$$

where $\omega = cK/\mu$.

As an example of an application of the transport equation, (53), we consider the energy spectrum of the photons emerging from a sphere of radius R containing a fully ionized gas in kinetic equilibrium at temperature, T. Since this is a steady state situation, Ψ_1, is, of course, not dependent upon time. It is a straight forward matter to obtain the solution to equation (53) for this case, and we find for the total number of photons emerging from the surface of the sphere per sec with wave vectors (magnitudes) in dK about K,

$$J(K)\,dK = \frac{2uc\epsilon\rho_K\,dK}{\alpha - \epsilon}(4\pi R^2)\left[\frac{1}{2} + \frac{e^{-\zeta}}{\zeta} - \frac{1 - e^{-\zeta}}{\zeta^2}\right], \tag{55}$$

where $\rho_K = K^2/4\pi^3$ is the density of states and $\zeta = 2R(\alpha - \epsilon)/\mu c$ is the optical radius of the sphere.

Two limiting cases of this result are of some interest. The first is the "transparent sphere" limit, which is obtained as $\zeta \to 0$. In this instance the bracketed factor becomes just $\zeta/3$ and equation (55) reduces to

$$J(K)\,dK \underset{\zeta\to 0}{\longrightarrow} \left(\frac{4\pi R^3}{3}\right)(4\pi\epsilon\rho_K\,dK), \tag{56}$$

i.e. to the expected result that the total emission rate is just the volume of the sphere times the emission rate per cm³. The "black sphere" limit, obtained as $\zeta \to \infty$, is more interesting. We find, using equation (54),

$$J(K)\,dK \underset{\zeta\to \infty}{\longrightarrow} (4\pi R^2)\frac{\pi\mu c\rho_K\,dK}{\exp(\hbar\omega/\theta) - 1}. \tag{57}$$

In this case, the energy per unit frequency per cm² is

$$I(\omega) = \mu P(\omega), \tag{58}$$

where

$$P(\omega) = \frac{\hbar\omega^3/4\pi^2c^2}{\exp(\hbar\omega/\theta) - 1} \tag{59}$$

is the Planck spectrum. We note that the emergent frequency distribution is not Planckian in the frequency range for which μ, the index of refraction, differs from unity and is frequency dependent. For this case, in which we have assumed an ionized gas, we have roughly

$$\mu \simeq (1 + \omega_p^2/\omega^2)^{-1/2}, \tag{60}$$

where $\omega_p^2 = 4\pi ne^2/m$ is the electron plasma frequency. Thus, for frequencies below the plasma frequency,

$$I(\omega) \simeq (\omega/\omega_p)P(\omega). \tag{61}$$

REFERENCES

1. S. Ono, *The quantum-statistical theory of transport phenomena—III: Coarse-grained phase space distribution functions*, Prog. Theor. Phys., Kyoto **12** (1954), 113.

2. R. K. Osborn, *Kinetic equations for fully ionized, inhomogeneous plasmas*, Phys. Rev. **130** (1963), 2142.

3. L. P. Kadanoff and P. C. Martin, *Hydrodynamic equations and correlation functions*, Ann. Phys. **24** (1965), 419.

4. B. V. Felderhof, *Onsager relations and the spectrum of critical opalescence*, J. Chem. Phys. **44** (1966), 602.

5. M. Lax, *Classical noise. IV: Langevin methods*, Rev. Mod. Phys. **38** (1966), 541.

6. A. Z. Akcasu and R. K. Osborn, *Application of Langevin's technique to space-and-energy-dependent noise analysis*, Nuclear Sci. and Engr. **26** (1966), 13.

7. I. L. Klimontovich, *On the method of second quantization in phase space*, Dokl. Acad. Nauk SSSR **96** (1954), 43.

8. M. N. Rosenbluth and N. Rostoker, *Scattering of electromagnetic waves by a nonequilibrium plasma*, Phys. Fluids **5** (1962), 776.

9. T. H. Dupree, *Kinetic theory of plasma and the electromagnetic field*, Phys. Fluids **6** (1963) 1714.

10. R. K. Osborn and S. Yip, *The foundations of neutron transport theory*, Gordon & Breach New York, 1967

11. M. Natelson, R. K. Osborn and F. Shure, *Space and energy effects in reactor fluctuation experiments*, J. Nuclear Energy Parts A/B, **20** (1966), 557.

12. R. K. Osborn and M. Natelson, *Kinetic equations for neutron distributions*, J. Nuclear Energy Parts A/B, **19** (1965), 619.

13. P. W. Chan and R. A. Nodwell, *Collective scattering of laser light by a plasma*, Phys. Rev. Letters **16** (1966), 122.

14. S. A. Ramsden and W. E. R. Davies, *Observation of cooperative effects in the scattering of a laser beam from a plasma*, Phys. Rev. Letters **16** (1966), 303.

15. O. A. Anderson, *Measurement of the electron correlation spectrum in a plasma*, Phys. Rev. Letters **16** (1966), 978.

16. D. F. Dubois and V. Gilinsky, *Incoherent scattering of radiation by plasmas. I. Quantum mechanical calculation of scattering cross-sections*, Phys. Rev. **133** (1964), A1308.

17. E. E. Salpeter, *Electron density fluctuations in a plasma*, Phys. Rev. **120** (1960), 1528.

18. I. B. Bernstein, S. K. Trehan and M. P. H. Weenink, *Plasma oscillations. II. Kinetic theory of waves in plasmas*, Nuclear Fusion **4** (1964), 61.

19. L. L. Foldy, *Diffusion of high energy gamma rays through matter. I. Fundamental equations*, Phys. Rev. **81** (1951), 395.

20. H. Goldstein, *The attenuation of gamma rays in reactor shields*, Addison-Wesley, Reading, Mass., 1959.

21. E. H. Klevans, *A theory of photon transport in dispersive media*, Univ. of Mich. Rad. Lab. Tech. Report 2764-12-T, 1962.

22. C. A. Mead, *Quantum theory of the refractive index*, Phys. Rev. **110** (1958), 359.

23. E. A. Milne, *Thermodynamics of the stars*, *Handbuch der Astrophysik* **3**, (1930) Chapter 2.

24. S. Chandrasekhar, *Radiative transfer*, Dover, New York, 1960.

25. D. C. Montgomery and D. A. Tidman, *Plasma kinetic theory*, McGraw-Hill, New York, 1964.

UNIVERSITY OF MICHIGAN

Monte Carlo solutions of linear transport problems[1]

M. H. Kalos

It has been known for many years that Monte Carlo methods permit the numerical solution of problems of linear transport in the face of arbitrary complications in the geometrical arrangement of a medium and of the interactions of the radiation with matter. This is because these interactions are naturally stochastic and lend themselves readily to simulation.

While a more or less direct simulation works satisfactorily in some applications, there are many others in which the significant events are too rare for it to be efficient. I shall discuss several kinds of problems for which this is true and some of the techniques that have been used to deal with them.

As we shall see, some of these techniques permit results to be obtained with an arbitrarily small statistical error; all require varying departures from simulation. Another general character of these methods is that they work best if a single answer or a few closely related parameters are calculated.

Although not all applications of Monte Carlo have this limitation, we consider only the calculation of average radiation intensity of neutrons or gamma rays. For compactness let

$$\boldsymbol{P} = \{\boldsymbol{x}, E, \boldsymbol{\Omega}\}$$

where $\boldsymbol{x}$ is a position, E an energy variable, and $\boldsymbol{\Omega}$ gives the direction of a particle. If necessary, time and variables describing polarization may be included as well. Let $\Psi(\boldsymbol{P})$ denote the average density in phase space with which particles emerge from the source or from collisions. Because the transport process is Markovian

[1] The writing of this paper was supported by the AEC Computing and Applied Mathematics Center, Courant Institute of Mathematical Sciences, New York University, under Contract AT(30-1)-1480 with the U.S. Atomic Energy Commission.

in character, no distinction need be made between source particles and others with the same $\boldsymbol{P}$. Hence

$$\Psi(\boldsymbol{P}) = S(\boldsymbol{P}) + \int K(\boldsymbol{P}' \to \boldsymbol{P})\Psi(\boldsymbol{P}')\, d\boldsymbol{P}',$$

where $S(\boldsymbol{P})$ is the source density and $K(\boldsymbol{P}' \to \boldsymbol{P})$ gives, for a particle leaving a collision at $\boldsymbol{P}'$, the density at $\boldsymbol{P}$ of the next collision. K may be factored as

$$K(\boldsymbol{P}' \to \boldsymbol{P}) = T(\boldsymbol{x}' \to \boldsymbol{x} \mid E', \boldsymbol{\Omega}')C(E' \to E, \boldsymbol{\Omega}' \to \boldsymbol{\Omega} \mid \boldsymbol{x}),$$

where

$$T(\boldsymbol{x}' \to \boldsymbol{x} \mid E', \boldsymbol{\Omega}') = \sigma(\boldsymbol{x}, E') \exp\left\{-\int_0^{|\boldsymbol{x}-\boldsymbol{x}'|} \sigma(\boldsymbol{x}' + \boldsymbol{\Omega}'s, E')\, ds\right\}$$
$$\times |\boldsymbol{x} - \boldsymbol{x}'|^{-2} \delta\left(\frac{\boldsymbol{x} - \boldsymbol{x}'}{|\boldsymbol{x} - \boldsymbol{x}'|} - \boldsymbol{\Omega}'\right)$$

is the density of next collisions at $\boldsymbol{x}$ and so describes the flight from one collision to the next. In the last formula $\sigma(\boldsymbol{x}, E)$ is the macroscopic attenuation coefficient (or "cross-section") at position $\boldsymbol{x}$ and energy E. C gives all details of the scattering process and so depends upon the kind of atomic or nuclear interactions which may take place at $\boldsymbol{x}$. $\Psi(\boldsymbol{P})$ is a density averaged over all particle paths which may occur in this way.

The problems of average linear transport reduce to the evaluation of linear functionals of the type

$$F = \int f(\boldsymbol{P})\Psi(\boldsymbol{P})\, d\boldsymbol{P}.$$

A possible Monte Carlo treatment is one in which numerical random variables are generated so as to have the distribution Ψ. To do this, points $\boldsymbol{P}_0$ may be selected from $S(\boldsymbol{P}_0)$. Then a random walk in phase space is carried out by choosing points $\boldsymbol{P}_{n+1}$ from $K(\boldsymbol{P}_n \to \boldsymbol{P}_{n+1})$. If the function $f(\boldsymbol{P})$—or a random variable having f as expected value—is evaluated at each $\boldsymbol{P}_n$, the sum has expected value F. The sample average of such sums over independent walks is an approximation to F.

The first problem to be discussed is a classic for Monte Carlo, namely, to calculate the penetration of radiation through a thick barrier, i.e. one for which the a priori chance of emergence is very small. To do this with reasonable efficiency it is necessary to change the rules of the random walk so as to make success more likely. One way to do this is to multiply the Ψ equation by some $J(\boldsymbol{P})$ to get

$$\Psi'(\boldsymbol{P}) = J(\boldsymbol{P})\Psi(\boldsymbol{P}) = S(\boldsymbol{P})J(\boldsymbol{P}) + \int \frac{J(\boldsymbol{P})}{J(\boldsymbol{P}')} K(\boldsymbol{P}' \to \boldsymbol{P})\Psi(\boldsymbol{P}')\, d\boldsymbol{P}'.$$

Aside from normalization of the source, this may be solved, as was the original equation, provided the new kernel

$$K'(\boldsymbol{P}' \to \boldsymbol{P}) = \frac{J(\boldsymbol{P})}{J(\boldsymbol{P}')} K(\boldsymbol{P}' \to \boldsymbol{P})$$

is not made pathological by the choice of J. Then

$$F = \int (\Psi'(\boldsymbol{P}) f(\boldsymbol{P}) / J(\boldsymbol{P}))\, d\boldsymbol{P}$$

so that averages of f/J may be used in place of f alone.

$J(\boldsymbol{P})$ may be chosen so as to minimize the statistical error. Goertzel [**1**], [**2**, pp. 343 ff.] has shown that the proper choice is J which obeys

$$J(\boldsymbol{P}) = f(\boldsymbol{P}) + \int K(\boldsymbol{P} \to \boldsymbol{P}') J(\boldsymbol{P}')\, d\boldsymbol{P}'$$

from which it follows that

$$F = \int J(\boldsymbol{P}) S(\boldsymbol{P})\, d\boldsymbol{P}.$$

If a modified kernel is derived from this J and if an estimate is obtained only when the random walk stops, then every walk, whatever its path, gives the same result. That is, the variance is zero. If in the last equation we use $S = \delta(\boldsymbol{P} - \boldsymbol{P}_0)$, there follows the interesting observation that $J(\boldsymbol{P}_0)$ is the result which follows the introduction of a particle at $\boldsymbol{P}_0$.

It seems plausible that a zero variance scheme should be a very good point of departure for making efficient sampling schemes. One does not know the solution of the J equation, but we conjecture that use of an approximation to it which is too crude for direct calculation of F may yet offer great reductions in variance. Experience has borne out this expectation. Indeed, many significant and useful calculations have been made possible by the use of extremely simple assumptions about J. The old device of "splitting" [**2**, p. 347] may be regarded as the result of using J which is a piecewise constant function of position alone. The "exponential transformation" [**3**], [**2**, p. 346] similarly follows $J(\boldsymbol{P}) \propto \exp(\sigma_0 x)$ in slab geometry. The understanding of the meaning of J and of K' has led to more efficient uses of these simple forms for J than the original devices [**4**], [**5**], [**6**].

Several years ago I tried to carry through Goertzel's scheme more completely [**7**]. A semiquantitative analytical form for J was found (emphasizing, but not limited to, the first terms of its Neumann series). Then analytical but easily sampled approximations to the significant features of the modified kernel were found. In spite of the crude approximations, penetrations through very deep layers were calculated. The experience and confidence gained in that experiment have been used in designing sampling methods based on simple transport models. Recently, Coveyou and his colleagues at Oak Ridge have made further advances in the understanding and practice of this technique [**8**], [**9**].

$J(\boldsymbol{P})$ has the significance of the expected answer for F which results from a source at $\boldsymbol{P}$. Estimates of this can therefore be obtained while solving the Ψ' equation. This suggests the possibility of "learning" a satisfactory sampling scheme while following paths in the ordinary way. But whether this process can be started with no a priori guesses or when and how likely the efficiency would grow worse rather than better are questions that remain to be answered. Mathematical results to guide experimentation are needed.

Goertzel's scheme is not the only one which permits Monte Carlo results to be obtained with zero error. Nakache and Kalos [**10**] have given another based upon the following variational expression for F:

$$F = \frac{\{\int J(\boldsymbol{P})S(\boldsymbol{P})\,d\boldsymbol{P}\}\{\int \Psi(\boldsymbol{P})f(\boldsymbol{P})\,d\boldsymbol{P}\}}{\int J(\boldsymbol{P})\{\Psi(\boldsymbol{P}) - \int K(\boldsymbol{P}' \to \boldsymbol{P})\Psi(\boldsymbol{P}')\,d\boldsymbol{P}'\}\,d\boldsymbol{P}}.$$

Nakache used the method with an approximate J and found significant improvements in efficiency without the necessity of altering the paths. This too seems a promising technique but its potential remains largely unexplored. Perhaps there are still other formulations from which efficient sampling schemes may be devised.

There is an entirely different kind of problem than deep penetration that presents the difficulty associated with rare events. In its extreme form it is the estimation of average radiation flux at a point—through which no simulated history can be expected to pass.

To give such an estimate one can carry out a Monte Carlo procedure designed to solve the J equation. This uses the transposed kernel so that paths are followed backwards. These may be started at any point and followed to a dispersed source distribution. In spite of some technical problems this approach has been used with success at Westinghouse, by Maynard, Spanier and co-workers [**11**].

But if the source is also a point, the backwards Monte Carlo offers no advantages. There is an estimation procedure which may be used which directs a last flight toward any given point, $\boldsymbol{x}_0$. For a homogeneous medium with isotropic scattering, evaluating

$$f(\boldsymbol{x}, \boldsymbol{x}_0) = C \exp(-\sigma r)/4\pi r^2, \quad \begin{aligned} C &= \text{scattering probability}, \\ \sigma &= \text{attenuation coefficient}, \\ r &= |\boldsymbol{x} - \boldsymbol{x}_0|, \end{aligned}$$

at each collision gives an unbiased estimate of flux at $\boldsymbol{x}_0$. This is still very inefficient if the point $\boldsymbol{x}_0$ lies within the scattering medium, for then the r^{-2} singularity makes the variance diverge. The distribution of the mean obeys a central limit law whose width decreases only as $N^{-1/3}$ for N independent samples [**12**]. Thus an estimator with finite variance is useful. Because the once collided flux has a weaker singularity $\sim r^{-1}$, the average of this quantity for each $\boldsymbol{x}$ will have a finite variance. The integral for the once collided flux can itself be done by Monte Carlo, and providing that an appropriate singular kernel is used to select the intermediate point of collision, its variance will also exist.

It is interesting to note in passing that estimation of average flux on a surface presents similar, though less severe problems. If θ is the angle of particle direction with the normal, two estimators are $|\cos\theta|^{-1}$ for rays that cross the surface and $|\cos\theta|^{-1}\exp(-\sigma s)$ for rays that do if extended. Each of these has infinite variance if the surface lies within a scattering medium, but in each case the mean is distributed in an asymptotically normal distribution whose width is proportional to $(\log N/N)^{1/2}$. A bounded estimator is easy to devise.

These estimators call attention to some unusual problems in the calculation of confidence limits. Even where the parent populations do not have these pathological features, the distribution of the mean in a practical problem frequently is far from its limiting behavior. It would be most useful to have practical methods for assessing the validity of sample means. Such methods might use the analytical and physical character of the problem together with the power of the digital computers with which these problems are usually done.

There are of course many other interesting practical problems associated with estimation, such as finding new and eventually optimum procedures [**4**], [**13**], [**14**]. In such work numerical experiments play a significant role.

A last key problem that I will touch on is that of the estimation of criticality of a fissile assembly. This is described most succinctly as follows. There is a kernel $M(\mathbf{x}, \mathbf{x}')$ which gives the expected density of fissions at $\mathbf{x}$ which follow a fission earlier at $\mathbf{x}'$. It is then necessary to find the maximum eigenvalue λ in

$$\mathscr{M}\phi \equiv \int M(\mathbf{x}, \mathbf{x}')\phi(\mathbf{x}')\, d\mathbf{x}' = \lambda\phi(\mathbf{x}).$$

In all but the simplest model problems, M is not known explicitly but may be sampled by following Monte Carlo paths in the usual way. The whole process may be iterated from some ϕ_0 to generate

$$\phi_n = \mathscr{M}^n\phi_0$$

and, eventually, the equilibrium distribution from which the eigenvalue may be obtained. For example, one may multiply the $\mathscr{M}$ equation by any function $\psi(x)$ and integrate

$$\lambda = \frac{\int \psi(\mathbf{x})\mathscr{M}\phi(\mathbf{x})\, d\mathbf{x}}{\int \psi(\mathbf{x})\phi(\mathbf{x})\, d\mathbf{x}} = \frac{\Sigma \int \psi(\mathbf{x})\phi_{n+1}(\mathbf{x})\, d\mathbf{x}}{\Sigma \int \psi(\mathbf{x})\phi_n(\mathbf{x})\, d\mathbf{x}}.$$

The sums extend over those iterations which are judged to be close to the equilibrium distribution. Setting $\psi = 1$ means simply counting the ratio of fissions in successive generations.

This is very straightforward but is used less than might be expected, partly because in eigenvalue problems high precision is generally required and partly because of the labor of the preliminary iterations.

The acceleration of this preliminary convergence is more difficult than in finite difference calculations. A method that has been used with success [**15**], [**16**] is to divide the fuel region into zones and, in the course of following paths, estimate M_{ij}, the expected number of fissions in zone j which follow one in zone i. Then a numerical calculation of the eigenvalue of this discretized operator can be obtained well before the iterated distribution settles down. It should be possible to accelerate further by computing a discretized eigenvector and adjusting the population to conform to it.

This suggests again the possibility of using variational expressions to reduce the statistical error. A serious difficulty is of course that M is not known and cannot be manipulated analytically. Thus a procedure which can be made arbitrarily close to zero variance cannot be carried out on this basis. A new approach seems to be needed.

In conclusion, it is worth stressing that many practical technical problems dominate the practice of Monte Carlo methods. This is particularly true for problems having complex geometry and interactions, for which Monte Carlo is well suited. Then the manipulation of data and the convenient but flexible and computationally efficient specification of the geometry become organizational problems which are not at all trivial. The geometrical problem is simple enough to state: to calculate the intersection of a given ray with a set of prescribed surfaces. Several codes exist in which quadric surfaces may be specified, but the reduction to the required format in error free form can be a tedious and thankless job. Perhaps there is a logical language like a computer language to make it more amenable. One tack that has been tried **[17]** is to allow configurations to be built up as the operations of union, intersection and negation of the sets of points defined by simple figures.

I hope that the flavor of some current questions in Monte Carlo has been conveyed. Perhaps it is too bad that the elegant simplicity of the idea of simulating particle histories has been corrupted, but it has generally been true that man's reach expands so as to exceed his grasp.

Bibliography

1. G. Goertzel, *Quota sampling and importance functions in stochastic solutions of particle problems*, ORNL 434 (unpublished).

2. G. Goertzel and M. H. Kalos, *Monte Carlo methods in transport problems*, Ser. I, Vol. 2, Progress in Nuclear Energy, Pergamon Press, New York, 1958, pp. 315–369.

3. H. Kahn, *Modification of the Monte Carlo method*, Rand Report P132, 1949 (unpublished).

4. M. H. Kalos, F. R. Nakache, and J. Celnik, "Monte Carlo methods in reactor computations" in Chapter V of *Computing methods in reactor physics*, edited by Kelber, Greenspan, and Okrent, Gordon and Breach, New York (in press).

5. M. Leimdorfer, *On the use of Monte Carlo methods for calculating the deep penetration of neutrons in shields*, Trans. Chalmers Univ., No. 287 (1964).

6. E. S. Troubetzkoy, UNC SAM II, *A Fortran Monte Carlo program treating time dependent neutron and photon transport through matter*, UNC 5157, Sept. 1966 (unpublished).

7. M. H. Kalos, *Importance sampling in Monte Carlo shielding calculations—neutron penetration through thick hydrogen slabs*, Nucl. Sci. and Engrg. **16** (1963), 227.

8. R. R. Coveyou, V. R. Cain, and K. J. Yost, *Adjoint and importance in Monte Carlo applications*, Nucl. Sci. and Engrg. **27** (1967), 219.

9. V. R. Cain, Private communication, October 1966.

10. F. R. Nakache, *Variational Monte Carlo calculation of thermal utilization factors and neutron density depressions in foils*, Doctoral Dissertation, Columbia Univ., 1962 (unpublished).

11. C. W. Maynard, *An application of the reciprocity theorem to the acceleration of Monte Carlo calculations*, Nucl. Sci. and Engrg. **10** (1961), 97.

12. M. H. Kalos, *On the estimation of flux at a point by Monte Carlo*, Nucl. Sci. and Engrg. **16** (1963), 111.

13. D. B. MacMillan, *Comparison of statistical estimators for neutron Monte Carlo calculations*, Nucl. Sci. and Engrg. **26** (1966), 366.

14. J. Spanier, *Two pairs of families of estimators for transport problems*, SIAM J. Appl. Math. **14** (1966), 702.

15. K. W. Morton, *Criticality calculations by Monte Carlo methods*, AERE T/R 1903, Harwell, 1956 (unpublished).

16. E. L. Kaplan, *Monte Carlo methods for equilibrium problems in neutron multiplication*, UCRL 5275-T, 1958 (unpublished).

17. W. Guber and P. S. Mittelman, *A new technique for the computer representation of the human body*, Trans. Amer. Nucl. Soc. **9** (1966), 358.

Courant Institute of Mathematical Sciences
New York University
New York, New York

IV. Kinetic Theory and Plasma Transport

Radiative transfer in fluctuating media

Max Krook[1] and G. B. Rybicki[2]

1. **Introduction.** Radiative energy transport plays an essential role in the theory of stellar atmospheres. In principle, the structure of an atmosphere is completely determined when we specify its chemical composition, its surface gravity g, and the net energy flux that it is required to transport outward; the latter two parameters are supposed to have been obtained beforehand by a solution of the internal structure problem.

Physically, a stellar atmosphere is composed of molecules (atoms in various stages of ionization and in various internal energy states) and of photons. The state of the system can be described rather generally by molecular distribution functions for the matter and by photon distribution functions. The distribution functions satisfy kinetic equations in which all relevant interaction processes of molecules with molecules and molecules with photons are represented. However, because of the enormous complexity of these equations, approximation procedures have to be invoked in order to render them mathematically tractable.

Over the greater part of an atmosphere, the mean free path for molecular collisions is very small compared to macroscopic characteristic lengths. A reasonable zero-order approximation then consists in the treatment of the matter as being in a state of local thermodynamic equilibrium (LTE). In this approximation, the local microscopic state of the matter is determined (for given chemical composition) by its density ρ (or pressure p) and its temperature T; as a consequence, the interaction coefficients for matter with the radiation are functions of ρ and T. On

[1] Department of Astronomy and Division of Engineering and Applied Physics, Harvard University, Cambridge, Mass.

[2] Smithsonian Astrophysical Observatory and Department of Astronomy, Harvard University, Cambridge, Mass.

the other hand, because of the condition that at the top of the atmosphere no radiation is incident from outside, the radiation field may not be regarded as corresponding to local thermodynamic equilibrium.

The radiation field is described by a photon distribution function $F_\nu(c\mathbf{\Omega}, \mathbf{r}, t)$ with the interpretation that $F_\nu(c\mathbf{\Omega}, \mathbf{r}, t)\, d\mathbf{r}\, d\mathbf{\Omega}\, d\nu$ is the probability at time t of finding a photon in volume element $d\mathbf{r}$ about position $\mathbf{r}$, with frequency in range $d\nu$ about ν, and with direction of motion in solid angle $d\mathbf{\Omega}$ about direction $\mathbf{\Omega}$; c is the speed of light. In radiative transport theory, it is customary to use the specific intensity defined by $f(\mathbf{r}, \mathbf{\Omega}, \nu, t) = h\nu c F_\nu(c\mathbf{\Omega}, \mathbf{r}, t)$, where h is Planck's constant.

For an atmosphere in a steady state without mass motions of material, the basic equations are

(i) The equation of radiative transfer

$$\mathbf{\Omega} \cdot \nabla f(\mathbf{r}, \mathbf{\Omega}, \nu) = -\kappa_\nu(\rho, T) f + \varepsilon_\nu(\mathbf{r}, \mathbf{\Omega}), \tag{1}$$

where κ_ν is a general absorption coefficient for radiation of frequency ν, and $\varepsilon_\nu(\mathbf{r}, \mathbf{\Omega})$ is the emissivity of the material for ν radiation in the direction $\mathbf{\Omega}$.

(ii) The equation of hydrostatic equilibrium

$$\nabla p = -g\rho. \tag{2}$$

When the effective depth of the atmosphere is very small compared to the radius of the star, the atmosphere may be regarded as stratified in parallel planes.

The photon processes described in equation (1) are conveniently classified into continuum processes and line processes. In many cases of interest it suffices, in a first approximation, to neglect line processes in the evaluation of the thermal structure of the atmosphere. The transfer equation (1) then reduces to the form

$$\mathbf{\Omega} \cdot \nabla f(\mathbf{r}, \mathbf{\Omega}, \nu) = \kappa_\nu(\rho, T)\{-f + B_\nu[T(\mathbf{r})]\}, \tag{3}$$

where B_ν is the Planck function corresponding to the local matter temperature $T(\mathbf{r})$;

$$B_\nu(T) = 2h\nu^3 c^{-2}(e^{h\nu/kT} - 1)^{-1}, \tag{4}$$

where k is Boltzmann's constant. The replacement of the source function $\varepsilon_\nu/\kappa_\nu$ by the Planck function is a consequence of treating the matter as being in LTE.

When the atmosphere can be regarded as stratified in parallel planes, all quantities depend only on depth z in the atmosphere, and the basic equations become

$$\mu\, \partial f(z, \mu, \nu)/\partial z = \kappa_\nu\{-f + B_\nu[T(z)]\}, \tag{5}$$

$$dp/dz = -g\rho, \tag{6}$$

together with the equation of state $p = p(\rho, T)$; μ is the cosine of the angle between the direction of the intensity and the outward normal. The solution of equations (5) and (6) for the special degenerate case of gray absorption coefficient, i.e. $\kappa_\nu \equiv \kappa$ independent of ν, is of course well known, since it is reducible to the Milne problem.

In practice, κ_ν exhibits large variations in its dependence on ν. The problem posed by equations (5) and (6) is then highly nonlinear. However, satisfactory perturbation-iteration procedures have been devised for handling the problem numerically.

At present, considerable attention is being focussed on transfer problems involving line processes; the solution of these problems permits detailed and extensive comparison of models with observational results. In particular, there has been significant progress in the handling of the line transfer problem for multi-level atoms, including the effect of noncoherent scattering and departures from local thermodynamic equilibrium.

Another important class of problems is posed by extended stellar atmospheres, in which the assumption of stratification in parallel planes is not valid, and the effect of curvature must be taken into account.

In certain classes of atmosphere models, the assumption that the material is everywhere at rest is found not to be consistent in the sense that the resulting model would be dynamically unstable. In such atmospheres, energy is transported both by radiation and by (turbulent) convection, as in the solar atmosphere. The local physical state is then a fluctuating function of position and time. A proper dynamical treatment of the structure problem for such atmospheres is not possible at the present time. However, important insights into the problem can be gained by studying a sequence of problems in which the material fluctuations are supposed given, and one then inquires into the properties of the emergent radiation from such an atmosphere. It is with problems of this kind that the rest of this paper is concerned.

2. **The transfer equation.** In this section, we derive the basic equations for three-dimensional transfer that we shall need subsequently. We shall make the assumption of LTE and the gray absorption coefficient. Furthermore, we shall assume that the absorption coefficient is constant in space. By measuring all lengths in units of a mean free path, we can then set

$$\kappa_\nu(\boldsymbol{r}) \equiv 1, \tag{7}$$

and integrate equation (3) over frequency, which gives

$$\boldsymbol{\Omega} \cdot \nabla I(\boldsymbol{r}, \boldsymbol{\Omega}) = -I(\boldsymbol{r}, \boldsymbol{\Omega}) + B(\boldsymbol{r}). \tag{8}$$

The quantities

$$I(\boldsymbol{r}, \boldsymbol{\Omega}) = \int_0^\infty d\nu f(\boldsymbol{r}, \boldsymbol{\Omega}, \nu), \qquad B(\boldsymbol{r}) = \int_0^\infty d\nu B_\nu(T(\boldsymbol{r})) = \frac{\sigma}{\pi} T^4(\boldsymbol{r}) \tag{9}$$

are the *integrated intensity* and *integrated Planck function*, respectively; the σ appearing here is the Stefan-Boltzmann constant.

The radiation flux vector is defined by

$$\boldsymbol{F}(\boldsymbol{r}) = \int_0^\infty d\nu \int_{4\pi} d\boldsymbol{\Omega}\boldsymbol{\Omega} f(\boldsymbol{r}, \boldsymbol{\Omega}, \nu) = \int_{4\pi} d\boldsymbol{\Omega}\boldsymbol{\Omega} I(\boldsymbol{r}, \boldsymbol{\Omega}). \tag{10}$$

An expression for the divergence of this vector can be found by integration of the transfer equation (8) over the entire solid angle:

$$\nabla \cdot \boldsymbol{F}(\boldsymbol{r}) = 4\pi[-\phi(\boldsymbol{r}) + B(\boldsymbol{r})], \tag{11}$$

where the *mean intensity* $\phi(\boldsymbol{r})$ is defined by

$$\phi(\boldsymbol{r}) = (4\pi)^{-1}\int_{4\pi} d\boldsymbol{\Omega} I(\boldsymbol{r}, \boldsymbol{\Omega}). \tag{12}$$

We shall now define a function $e(\boldsymbol{r})$ by the equation

$$\nabla \cdot \boldsymbol{F}(\boldsymbol{r}) = 4\pi e(\boldsymbol{r}). \tag{13}$$

This function measures the deviation from radiative equilibrium at each point $\boldsymbol{r}$. The classic case of an atmosphere in radiative equilibrium would correspond to the case $e(\boldsymbol{r}) \equiv 0$. From equations (11) and (13) it follows that

$$B(\boldsymbol{r}) = \phi(\boldsymbol{r}) + e(\boldsymbol{r}). \tag{14}$$

With the assumption that no radiation is incident on the system from outside, formal integration of the transfer equation (8) along rays to find the specific intensity, and subsequent integration over the solid angle yield the result

$$\phi(\boldsymbol{r}) = \int L(|\boldsymbol{r} - \boldsymbol{r}'|)B(\boldsymbol{r}')\, d\boldsymbol{r}', \tag{15}$$

where

$$L(r) = e^{-r}/4\pi r^2 \tag{16}$$

(see, e.g. [2]). Combined with equation (14) this gives the following linear integral equation for $B(\boldsymbol{r})$:

$$B(\boldsymbol{r}) = \int L(|\boldsymbol{r} - \boldsymbol{r}'|)B(\boldsymbol{r}')\, d\boldsymbol{r}' + e(\boldsymbol{r}). \tag{17}$$

With $e(\boldsymbol{r})$ specified, this equation can be solved for $B(\boldsymbol{r})$. Then the specific intensity $I(\boldsymbol{r}, \boldsymbol{\Omega})$ can be found easily by integration of equation (8).

This solution for $B(\boldsymbol{r})$ can be conveniently expressed in terms of the *Green's function* of the problem, namely, the function $G(\boldsymbol{r}, \boldsymbol{r}')$ satisfying

$$G(\boldsymbol{r}, \boldsymbol{r}') = \int L(|\boldsymbol{r} - \boldsymbol{r}''|)G(\boldsymbol{r}'', \boldsymbol{r}')\, d\boldsymbol{r}'' + \delta(\boldsymbol{r} - \boldsymbol{r}'), \tag{18}$$

where $\delta(\boldsymbol{r} - \boldsymbol{r}')$ is a three-dimensional δ-function. In terms of the Green's function, the solution to equation (17) has the form

$$B(\boldsymbol{r}) = \int G(\boldsymbol{r}, \boldsymbol{r}'')e(\boldsymbol{r}'')\, d\boldsymbol{r}''. \tag{19}$$

This can be proved by multiplying (18) by $e(\boldsymbol{r}')$ and integrating with respect to $\boldsymbol{r}'$. It will then be noted that each side of equation (19) satisfies the same equation, namely, equation (17).

3. **Fluctuations.** Let us first specify the geometry of the problem. In terms of the cartesian coordinates (x, y, z) we define the atmosphere to be the half-space $z \geqq 0$. In the usual theory of stellar atmospheres, the assumption is made that all physical quantities depend only on the single space variable z. We now wish to eliminate this assumption and allow variations in x and y as well. However, it is convenient to imagine that there is some mean stratified atmosphere about which fluctuations occur. Thus, we write

$$\begin{aligned} B(\mathbf{r}) &= B^{(0)}(z) + \Delta B(\mathbf{r}), \\ e(\mathbf{r}) &= e^{(0)}(z) + \Delta B(\mathbf{r}), \\ I(\mathbf{r}, \mathbf{\Omega}) &= e^{(0)}(z, \mathbf{\Omega}) + \Delta I(\mathbf{r}, \mathbf{\Omega}), \end{aligned} \tag{20}$$

and similarly for all other quantities. The superscript (0) refers to the mean atmosphere, which does not fluctuate, and the prefix Δ refers to the fluctuations. We shall assume that these fluctuations are statistical in nature and that they have zero mean,

$$\langle \Delta B(\mathbf{r}) \rangle = 0, \qquad \langle \Delta e(\mathbf{r}) \rangle = 0, \qquad \langle \Delta I(\mathbf{r}, \mathbf{\Omega}) \rangle = 0, \tag{21}$$

etc. We also define the correlation functions:

$$C_{BB}(\mathbf{r}, \mathbf{r}') = \langle \Delta B(\mathbf{r})\, \Delta B(\mathbf{r}') \rangle, \qquad C_{ee}(\mathbf{r}, \mathbf{r}') = \langle \Delta e(\mathbf{r})\, \Delta e(\mathbf{r}') \rangle. \tag{22}$$

The correlation function of emergent intensity will be treated in a later section.

One important type of problem in a fluctuating atmosphere is to compute the correlation function of the source function C_{BB} when the correlation function C_{ee} is given. First, let us note that the fluctuations obey the same equation (19) as do the total quantities because of the linearity of the problem. Thus we can write

$$\Delta B(\mathbf{r}) = \int G(\mathbf{r}, \mathbf{r}'')\, \Delta e(\mathbf{r}'')\, d\mathbf{r}'', \qquad \Delta B(\mathbf{r}') = \int G(\mathbf{r}', \mathbf{r}''')\, \Delta e(\mathbf{r}''')\, d\mathbf{r}'''. \tag{23}$$

Multiplying these two equations together and averaging, we obtain

$$C_{BB}(\mathbf{r}, \mathbf{r}') = \int d\mathbf{r}'' \int d\mathbf{r}''' G(\mathbf{r}, \mathbf{r}'') G(\mathbf{r}', \mathbf{r}''') C_{ee}(\mathbf{r}'', \mathbf{r}'''). \tag{24}$$

This is the formal solution to the problem. In the present form, the solution for C_{BB} requires a six-fold integration involving the auxiliary function G of six variables. Furthermore, in order to determine the correlation function of emergent intensity, we require two more integrations. The reduction of these expressions to tractable form is accomplished in the following section.

4. **Invariance principles and symmetries.** There are, fortunately, several simplifying features in these problems. First of all, there are two types of symmetries:

(i) The medium is intrinsically invariant under translations and rotations in the x-y plane. There is also an "invariance principle," which applies in the z-direction in analogy to the invariance principles of the plane-parallel case. These invariances lead to a great reduction in the complexity of the Green's function; we shall show that it can be derived from a single function of two variables, rather than the apparent six variables.

(ii) On physical grounds, the fluctuations themselves may be assumed to be statistically uniform in the x-y plane. This implies the rotational and translational invariance in the x-y plane of the correlation functions, which become functions of three variables instead of six.

There is a third simplification that comes about from a mathematical coincidence, namely, Laplace transforms are used in the z-direction in order to give concise expression to the invariance principles, and Laplace transforms of the correlation function C_{BB} are needed to find the correlation function of emergent intensity. Therefore, the emergent intensity correlations are found more or less automatically, in close analogy to the corresponding plane-parallel case.

These three simplifications lead to the possibility of our evaluating the correlation function of emergent intensity from a given C_{ee} with two integrations instead of eight, and with the introduction of auxiliary functions of only two variables.

Experience has shown that the invariance in the x-y plane is best taken into account by use of Fourier transforms. Let us define a two dimensional vector $\boldsymbol{\rho}$ with components (x, y): this is the projection of $\boldsymbol{r}$ onto the x-y plane. The symmetry of any correlation function demands that

$$C(\boldsymbol{r}, \boldsymbol{r}') = C(z, z', |\boldsymbol{\rho} - \boldsymbol{\rho}'|), \tag{25}$$

and similarly for $G(\boldsymbol{r}, \boldsymbol{r}')$. Let us define the Fourier transform of any of these functions by

$$\bar{C}(z, z', |\boldsymbol{k}|) = \int d\boldsymbol{\rho} e^{i\boldsymbol{k}\cdot\boldsymbol{\rho}} C(z, z', |\boldsymbol{\rho}|), \tag{26}$$

which clearly depends only on the magnitude of the wave vector $\boldsymbol{k}$, which we denote by $k \equiv |\boldsymbol{k}|$. Applying the convolution theorem for Fourier transforms to equation (24), we obtain

$$\bar{C}_{BB}(z, z', k) = \int_0^\infty dz'' \int_0^\infty dz''' \bar{G}(z, z'', k) \bar{G}(z', z''', k) \bar{C}_{ee}(z'', z''', k). \tag{27}$$

Fourier transformation of equation (18) similarly yields

$$\bar{G}(z, z', k) = \int_0^\infty \bar{K}(|z - z''|, k) \bar{G}(z'', z', k)\, dz'' + \delta(z - z'), \tag{28}$$

where $\bar{K}(z, k)$ is the Fourier transform of the kernel function of equation (16), namely,

$$\bar{K}(z, k) = \int d^2\boldsymbol{\rho} e^{i\boldsymbol{k}\cdot\boldsymbol{\rho}} \frac{\exp[-(z^2 + \rho^2)^{1/2}]}{4\pi(z^2 + \rho^2)}. \tag{29}$$

Elliott [3] has given a solution of equation (28) by means of an invariance principle combined with the Wiener-Hopf technique. While we shall essentially use Elliott's invariance technique, which finds more general expression in the work of Sobolev

[7], we depart somewhat from his derivation in that we determine the structure of the Green's function in its Laplace-transformed form. Let us define the resolvent function by

$$\bar{\Gamma}(z, z', k) = \bar{G}(z, z', k) - \delta(z - z'). \tag{30}$$

Sobolev [7] shows that $\bar{\Gamma}$ is symmetric in z and z', and that it satisfies the differential relation

$$\begin{aligned} \frac{\partial \bar{\Gamma}}{\partial z} + \frac{\partial \bar{\Gamma}}{\partial z'} &= \bar{\Gamma}(z, 0, k)\bar{\Gamma}(0, z', k) \\ &= \Phi(z, k)\Phi(z', k), \end{aligned} \tag{31}$$

where we have set

$$\Phi(z, k) = \bar{\Gamma}(z, 0, k) = \bar{\Gamma}(0, z, k). \tag{32}$$

By taking Laplace transforms of equation (31) with respect to z and using integration by parts, we obtain

$$s\bar{\Gamma}(s, z', k) - \Phi(z', k) + \partial\bar{\Gamma}(s, z', k)/\partial z' = \Phi(s, k)\Phi(z', k).$$

Laplace transformation with respect to z' then gives

$$(s + s')\bar{\Gamma}(s, s', k) = \bar{\Phi}(s, k)\bar{\Phi}(s', k) + \bar{\Phi}(s', k) + \bar{\Phi}(s, k). \tag{33}$$

We have used s and s' as the transform variables for the transforms with respect to z and z', respectively. To avoid complicated notation, we have depended somewhat on the variable names to indicate what transform has been taken. Since the Laplace transform of $\delta(z)$ is 1, and the double Laplace transform of $\delta(z - z')$ is $1/(s + s')$, we can write

$$\bar{\bar{G}}(s, s', k) = \mathscr{H}(s, k)\mathscr{H}(s', k)/(s + s'), \tag{34}$$

where $\bar{\bar{G}}(s, s', k)$ is the double Laplace transform of the Green's function $\bar{G}(z, z', k)$, and

$$\mathscr{H}(s, k) \equiv \bar{\Phi}(s, k) + 1. \tag{35}$$

From equations (27), (29), (31), and (35), we see that

$$\mathscr{H}(s, k) = \int_0^\infty e^{-sz}h(z, k)\, dz, \tag{36}$$

where $h(z, k)$ is the solution of

$$h(z, k) = \int_0^\infty \bar{K}(|z - z'|, k)h(z', k)\, dz' + \delta(z). \tag{37}$$

That is, $h(z, k)$ is the special Green's function that results when the variable z' is set equal to zero. The function $\mathscr{H}(s, k)$ is very closely related to the usual H-function of radiative transfer. In fact, if we used the function

$$H(\mu, k) = \mathscr{H}(1/\mu, k) \tag{38}$$

instead, the correspondence would be complete (see [6]). However, it is more convenient here to use $\mathscr{H}(s, k)$.

Taking Laplace transforms of equation (27) with respect to z and z', and using Parseval's theorem to convert the integrations over z'' and z''' to complex integrals over s'' and s''', we obtain

$$(39)\quad \bar{\bar{C}}_{BB}(s, s', k) = \frac{1}{(2\pi i)^2} \int_c ds'' \int_c ds'' \bar{\bar{G}}(s, s'', k) \bar{\bar{G}}(s', s''', k) \bar{\bar{C}}_{ee}(-s'', s''', k),$$

where the contour c can be taken from $-i\infty$ to $+i\infty$ along the imaginary axis. Use of equation (34) now yields

$$(40)\quad \begin{aligned} (2\pi i)^2 \bar{\bar{C}}_{BB}(s, s', k) = \mathscr{H}(s, k)\mathscr{H}(s', k) \int_c ds'' \int_c ds''' \frac{\mathscr{H}(s'', k)\mathscr{H}(s''', k)}{(s + s'')(s' + s''')} \\ \times \bar{\bar{C}}_{ee}(-s'', -s''', k). \end{aligned}$$

This equation represents the maximum simplifications apparently allowed by the symmetries of the problem.

5. **Correlation function of emergent intensity; the inverse problem.** In the usual plane-parallel case, there is a simple connection between the Laplace transform of the source function and the emergent intensity $I(\mu) = I(0, \mu)$ at angle $\theta = \cos^{-1} \mu$, namely,

$$(41)\quad I(\mu) = \frac{1}{\mu} \mathscr{L}_{1/\mu} B(z).$$

(See, e.g. [1].) In addition to the usefulness of this relation for computing $I(\mu)$, it is essential in the inverse problem—that of determining $B(z)$ from the observed values of the so-called limb-darkening law $I(\mu)$.

We shall show that there exists an analogous relation between a certain observable correlation function of emergent intensity C_{II} and the correlation function of the source function C_{BB}. The inverse problem is likewise capable of solution, and the function C_{BB} is in principle determinable from the observed values of C_{II}. The development given here supersedes an earlier, less concise one [6].

There is a restriction on the type of correlation functions of emergent intensity that we shall consider here, namely, that the two intensities have the same direction. (This restriction is motivated by the type of observations of the sun's surface that are actually possible from the earth.) Let $\theta = \cos^{-1} \mu$ be the usual angle between each emergent intensity and the outward normal to the surface, and let $\boldsymbol{\nu}$ be the two-dimensional projection of the unit vector $\boldsymbol{\Omega}$ onto the x-y plane. Then formal integration of the transfer equation yields the following results for the fluctuations in the emergent intensity at two surface points $\boldsymbol{\rho}$ and $\boldsymbol{\rho}'$:

$$(42)\quad \begin{aligned} \Delta I(\mu, \boldsymbol{\nu}, \boldsymbol{\rho}) &= \frac{1}{\mu} \int_0^\infty dz e^{-z/\mu} \Delta B\left(z, \boldsymbol{\rho} - \frac{\boldsymbol{\nu}}{\mu} z\right), \\ \Delta I(\mu, \boldsymbol{\nu}, \boldsymbol{\rho}') &= \frac{1}{\mu} \int_0^\infty dz' e^{-z'/\mu} \Delta B\left(z', \boldsymbol{\rho}' - \frac{\boldsymbol{\nu}}{\mu} z'\right). \end{aligned}$$

Multiplying these two equations together and averaging, we obtain

$$C_{II}(\mu, \boldsymbol{\nu}, \boldsymbol{\rho} - \boldsymbol{\rho}') \equiv \langle \Delta I(\mu, \boldsymbol{\nu}, \boldsymbol{\rho})\, \Delta I(\mu, \boldsymbol{\nu}, \boldsymbol{\rho}') \rangle$$

$$= \frac{1}{\mu^2} \int_0^\infty dz \int_0^\infty dz' e^{-(z+z')/\mu} \tag{43}$$

$$\times\, C_{BB}\left(z, z', \left| \boldsymbol{\rho} - \boldsymbol{\rho}' - \frac{\boldsymbol{\nu}}{\mu}(z - z') \right|\right),$$

which defines C_{II}. A Fourier transform with respect to $\boldsymbol{\rho} - \boldsymbol{\rho}'$ then yields

$$\bar{C}_{II}(\mu, \boldsymbol{\nu}, k) = \frac{1}{\mu^2} \int_0^\infty dz \int_0^\infty dz' \exp\,[-(z + z')/\mu] \tag{44}$$

$$\exp\,[i\boldsymbol{k} \cdot \boldsymbol{\nu}(z - z')/\mu] \bar{C}_{BB}(z, z', k),$$

where the shift theorem for Fourier transforms has been used. Therefore, we have the desired result

$$\bar{C}_{II}(\mu, \boldsymbol{\nu}, \boldsymbol{k}) = (1/\mu^2) \mathscr{L}_{(1-ik\cdot\nu)/\mu} \mathscr{L}'_{(1+ik\cdot\nu)/\mu} \bar{C}_{BB}(z, z', k), \tag{45}$$

which is analogous to the usual plane-parallel result, equation (41). Here $\mathscr{L}$ acts on z and $\mathscr{L}'$ acts on z'. It should be noted that $\bar{C}_{II}$ depends on its arguments only through the quantities μ, k, and

$$p \equiv \boldsymbol{k} \cdot \boldsymbol{\nu}/\mu, \tag{46}$$

so that we can define

$$l(\mu, k, p) \equiv C_{II}(\mu, \boldsymbol{\nu}, \boldsymbol{k}). \tag{47}$$

The inverse problem can now be solved by changing the variables of integration in equation (44) to

$$\xi = z + z', \qquad \eta = z - z'. \tag{48}$$

Then

$$l(\mu, k, p) = \int_{-\infty}^\infty d\eta e^{i\eta p} \frac{2}{\mu^2} \int_{|\eta|}^\infty d\xi e^{-\xi/\mu} \bar{C}_{BB}(z, z', k). \tag{49}$$

The dependence on the variable p can be inverted by Fourier transformation, which reduces the problem to

$$g(\mu, k, \eta) = \frac{2}{\mu^2} \int_{|\eta|}^\infty d\xi e^{-\xi/\mu} \bar{C}_{BB}(z, z', k), \tag{50}$$

or

$$\frac{\mu}{2} e^{|\eta|/\mu} g(\mu, k, \eta) = \frac{1}{\mu} \int_0^\infty d\xi' e^{-\xi'/\mu} \bar{C}_{BB}(z, z', k), \tag{51}$$

where

$$\xi' = \xi - |\eta|. \tag{52}$$

In its dependence on μ, equation (51) is the usual limb-darkening law, which can be inverted in a number of ways to give $\bar{C}_{BB}$ as a function of k, η, and ξ'. The algebraic solution of equations (48) and (52) then gives $\bar{C}_{BB} = \bar{C}_{BB}(z, z', k)$.

One point that should be mentioned in connection with the inversion of the Fourier transform in equation (49) is that the physical range of the variable p is finite, namely, $|p| \leq k[(1/\mu^2) - 1]^{1/2}$. Therefore, the usual integral inversion formula is not directly applicable. However, it can be seen from equation (45) that the function $I(\mu, k, p)$ is analytic in p within a strip about the real p axis, since values in this strip correspond to values of the Laplace transforms within their half-planes of regularity. Thus, in principle, $I(\mu, k, p)$ may be continued analytically to all real values of p, and then the inversion formula applied.

6. **An approximation for the function $\mathscr{H}(s, k)$.** The general theory of the function $\mathscr{H}(s, k)$ is treated by Rybicki [6] (where it is denoted $\bar{G}(s, k)$). In this reference, it is shown that the transform of the kernel function can be expressed as

$$\bar{K}(z, k) = \frac{1}{2}\int_0^1 \frac{d\mu}{\mu(1 + k^2\mu^2)^{1/2}} \exp\left[-z\mu^{-1}(1 + k^2\mu^2)^{1/2}\right]. \tag{53}$$

When $k = 0$, this reduces to the usual kernel of the plane-parallel case, which is one-half the exponential integral function of z. The special advantage of this form of the kernel is that all the z-dependence is in the exponent. It is, therefore, a suitable starting point for the so-called *kernel approximation method* [5], in which a Gaussian quadrature formula is used to convert the integral over μ into a finite sum. Thus,

$$\bar{K}(z, k) \approx \sum_{i=1}^{N} a_i(k) e^{-b_i(k)z}. \tag{54}$$

Krook [5] showed in the context of the plane-parallel case that this procedure is equivalent to the Sykes-Yvon version of the discrete ordinate method. In the present three-dimensional problem, where it is far from obvious how "discrete ordinates" can be easily introduced, the kernel approximation is the simplest method of gaining the usual advantages of the discrete-ordinate method.

With this kernel approximation, the solution for $\mathscr{H}(s, k)$ is given as a rational function of s

$$\mathscr{H}(s, k) = \prod_{i=1}^{N} \frac{s + (\lambda_i^2 + k^2)^{1/2}}{s + (\gamma_i^2 + k^2)^{1/2}}, \tag{55}$$

[6], where $\lambda_i = 1/\mu_i$, the μ_i being the division points of the Gaussian formula, and where the γ_i^2 are the roots of the characteristic equation

$$\sum_{j=1}^{N} \frac{w_j}{1 - \mu_j^2\gamma_i^2} = 1. \tag{56}$$

The w_j are the weights of the quadrature formula. Equation (56) is recognized as the characteristic equation of the *plane-parallel* case [1] and [4]. It is one of the special bonuses of the kernel approximation method that it yields a form for

$\mathscr{H}(s, k)$ that depends only on parameters of the plane-parallel case. Also, with special forms of the function $\bar{\bar{C}}_{ee}$, the complex integrals in equation (40) can be explicitly evaluated because of the rational form of $\mathscr{H}(s, k)$.

It should be pointed out that when $N = 1$ in equation (54) the resulting approximation is equivalent to a diffusion approximation, and in this case

$$\mathscr{H}(s, k) \approx [s + (k^2 + 3)^{1/2}]/(s + k). \tag{57}$$

See Rybicki [6] for details.

7. **An example.** It is possible to work out in more detail a simple problem defined by the correlation function

$$C_{ee}(\boldsymbol{r}, \boldsymbol{r}') = \beta\, \delta(\boldsymbol{r} - \boldsymbol{r}'). \tag{58}$$

Then

$$\bar{C}_{ee}(z, z', k) = \beta\, \delta(z - z'), \tag{59}$$

and substitution into equation (27) gives

$$\bar{C}_{BB}(z, z', k) = \beta \int_0^\infty dz'' \bar{G}(z, z'', k)\bar{G}(z'', z', k). \tag{60}$$

Laplace transformations with respect to z and z' and use of Parseval's theorem then yield

$$\begin{aligned} \bar{\bar{C}}_{BB}(s, s', k) &= \frac{\beta}{2\pi i} \int_c ds'' \bar{\bar{G}}(s, -s'', k)\bar{\bar{G}}(s'', s', k) \\ &= \frac{\beta}{2\pi i} \mathscr{H}(s, k)\mathscr{H}(s', k) \int_c ds'' \frac{\mathscr{H}(-s'')\mathscr{H}(s'')}{(s - s'')(s'' + s')} . \end{aligned} \tag{61}$$

The use of the rational approximation of equation (55) now allows this integral to be explicitly evaluated. For simplicity, we shall use the diffusion approximation of equation (57), since it serves to illustrate the basic ideas. Also, we shall evaluate equation (61) for the case $s = s' = 1$, so that we obtain the correlation function of emergent intensity for normal emergence ($\mu = 1, \boldsymbol{\nu} = 0$). Closing the contour in the left half-plane, we obtain, after some elementary manipulations,

$$\bar{C}_{II}(\mu = 1, \boldsymbol{\nu} = 0, \boldsymbol{k}) = \beta \frac{k^2 + k + 3}{2k(k + 1)} \left(\frac{1 + \sqrt{k^2 + 3}}{1 + k}\right)^2. \tag{62}$$

The most interesting property of this result is the $1/k$ singularity near $k = 0$, which persists, incidentally, even when more accurate approximations for $\mathscr{H}$ are used. This singularity is removed if the fluctuations become zero when $z \geqq z_0$; then there is a cutoff of the singularity at wave numbers $k \leqslant z_0^{-1}$. Possible application of this result for investigating the convection zone of the sun are discussed by Rybicki [6].

References

1. S. Chandrasekhar, *Radiative transfer*, Clarendon Press, Oxford, 1950.

2. B. Davison and J. B. Sykes, *Neutron transport theory*, Clarendon Press, Oxford, 1957.

3. J. P. Elliott, *Milne's problem with a point-source*, Proc. Roy. Soc. London Ser. A **228** (1955), 424–433.

4. V. Kourganoff and I. W. Busbridge, *Basic methods in transfer problems*, Clarendon Press, Oxford, 1952.

5. Max Krook, *On the solution of equations of transfer*, I, Astrophys. J. **122** (1955), 488–497.

6. G. B. Rybicki, *Transfer of radiation in stochastic media*, Ph.D. Thesis, Harvard University; also in Smithsonian Astrophys. Obs. Spec. Rep. No. 180, 1965.

7. V. V. Sobolev, *A treatise on radiative transfer*, Van Nostrand, Princeton, N.J., 1963.

Harvard University

Boundary value problems in linearized kinetic theory

Carlo Cercignani[1]

Abstract. A "state-of-the-art" review on boundary value problems in linearized kinetic theory is presented. After a detailed discussion of the often neglected subject of boundary conditions, the available theorems of existence, uniqueness and convergence of model solutions are stated. A list of solved problems is given with brief comments on the methods of solution and particular reference to the recently proposed variational technique. Relations of the Chapman-Enskog and Hilbert expansions with boundary value problems are also discussed. Finally the subject of external flows is briefly reviewed.

1. **Introduction.** Boundary value problems for the linearized gas-dynamics Boltzmann equation have been largely investigated in the last ten years, with an evolution characterized by a trend away from "moment methods" (i.e. methods based upon approximating the distribution function by a finite sum of selected functions of the molecular velocity, containing undetermined "macroscopic" parameters depending on the position vector x and time t). This evolution has been suggested by the discovery of mathematical facts, such as the presence of various continuous spectra **[1]**–**[4]**, and has been largely made possible by the use of kinetic models **[5]**–**[8]**.

The aim of an applied mathematician working on the boundary value problem for the linearized Boltzmann equation is to build up a theory of the subject which is at the same time rigorous, systematic and useful. The ambitious plan of such a theory is to start with a theorem which insures existence and uniqueness of the solution for any physically reasonable problem; then to show that, if a sequence of models approximates a true Boltzmann equation in some sense, the sequence of the corresponding solutions approximates the true solution in some related

[1] On leave of absence from Applicazioni e Ricerche Scientifiche—Milano—and Universitá di Milano—Milano (Italy).

sense; finally, model equations of different accuracy have to be solved in correspondence with specific problems, not only in order to have the solutions but also to establish the practical agreement of the solutions of sufficiently simple models with the corresponding solutions of the Boltzmann equation.

The aim of this paper is to provide an up-to-date survey of the present state of the subject. It is hoped that this survey will show that the basic steps of the above outlined task have been largely accomplished when two basic restrictions are introduced:

(1) One considers only domains which are n-dimensionally bounded ($n = 1, 2, 3$). Therefore any flow or heat transfer between parallel plates, cylinders, or closed surfaces is included, while, e.g. flow past a solid body is not.

(2) One introduces a cutoff in the collision operator or considers only rigid sphere molecules, in such a way that the collision operator L can be split into two parts

$$Lh = Kh - \nu(v)h \tag{1.1}$$

where $\nu(v)$ is a function of the magnitude of the molecular velocity and K has the aspect of an integral operator; h denotes the perturbation of a basic Maxwellian $F(\mathbf{v})$ and is the unknown of the linearized Boltzmann equation (see §3). Different kinds of cutoff are possible; the most easy to handle mathematically seems to be the angular cutoff introduced by Grad **[2]**, but the recently studied radial cutoff **[9]** seems also to deserve attention.

One could ask to what extent the two above-mentioned restrictions are required by mathematical convenience rather than being dictated by basic features or intrinsic difficulties. Concerning the first assumption, we note that the restriction to bounded regions is vital to any approach based on Hilbert space, unless the norm is either weighted with a space dependent weight function or based on space derivatives of h rather than h itself; both these modifications seem to imply a considerable complication and do not present themselves as spontaneous. The troubles with an L^2 theory of the linearized Boltzmann equation in unbounded domains can be traced to the nonuniform validity of the linearization, to be described in §7. Concerning the second restriction, one has to distinguish between angular and radial cutoff. The former cuts all the grazing collisions such that the deflection angle is smaller than some given ϵ, the latter the interactions at distances larger than some σ_0. The angular cutoff was introduced by Grad **[2]**, the choice being admittedly dictated by mathematical convenience. A cutoff of this kind does not seem to have any basic physical meaning and would appear completely meaningful only if one could pass to the limit of a vanishing cutoff angle.

On the other hand, the radial cutoff can be regarded as physically meaningful **[9]** since self-consistency implies that the Boltzmann equation cannot describe the interactions of molecules separated by distances larger than the intermolecular distance.

The plan of this paper is as follows: we first discuss boundary conditions, a subject often neglected in spite of its basic importance in any boundary value problem (§2). The properties of the free-molecular and collision operators, the available theorems of existence and uniqueness, the convergence properties of the solutions of models are then reviewed (§3). §4 contains a commented list of the specific boundary value problems which have been solved either analytically or numerically. §5 briefly reviews a very efficient variational procedure for extracting information from the linearized Boltzmann equation. §6 deals with the relations between the boundary value problem and the Hilbert and Chapman-Enskog series. §7 finally discusses the nonuniform validity of linearization in external problems.

2. **Boundary conditions.** The boundary conditions to be matched with the Boltzmann equation are as important as the Boltzmann equation itself, although this does not seem to be always realized. The importance of the boundary conditions comes from the fact that they describe the interaction of the gas molecules with the molecules of the solid bodies which are in contact with the gas; it is to this interaction that one can trace the origin of the drag exerted by the gas upon the body and the heat transfer between the gas and the solid boundary. That the choice of boundary conditions is very important is clearly proved by the case of a specularly reflecting boundary. This assumption was rejected by Maxwell **[10]** in 1879, on the basis of the fact that a gas could not exert oblique stresses against such specularly reflecting surfaces. We can add more, i.e. that as a consequence of the *H*-theorem, the only steady solutions of the Boltzmann equation in presence of specularly reflecting walls are Maxwellians: no Couette flow, no heat transfer, etc. could exist between such surfaces!

In agreement with the statistical standpoint which underlies the Boltzmann equation, what we require is the probability density $R(\boldsymbol{x}; \boldsymbol{v}' \to \boldsymbol{v})$ that a molecule hitting the solid boundary ∂R at some point $\boldsymbol{x}$ with some velocity $\boldsymbol{v}'$, re-emerges at practically the same point with some other velocity $\boldsymbol{v}$; accordingly if $\boldsymbol{n}$ is the inward normal and $\boldsymbol{u}_0$ the velocity of the wall, we shall have $\boldsymbol{v}' \cdot \boldsymbol{n} < \boldsymbol{u}_0 \cdot \boldsymbol{n}$, $\boldsymbol{v} \cdot \boldsymbol{n} > \boldsymbol{u}_0 \cdot \boldsymbol{n}$. Accordingly, our boundary conditions for the full distribution f will have the form

$$|(\boldsymbol{v} - \boldsymbol{u}_0) \cdot \boldsymbol{n}| f(\boldsymbol{x}, \boldsymbol{v}) = \int_{\boldsymbol{v}' \cdot \boldsymbol{n} < \boldsymbol{u}_0 \cdot \boldsymbol{n}} R(\boldsymbol{x}; \boldsymbol{v}' \to \boldsymbol{v}) \, |(\boldsymbol{v}' - \boldsymbol{u}_0) \cdot \mathbf{n}| f(\boldsymbol{x}, \boldsymbol{v}') \, d\boldsymbol{v}' \tag{2.1}$$

$$(\boldsymbol{v} \cdot \boldsymbol{n} > \boldsymbol{u}_0 \cdot \boldsymbol{n}; \quad \boldsymbol{x} \in \partial R);$$

in fact we have to use the mass flow $|(\boldsymbol{v} - \boldsymbol{u}_0) \cdot \boldsymbol{n}| f(\boldsymbol{x}, \boldsymbol{v})$ in doing a surface balance. $R(\boldsymbol{x}; \boldsymbol{v}' \to \boldsymbol{v})$ being a probability density is normalized to unity

$$\int_{\boldsymbol{v} \cdot \boldsymbol{n} > \boldsymbol{u}_0 \cdot \boldsymbol{n}} R(\boldsymbol{x}; \boldsymbol{v}' \to \boldsymbol{v}) \, d\boldsymbol{v} = 1 \qquad (\boldsymbol{v}' \cdot \boldsymbol{n} < \boldsymbol{u}_0 \cdot \boldsymbol{n}; \boldsymbol{x} \in \partial R). \tag{2.2}$$

The linearity of equation (2.1) means that the gas molecules do not change appreciably the state of the wall molecules and do not interact with each other during the time of their adsorption. Equation (2.2) holds if the boundary does not capture molecules and the average time between an adsorption and an evaporation is negligible. Equation (2.2) implies that the normal component of the mass velocity of the gas at the wall ($\boldsymbol{v}_0 \cdot \boldsymbol{n}$) is the same as the normal velocity of the wall ($\boldsymbol{u}_0 \cdot \boldsymbol{n}$); accordingly, $(\boldsymbol{v} - \boldsymbol{u}_0) \cdot \boldsymbol{n} = \boldsymbol{c} \cdot \boldsymbol{n}$ where $\boldsymbol{c} = \boldsymbol{v} - \boldsymbol{v}_0$ is the peculiar velocity of the molecules.

We note here that, if the boundary conditions depend only upon local properties of the boundary (as, e.g. the velocity $\boldsymbol{u}_0$ and the temperature T_0 of the wall), then two additional relations are to be satisfied by the kernel $R(\boldsymbol{x}; \boldsymbol{v}' \to \boldsymbol{v})$. To see this, we note that, if we take the gas to have a Maxwellian distribution with temperature and mass velocity equal to the temperature and mass velocity of the wall, this gas is in thermal and dynamical equilibrium with the wall (at least locally); therefore, the number of molecules which change their velocity from $\boldsymbol{v}' - \boldsymbol{u}_0$ to $\boldsymbol{v} - \boldsymbol{u}_0$ because of the interaction with the walls must be equal to those which change their velocity from $\boldsymbol{v} - \boldsymbol{u}_0 - 2\boldsymbol{n}(\boldsymbol{n} \cdot \boldsymbol{c}) = \boldsymbol{v}_R$ to $\boldsymbol{v}' - \boldsymbol{u}_0 - 2\boldsymbol{n}(\boldsymbol{n} \cdot \boldsymbol{c}') = \boldsymbol{v}'_R$ (in a state of stable equilibrium specular symmetry obviously applies). Accordingly,

(2.3)

$$|\boldsymbol{c}' \cdot \boldsymbol{n}|\, R(\boldsymbol{x}; \boldsymbol{v}' \to \boldsymbol{v}) F_0(\boldsymbol{v}') = |\boldsymbol{c} \cdot \boldsymbol{n}|\, R(\boldsymbol{x}; \boldsymbol{v}_R \to \boldsymbol{v}'_R) F_0(\boldsymbol{v}) \qquad (\boldsymbol{c}' \cdot \boldsymbol{n} < 0, \quad \boldsymbol{c} \cdot \boldsymbol{n} > 0)$$

where $F_0(\boldsymbol{v})$ denotes a Maxwellian with mass velocity $\boldsymbol{u}_0$, temperature T_0 and arbitrary density, and we have used the fact that $F_0(\boldsymbol{v}'_R) = F_0(\boldsymbol{v}')$. Equation (2.3) expresses what is called sometimes "detailed balancing." Since $R(\boldsymbol{x}; \boldsymbol{v}' \to \boldsymbol{v})$ does not depend upon f (according to our assumptions), equation (2.3) must be identically satisfied.

Also, if we consider the same equilibrium situation, the corresponding Maxwellian must satisfy the boundary conditions, equation (2.1), i.e.

$$(2.4) \quad \int_{\boldsymbol{c}' \cdot \boldsymbol{n} < 0} |\boldsymbol{c}' \cdot \boldsymbol{n}|\, R(\boldsymbol{x}; \boldsymbol{v}' \to \boldsymbol{v}) F_0(\boldsymbol{v}')\, d\boldsymbol{v}' = |\boldsymbol{c} \cdot \boldsymbol{n}|\, F_0(\boldsymbol{v}') \qquad (\boldsymbol{c} \cdot \boldsymbol{n} > 0; \boldsymbol{x} \in \partial R).$$

This equation also follows from equation (2.3) plus equation (2.2). If the boundary conditions do not contain the temperature of the wall (as, e.g. in the case of specular reflection), then a Maxwellian with any temperature satisfies both the detailed balancing and equation (2.4). If this is true for any point of the boundary (including possible boundaries at infinity), it is clear that the above-mentioned boundary conditions are quite unrealistic in general, since they would allow the gas to stay in thermal equilibrium at any given temperature, irrespective of the surrounding bodies. This fact, in general, rules out these boundary conditions (adiabatic walls), which can be retained, however, in particular cases.

Equations (2.2), (2.3) and (2.4) are by no means sufficient to determine $R(\boldsymbol{x}; \boldsymbol{v}' \to \boldsymbol{v})$; to do this, we have to construct a model of the adsorption and re-evaporation of the molecules at the wall. It seems that, from this point of view, the situation has not improved since Maxwell's classical discussion **[10]** of the boundary conditions for the Boltzmann equation. His conclusions can be summarized by the following choice of $R(\boldsymbol{x}; \boldsymbol{v}' \to \boldsymbol{v})$:

$$(2.5) \quad \begin{aligned} R(\boldsymbol{x}; \boldsymbol{v}' \to \boldsymbol{v}) &= (1 - \alpha)\, \delta(\boldsymbol{v}' - \boldsymbol{v} + 2\boldsymbol{n}(\boldsymbol{c} \cdot \boldsymbol{n})) \\ &\quad + \alpha[2\pi(RT_0)^2]^{-1} \exp\, [-(\boldsymbol{v} - \boldsymbol{u}_0)^2/2RT_0] \end{aligned}$$

where α is any number between 0 and 1 (to be eventually determined by experimental data).

Maxwell's $R(\boldsymbol{x}; \boldsymbol{v}' \to \boldsymbol{v})$ satisfies equations (2.2), (2.3), (2.4), is simple enough, and is used by almost everybody; this does not mean, however, that Maxwell's choice is the only possible one. A more satisfactory boundary condition could, e.g. be based on the following choice:

$$R(\boldsymbol{x}; \boldsymbol{v}' \to \boldsymbol{v}) = (1 - \alpha)\int_{\boldsymbol{\beta}\cdot\boldsymbol{n}>0} \delta(\boldsymbol{v}' - \boldsymbol{v} + 2\boldsymbol{\beta}[\boldsymbol{\beta}\cdot(\boldsymbol{v} - \boldsymbol{u}_0)])H(\boldsymbol{\beta}, \boldsymbol{n}, \boldsymbol{c}/c)\, d\boldsymbol{\beta} + \alpha[2\pi(RT_0)^2]^{-1} \exp\,[-(\boldsymbol{v} - \boldsymbol{u}_0)^2(2RT_0)^{-1}] \tag{2.6}$$

where $\boldsymbol{\beta}$ is a unit vector and $H(\boldsymbol{\beta}, \boldsymbol{n}, \boldsymbol{c}/c)$ a function which describes the partial shielding of a molecule in the outer layer of the wall lattice by the molecules which lie next to it (compare with the discussion in Maxwell's paper [**10**]). Note that equation (2.6) reduces to equation (2.5) if we take $H(\boldsymbol{\beta}, \boldsymbol{n}, \boldsymbol{c}/c) = \delta(\boldsymbol{\beta} - \boldsymbol{n})$.

Another approach to the construction of reasonable boundary conditions can be based upon the following considerations.

Let us consider the following eigenvalue equation

$$A\psi = \int_{\boldsymbol{c}'\cdot\boldsymbol{n}>0} K(\boldsymbol{v}', \boldsymbol{v})\psi(\boldsymbol{v}')\, d\boldsymbol{v}' = \lambda\psi(\boldsymbol{v}) \qquad (\boldsymbol{c}\cdot\boldsymbol{n} > 0) \tag{2.7}$$

where, if $\boldsymbol{v}'_R = \boldsymbol{v}' - 2\boldsymbol{n}(\boldsymbol{n}\cdot\boldsymbol{c}')$ (hence $\boldsymbol{c}'_R\cdot\boldsymbol{n} < 0$), we have put

$$K(\boldsymbol{v}', \boldsymbol{v}) = [|\boldsymbol{c}'\cdot\boldsymbol{n}|\, F_0(\boldsymbol{v}')]^{1/2}[|\boldsymbol{c}\cdot\boldsymbol{n}|\, F_0(\boldsymbol{v})]^{-1/2}R(\boldsymbol{x}; \boldsymbol{v}'_R \to \boldsymbol{v}). \tag{2.8}$$

Thanks to equation (2.3), $K(\boldsymbol{v}', \boldsymbol{v})$ is symmetric; besides, equation (2.4) shows that

$$\psi_0(\boldsymbol{v}) = [2\pi(RT_0)^2]^{-1/2}\, |\boldsymbol{c}\cdot\boldsymbol{n}|^{1/2} \exp\,[-(\boldsymbol{v} - \boldsymbol{u}_0)^2/RT_0] \tag{2.9}$$

is an eigensolution of equation (2.7) corresponding to the eigenvalue $\lambda = 1$. If K has a purely discrete spectrum, then

$$K(\boldsymbol{v}', \boldsymbol{v}) = \sum_{n=0}^{\infty} \lambda_n\psi_n(\boldsymbol{v})\psi_n(\boldsymbol{v}'), \tag{2.10}$$

where ψ_n is the normalized eigenfunction corresponding to the eigenvalue $\lambda = \lambda_n$. The generalization to cases when the spectrum is not purely discrete is obvious and will not be considered explicitly. Equation (2.10) implies a similar relation for $R(\boldsymbol{x}; \boldsymbol{v}' \to \boldsymbol{v})$ which can be obtained from equation (2.8). We also assume that all the eigenvalues λ satisfy

$$0 \leqq \lambda_n \leqq 1 \qquad (n = 0, 1, 2, \ldots), \tag{2.11}$$

and, in general, $\lambda_n = 1$ only for $n = 0$; the fact that $\lambda_n < 1$ $(n \neq 0)$ depends upon the fact that the Maxwellian must be a stable solution, while $\lambda_n > 0$ is implied by the positive character of the kernel $R(\boldsymbol{x}; \boldsymbol{v}' \to \boldsymbol{v})$.

We can now use the above relations to construct models for the boundary conditions in the same way as it is usually done for the collision operator. I.e.

instead of assuming a detailed physical model, we can make simplifying assumptions on the eigenfunctions and eigenvalues appearing in equation (2.10).

The simplest assumption is complete degeneracy, i.e. $\lambda_n = 1$ for any n (note that this violates the general condition that $\lambda_n = 1$ can hold only for $n = 0$); then equation (2.10) shows that $K(\boldsymbol{v}', \boldsymbol{v}) = \delta(\boldsymbol{v}' - \boldsymbol{v})$ and we obtain the boundary conditions of purely specular reflection.

The next possible assumption (which takes into account the above-mentioned restriction on the eigenvalue $\lambda = 1$) is that we have two eigenvalues $\lambda_0 = 1$, $\lambda_n = 1 - \alpha$ $(0 < \alpha < 1; n = 1, 2, \ldots)$; then

$$
\begin{aligned}
K(\boldsymbol{v}, \boldsymbol{v}') &= \psi_0(\boldsymbol{v})\psi_0(\boldsymbol{v}') + (1 - \alpha) \sum_{n=1}^{\infty} \psi_n(\boldsymbol{v})\psi_n(\boldsymbol{v}') \\
&= \alpha\psi_0(\boldsymbol{v})\psi_0(\boldsymbol{v}') + (1 - \alpha) \sum_{n=0}^{\infty} \psi_n(\boldsymbol{v})\psi_n(\boldsymbol{v}') \qquad (2.12)\\
&= \alpha\psi_0(\boldsymbol{v})\psi_0(\boldsymbol{v}') + (1 - \alpha)\delta(\boldsymbol{v} - \boldsymbol{v}').
\end{aligned}
$$

Use of equations (2.8) and (2.9) shows that this assumption gives Maxwell boundary conditions, equation (2.5); a particularly simple choice is $\lambda_0 = 1$, $\lambda_n = 0$ $(n = 1, 2, \ldots)$, which corresponds to $\alpha = 1$ and gives boundary conditions of pure diffusion according to the wall Maxwellian.

It is clear that in order to introduce more sophisticated boundary conditions, we have either to do some specializing assumptions on the eigenfunctions or use a more general relation than equation (2.10). If we follow the first procedure, we can assume the ψ_n to have the following form:

$$
\psi_n(\boldsymbol{v}) = \psi_0(\boldsymbol{v})P_n(\boldsymbol{v}), \qquad (2.13)
$$

where the P_n are half-range polynomials chosen in such a way that the ψ_n are orthonormal according to the L norm in the Hilbert space of the functions of $\boldsymbol{v}(\boldsymbol{c} \cdot \boldsymbol{n} > 0)$. We can choose the P_n in two ways, by assuming that they are polynomials either in the three components of $\boldsymbol{c}$ or in the two tangential components of $\boldsymbol{c}$ and in $|\boldsymbol{c} \cdot \boldsymbol{n}|^2$. The first procedure is similar to the one used by Gross, Jackson and Ziering **[11]** for different purposes, the second one has the advantage of producing classical polynomials and seems also to be more convenient for the present purposes. With such a choice of the ψ_n, the eigenvalues λ_n receive a physical interpretation in terms of generalized accommodation coefficients $\alpha_n = 1 - \lambda_n$, each moment having a different accommodation coefficient. In the case of specular reflection, e.g. $\lambda_n = 1$ for any n; accordingly, there is no accommodation. In the case of Maxwell's boundary conditions, all the significant moments (except density, which never accommodates) have the accommodation coefficient equal to $1 - (1 - \alpha) = \alpha$.

The above procedure is analogous to the Maxwell modeling [7], **[8]** for the collision term of the Boltzmann equation (the main difference being that here we do not know whether a dynamical model does exist such that the corresponding ψ_n are exactly given by equation (2.13)). A more general (and justified) procedure is analogous to the non-Maxwell modelling proposed by Sirovich **[8]** for the collision term of the Boltzmann equation. I.e. we assume, as a vector basis in Hilbert space, the functions ψ_n given by equation (2.13), even if we do not regard them as eigenfunctions of A; in this case equation (2.10) must be written as follows:

$$K(\boldsymbol{v}, \boldsymbol{v}') = \sum_{m,n=0}^{\infty} \lambda_{mn}\psi_n(\boldsymbol{v})\psi_n(\boldsymbol{v}') \tag{2.14}$$

where $\|\lambda_{mn}\|$ is a symmetric matrix which expresses the nonaccommodation properties of the wall (λ_{00} is always equal to 1, because of equation (2.4)). $\|\alpha_{mn}\|$, where $\alpha_{mn} = \delta_{mn} - \lambda_{mn}$ can be regarded as an "accommodation matrix"; the meaning of the diagonal terms can be taken to be the same as the above α_n, while off diagonal terms measure how much memory of the moments of the arriving distribution is retained by the moments of the leaving distribution. It is clear that the present procedure reduces to the previous one when the nondiagonal terms are zero.

For example, one can take ψ_1 and ψ_2 to be the two tangential components of the molecular velocity and

$$\begin{gathered}\lambda_{00} = 1, \qquad \lambda_{11} = \lambda_{22} = 1 - \beta, \qquad \lambda_{33} = \lambda_{44} = \cdots = \lambda_{nn} = 1 - \alpha, \\ \lambda_{mn} = 0 \quad (m \neq n),\end{gathered} \tag{2.15}$$

where $0 < \alpha < 1$, $0 < \beta < 1$. Then

$$\begin{aligned} K(\boldsymbol{v}, \boldsymbol{v}') = {} & [2\pi(RT_0)^2]^{-1} \exp[-(\boldsymbol{v} - \boldsymbol{u}_0)^2(RT_0)^{-1}] \exp[-(\boldsymbol{v}' - \boldsymbol{u}_0)^2(RT_0)^{-1}] \\ & \times \left\{\alpha + \frac{\alpha - \beta}{RT_0}[\boldsymbol{v} - \boldsymbol{u}_0 - \boldsymbol{n}(\boldsymbol{n}\cdot\boldsymbol{c})]\cdot[\boldsymbol{v}' - \boldsymbol{u}_0 - \boldsymbol{n}(\boldsymbol{n}\cdot\boldsymbol{c}')]\right\} \\ & \times (|\boldsymbol{c}\cdot\boldsymbol{n}|\,|\boldsymbol{c}'\cdot\boldsymbol{n}|)^{1/2} + (1 - \alpha)\delta(\boldsymbol{v} - \boldsymbol{v}'). \end{aligned} \tag{2.16}$$

Let us now investigate what happens to the boundary conditions when we linearize the Boltzmann equation by putting

$$f = F(\boldsymbol{v})(1 + h) \tag{2.17}$$

where F is Maxwellian which will be assumed to have constant density, velocity and temperature (its mass velocity can be assumed to be zero without loss of generality). Equation (2.1) can now be rewritten

$$h(\boldsymbol{v}) = h_0(\boldsymbol{v}) + \int A(\boldsymbol{x}; \boldsymbol{v}' \to \boldsymbol{v})h(\boldsymbol{v}')\, d\boldsymbol{v}' \tag{2.18}$$

where

$$h_0(\boldsymbol{v}') = [F(\boldsymbol{v}')\,|\boldsymbol{c}\cdot\boldsymbol{n}|]^{-1}\int R(\boldsymbol{x}; \boldsymbol{v}' \to \boldsymbol{v})F(\boldsymbol{v}')\,|\boldsymbol{c}\cdot\boldsymbol{n}|\, d\boldsymbol{v}' - 1, \tag{2.19}$$

$$A(\boldsymbol{x}; \boldsymbol{v}' \to \boldsymbol{v}) = [F(\boldsymbol{v})\,|\boldsymbol{c}\cdot\boldsymbol{n}|]^{-1}R(\boldsymbol{x}; \boldsymbol{v}' \to \boldsymbol{v})F(\boldsymbol{v}')\,|\boldsymbol{c}'\cdot\boldsymbol{n}|. \tag{2.20}$$

Equations (2.18), (2.19), (2.20) are exact; it is usual and convenient, however, to take into account some simplifications, which can be introduced without altering the order of accuracy implied by the linearization of the Boltzmann equation. As a matter of fact, when one linearizes the Boltzmann equation, he makes the assumption that, in some sense, h^2 is small with respect to h; this can be true only if h_0 is small with respect to 1. This means that, if $F_0(\boldsymbol{v}; \boldsymbol{x})$ $(x \in \partial R)$ is the Maxwellian with the same density as $F(\boldsymbol{v})$ but with velocity and temperature equal to the wall velocity and temperature at $\boldsymbol{x}$ (in such a way that equation (2.4) is satisfied), then the difference $1 - F_0/F$ is small with respect to 1 and determines the order of magnitude of h; this implies that the velocity and temperature of the boundaries (including possible boundaries at infinity) must be considered as small perturbations, i.e. relative velocities and temperature differences must be small (with respect to the average temperature and thermal speed, respectively). These qualitative conditions for the validity of the linearization are well known. We can now linearize $R(\boldsymbol{x}; \boldsymbol{v}' \to \boldsymbol{v})$ in both equation (2.19) and (2.20): first order terms are retained in equation (2.19) (zero order terms cancel because of equation (2.4)), while only zero order terms are kept in equation (2.20) (since first order terms in $A(\boldsymbol{x}; \boldsymbol{v}' \to \boldsymbol{v})$ become second order terms when multiplied by h as in equation (2.18)). The zero order $R(\boldsymbol{x}; \boldsymbol{v}' \to \boldsymbol{v})$ is no longer dependent upon $\boldsymbol{x}$ and satisfies the same relations as the full $R(\boldsymbol{x}; \boldsymbol{v}' \to \boldsymbol{v})$, except for the fact that the basic Maxwellian $F(\boldsymbol{v})$ now substitutes the wall Maxwellian $F_0(\boldsymbol{v}; \boldsymbol{x})$; all the above treatment remains true with obvious modifications and, in particular, we can construct model boundary conditions for the linearized Boltzmann equation in the same way as we did above for the full equation. We note also that equation (2.11) implies that

$$\int_{\boldsymbol{v}\cdot\boldsymbol{n}>0} |\boldsymbol{v}\cdot\boldsymbol{n}|\, |Ah|^2 F(\boldsymbol{v})\, d\boldsymbol{v} \leqq \int_{\boldsymbol{v}\cdot\boldsymbol{n}<0} |\boldsymbol{v}\cdot\boldsymbol{n}|\, |h|^2 F(\boldsymbol{v})\, d\boldsymbol{v}, \tag{2.21}$$

$$Ah = \int_{\boldsymbol{v}'\cdot\boldsymbol{n}} A(\boldsymbol{x}; \boldsymbol{v}' \to \boldsymbol{v}) h(\boldsymbol{v}')\, d\boldsymbol{v}'. \tag{2.22}$$

3. **Basic theory.** Equation (2.21) was introduced **[12]** as a basic assumption in the proof of existence and uniqueness for bounded domains; all the properties of the free-streaming operator $\boldsymbol{v}\cdot\nabla_x$ essentially follow from elementary manipulations and equation (2.21). In particular, it has been shown **[12]** that, for the large class of boundary conditions considered in §2, the following inequality holds:

$$\int |\boldsymbol{v}\cdot\nabla_x h|^2 F(\boldsymbol{v})\beta(\boldsymbol{v})\, d\boldsymbol{v}\, d\boldsymbol{x} \geqq \frac{\gamma}{d^2}\int v^2\beta(\boldsymbol{v})^2 |h|^2 F(\boldsymbol{v})\, d\boldsymbol{v}\, d\boldsymbol{x} \qquad (\gamma > 0). \tag{3.1}$$

Here the space integrations are extended to the region R filled by gas, and $\beta(\boldsymbol{v})$ is any given function of the molecular speed such that the following integrals exist,

$$\int F(\boldsymbol{v})\beta(\boldsymbol{v})\, d\boldsymbol{v}, \qquad \int F(\boldsymbol{v})[\beta(\boldsymbol{v})]^{-1}\, d\boldsymbol{v}; \tag{3.2}$$

also, d is the maximum chord which can be drawn in R, γ a numerical constant (depending upon β and the shape of R).

We shall consider the boundary value problem for the separated time equation

$$sh + \boldsymbol{v}\cdot\nabla_x h = Lh \qquad (\operatorname{Re} s \geqq 0) \tag{3.3}$$

where L is given by equation (1.1) and the boundary conditions by equation (2.18) (with the linearized versions of $A(\boldsymbol{x}; \boldsymbol{v}' \to \boldsymbol{v})$ and $h_0(\boldsymbol{v})$, as discussed at the end of §2). As was recalled in §1, there are different procedures to put L into the form specified by equation (1.1); the important point **[9]**, **[12]**, **[13]** is that, for both angular and radial cutoff, one can find a suitable function $\mu(v) \geqq \nu(v)$ such that L can be split as follows:

(3.4) $$Lh = Hh - \mu h$$

where H satisfies

(3.5) $$0 \leqq (h, Hh) \leqq (h, \mu h),$$

(3.6) $$(\rho^{-1}Hh, Hh) \leqq \alpha(h, \rho h) \qquad (\alpha < 1).$$

Here (,) denotes the scalar product in the Hilbert space $\mathscr{H}$ of the function of $\boldsymbol{v}$ which are square integrable with respect to the weight $F(\boldsymbol{v})$, and

(3.7) $$\rho(v) = \{[\mu(v)]^2 + \eta^2 v_i^2/d^2\}^{1/2}$$

where η is any given numerical constant and v_i is simply v if the domain is three-dimensionally bounded, while it reduces to the absolute value of the projection of $\boldsymbol{v}$ into the relevant plane or straight line for two- and one-dimensionally bounded regions.

As a consequence of the splitting in equation (3.4), one can rewrite equation (3.3) as an "integral" equation

(3.8) $$h = \bar{h}_0 + UHh;$$

(3.9) $$\bar{h}_0 = h_0 \exp\{-[\mu(\boldsymbol{v}) + s](\boldsymbol{x} - \tilde{\boldsymbol{y}}) \cdot \boldsymbol{v}/v^2\}$$

($\tilde{\boldsymbol{y}}$ being the closest point of the boundary such that $(\boldsymbol{x} - \tilde{\boldsymbol{y}})$ is directed as $\boldsymbol{v}$ and $(\boldsymbol{x} - \tilde{\boldsymbol{y}}) \cdot \boldsymbol{v} > 0$) and U is the inverse of the operator

(3.10) $$\boldsymbol{v} \cdot \nabla_x + [\mu(v) + s]$$

plus the homogeneous boundary conditions corresponding to $h_0 = 0$. For a steady problem we have merely to set $s = 0$.

By means of equation (3.1) it is possible to show **[12]** that a numerical constant η exists, such that, if $\rho(\boldsymbol{v})$ is given by equation (3.7), then

(3.11) $$((\rho Ug, Ug)) \leqq ((\rho^{-1}g, g))$$

where the double parentheses denote the scalar product in the Hilbert space of the square integrable functions of both $\boldsymbol{x}$ and $\boldsymbol{v}$, with respect to the weight $F(\boldsymbol{v})$.

Equations (3.1) and (3.6) together give

(3.12) $$((\rho UHh, UHh)) \leqq ((\rho^{-1}Hh, Hh)) \leqq \alpha((\rho h, h)) \qquad (\alpha < 1),$$

i.e. UH is a contracting operator in the Hilbert space $\mathscr{K}$ of the square integrable functions of both $\boldsymbol{x}$ and $\boldsymbol{v}$, with respect to the weight $\rho(\boldsymbol{v})F(\boldsymbol{v})$. This result immediately implies the following

THEOREM OF EXISTENCE AND UNIQUENESS. *If $\bar{h}_0 \in \mathscr{H}$, then a unique solution $h \in \mathscr{H}$ of equation* (3.8) *exists, which can, in principle, be found by an iterative technique.*

The same contraction property can be used to establish a rigorous connection between the existence and uniqueness theory and the practical procedures of solution, based upon kinetic models. In fact, one can prove the following:

THEOREM ON THE SOLUTIONS OF MODELS. *If L is a collision operator which can be split as in equation* (3.4) *and L_N a sequence of operators which can be similarly split* (*with the same μ*), *and*

$$\|\rho^{-1/2}(L_N - L)\rho^{-1/2}\| \to 0 \quad \text{as} \quad N \to \infty, \tag{3.13}$$

then the solution h_N of the integral version of the model equation corresponding to L_N tends (*in the $\mathscr{H}$ norm*) *to the solution h of the analogous equation corresponding to L and the same boundary conditions.* (*The norm appearing in equation* (3.13) *is the operator norm in the $\mathscr{H}$ space.*)

It is believed that the above results can be used as starting points for investigations of the smoothness properties of the solutions. Research in this direction has begun in a very recent paper **[14]**, where, for the case of one-dimensional problems and simplified boundary conditions, existence and uniqueness have been proved for two function spaces E and V, different from $\mathscr{H}$ and endowed with the following norms:

$$\|h\|_V^2 = \max_x \int dv\rho(v)F(v)\,|h(v, x)|^2, \tag{3.14}$$

$$\|h\|_E^2 = \max_{x,v}\,(1 + v^2)\,|h(v, x)|\,[F(v)]^{1/2}. \tag{3.15}$$

One of the advantages in using the spaces E and V lies in the fact that one can prove the existence and uniqueness theorem for the nonlinear problem (full Boltzmann equation) under the condition that the boundary data are small in the E norm. One can also show **[14]** that if $F(1 + h^*)$ is the solution of the nonlinear Boltzmann equation corresponding to a given boundary problem (whose linearized solution is h), then

$$\|h - h^*\|_E/\|h\|_E \to 0 \tag{3.16}$$

as the function $\bar{h}_0$ goes to zero in the E norm. This result gives a precise meaning to the use of the linearized Boltzmann equation as a tool for finding approximate solutions of the nonlinear equation.

4. **Accurate solutions of specific problems.** The results reviewed in §3 show that the present state of the boundary value problem in a bounded domain is satisfactory insofar as the basic theory is concerned (and provided some kind of cutoff is accepted); the aim of this section is to briefly review the available solutions of specific problems, in order to show that the situation is rather satisfactory also from this point of view.

Practically all the one-dimensional problems (in plane, cylindrical and spherical geometry) have been solved, either analytically or numerically, with the Bhatnager, Gross and Krook (BGK) model [5], and some of them even with more sophisticated models; with few exceptions, the boundary conditions were based upon the assumption of complete diffusion with Maxwellian distribution. Accordingly (if the restrictions to bounded domains and linearized cutoff operators are accepted), there are essentially two points which still require a more thorough investigation: the first is the model dependence of the solutions, the second is the influence of the boundary conditions. Both topics have only begun to be investigated and more work is required in order to estimate the accuracy of the solutions of models with simple boundary conditions, when regarded as approximate solutions of the full linearized Boltzmann equation and more general boundary conditions; the matter is well understood conceptually but numerical estimates are also required. We stress the fact that it is not realistic to solve the exact linearized Boltzmann equation (instead of a model) with poor boundary conditions, since changing the latter introduces corrections of the order of those introduced by changing the former.

Our review will be restricted to accurate solutions of linearized models, obtained by either analytical or numerical procedures; accordingly, no mention will be made of moment methods, iteration procedures or nonlinear problems, with few exceptions.

Analytical procedures for solving linear model equations can be based on either integral transform techniques or the method of separation of variables (method of "elementary solutions"). The two procedures are essentially equivalent, as is familiar from classical problems of mathematical physics (e.g. the initial and boundary value problems for the vibrating string or heat conduction); the separation of variables seems to be more concise and can be used to write down general solutions in a straightforward way, but choosing one method rather than the other seems a matter of personal taste, since any problem which can be solved by one method can be solved also with the other.

An example of a problem which can be solved analytically is Kramers' problem: an infinite expanse of gas flows parallel to an infinite plate and the velocity gradient tends to a constant as $x \to \infty$. The problem was first treated analytically by Welander [**6**]: he used the BGK model (which he was proposing independently) and Fourier transform techniques, but his analysis was quite involved and contained a basic mistake, which was later corrected by Willis [**15**] (the analysis given by Welander and Willis can be much simplified [**16**]). Kramers' problem can be very simply solved by means of the method of elementary solutions [**1**], [**17**]: in particular, a very simple expression for the slip coefficient ζ can be obtained and its accurate numerical evaluation [**18**] is made possible.

The same problem can be solved analytically even with models different from the BGK model, such as the so-called Ellipsoidal Statistical (ES) model (which, in its linearized form, is a Gross-Jackson model [7], but was first exploited, independently, by Holway [**19**] and Cercignani and Tironi [**20**]) and a model with velocity-dependent collision frequency [**21**]; introducing the ES model does not change the slip coefficient from its BGK value, while the introduction of a velocity-dependent collision frequency seems to slightly reduce the value of ζ [**22**]. Numerical estimates of the slip coefficient for these and other models have been also obtained [**22**], [**23**] by the variational technique to be discussed in the next section.

A preliminary investigation of the dependence of ζ upon the boundary condition [**24**] resulted into various series expansions of ζ in the case of Maxwell's boundary conditions and an exact integral expression for ζ in the case of boundary conditions based on the kernel (2.16) with $\alpha = 1$.

Other problems which can be solved analytically through the method of elementary solutions are the Rayleigh problem and the problem of a wall oscillating in its own plane. The solutions have been explicitly obtained only for the BGK model [**25**], [**26**], [**17**] but the essential theory is available also in the case of the above mentioned model with velocity-dependent collision frequency [**4**], [**27**]. Other solutions have been obtained for the problem of sound propagation in a half-space through Fourier transform techniques [**28**]–[**31**]; it is to be noted, however, that the BGK model had to be simplified further, by artificially assuming either one-dimensional collisions [**28**] or isothermal waves [**30**], [**31**].

The assumption of one-dimensional collisions can also be used for an analytical evaluation of the temperature jump [**17**]; since, however, the influence of the assumption upon the result is likely to be important, one can improve it by considering three-dimensional collisions, where, however, two degrees of freedom are artificially decoupled from the third one [**17**]. The resulting expression has never been evaluated numerically. There are, however, accurate numerical estimates of the temperature jump coefficient [**6**], [**32**], obtained by numerical integration of the usual (three-dimensional) BGK model.

With regard to two-plates problems, both plane Couette and plane Poiseuille flow have been largely investigated [**15**], [**20**], [**33**]–[**37**]; besides accurate numerical solutions based on both the integral [**15**], [**33**] and the integrodifferential versions of the BGK model [**34**], series expansions based on the method of elementary solutions have been studied, and the corrections implied by the use of the ES model have been evaluated in a closed form [**20**].

The linearized heat transfer between parallel plates has also been solved for the BGK model [**32**] (the nonlinear solution had been previously calculated by Willis [**38**] and the ES model [**39**], both with numerical procedures).

Problems in cylindrical and spherical geometry have also been investigated: numerical solutions have been obtained for both Poiseuille [**40**] and Couette flow [**41**] again, as well as for Poiseuille flow in an annulus [**42**] and for the problem of a cylinder rotating in an infinite extent of gas [**43**] (which is not a problem in a bounded domain, but is free from the troubles of linearized two-dimensional flows in unbounded domains, to be discussed in §7). The problem of heat transfer between concentric cylinders has been solved only for the nonlinear BGK model [**44**], while the linearized heat transfer from a sphere has been treated by a variational procedure [**45**].

5. **The variational approach.** The material reviewed in §4 shows that almost all the simple linearized problems have been solved by means of the BGK model: accordingly, future research is likely to be devoted to solving more realistic but more difficult problems, or investigating the influence of the collision term and the boundary conditions. In the latter case, the overall behavior of the solution should not change and only modifications of numerical values are to be investigated, while in the former case, the increasing complexity is likely to prevent accurate numerical solutions; accordingly, need will probably be felt for a method which is able to give quick and accurate answers for overall quantities, even if it gives poor approximations to the details of the solutions. These features are present in the variational approach recently proposed [**46**] and used [**45**]–[**47**], [**32**], [**22**], [**23**] for solving linearized boundary value problems. Although the original treatment [**46**] considers only kinetic models of the Gross and Jackson [**7**] or Sirovich [**8**] type, it is easily extended to models with velocity dependent collision frequency [**22**], [**23**] and even to the exact linearized Boltzmann equation. The variational principle is easily constructed by noticing that the operators appearing in the linearized Boltzmann equation (both in its integral and integrodifferential versions) are formally selfadjoint in a function space, which, in general, is a pseudo-Hilbert space, because its "metric" is not positive definite [**48**]. The integral version is usually preferred, because it already contains the boundary conditions and, accordingly, does not require restrictions on the choice of the possible trial functions to those satisfying the boundary conditions. The fact that we have to use a pseudometric implies that, in general, the variational principle is neither a maximum nor a minimum principle; it is possible, however, to show that it becomes a maximum or minimum principle for particularly simple problems [**46**], [**45**], [**32**].

Constructing a variational principle would be a matter of idle curiosity or, at best, would give a criterion for a best-fitting of the trial solutions, if an additional feature was not present: it turns out that the value which the functional to be varied attains in correspondence with its stationary point is usually related to a physically significant overall quantity, such as drag, total heat flux, flow rate, etc. This feature has not been shown to be true in general but has been found to hold for all the problems so far investigated: one can, e.g. relate the basic functional to the drag in Couette flow [**46**] (and flow past a body), to the flow rate in Poiseuille flow [**45**]–[**47**], to the total heat flux in heat transfer problems [**32**], [**45**], to the slip coefficient in Kramers' problem [**45**], [**22**], [**23**].

The advantage of relating the basic functional to such overall quantities comes from the well-known fact that the relative error made in estimating the functional is of the order of the square of the relative error made in estimating the trial function: accordingly, if we use a trial function containing disposable constants in such a way that it can be adjusted to represent the solution of the problem with an average error smaller than 10%, the functional (and hence the basic quantity related to it) will be evaluated with an error smaller than 1%.

The variational procedure has been used to evaluate the above mentioned overall quantities for plane Couette flow [**46**], both plane [**46**] and cylindrical [**45**],

[47] Poiseuille flows, heat transfer between parallel plates **[32]** and from a sphere **[45]** and Kramers' problem **[46]**, **[22]**, **[23]**. Whenever an accurate numerical solution is available (i.e. in all the above mentioned cases, except heat transfer from a sphere), the agreement is spectacular, in spite of the fact that very simple trial functions have been used; sometimes the variational result is even more accurate than the numerical solution (this can be inferred from the minimum or maximum properties of the functional).

6. **The Chapman-Enskog and Hilbert series versus boundary value problems.** The Chapman-Enskog and Hilbert series constitute another topic which is likely to be the subject of future investigations, in spite of the fact that such expansions have been known for half a century and are the main subject of a widely quoted book **[49]**. The reason for expecting new investigations is that the boundary conditions are never mentioned throughout this book, except in the proof of the H-theorem where the assumption of specularly reflecting walls is made; if the same assumption was to be used throughout, almost all solutions discussed in the book would be ruled out by the boundary conditions! (See §2.)

A question frequently asked is: what is the validity of the Chapman-Enskog and Hilbert expansions? Although one can be sure that they represent asymptotic expansions of solutions of the Boltzmann equation when the mean free path goes to zero, it is very difficult to make more precise statements till a connection with initial and boundary data is made. The trouble is that the Hilbert and Chapman-Enskog expansions surely cease to hold **[50]** through initial and boundary layers (as well as shock layers), a mean free time or a mean free path thick: and we have to go through these layers in order to effect the desired connection between the available expansions and the given initial or boundary data. This situation is commonly met whenever singular perturbations are considered (see, e.g. **[51]**); then one has to resort to more refined methods, such as the multiple time-scales method or matched inner-outer expansions. In the case of both Hilbert and Chapman-Enskog expansions, the initial value problem has been thoroughly investigated **[52]**–**[54]**, but the treatment of the boundary value problem has been only sketched **[55]**. The situation is rather clear up to the Navier-Stokes level plus slip (and temperature jump) boundary conditions, but is far from being clear at the next level of the Chapman-Enskog expansion (the Hilbert expansion is less ambitious and presents the same troubles as an expansion of solutions of the Navier-Stokes equations into a series of powers of the viscosity coefficient): the reason for this difficulty is that the Chapman-Enskog expansion is a multiple time-scales expansion, devised for treating correctly the time derivative without caring about possible troubles with the space derivatives, and, as a consequence, the order of differentiation with respect to space variables grows at each step. If one tries to solve the resulting higher-order differential equations, he finds some corrections to Navier-Stokes to be of the same order of terms, which have been neglected because of the asymptotic nature of the expansion: these corrective terms are easily discarded if one is solving the initial value problem in absence of boundaries but can give serious troubles if one is treating boundary value problems. In fact, the inner-outer expansions procedure seems to yield [55] the same number of boundary conditions at each step, while the order of the differential equations increases: the only conclusion which can be drawn is that either the Chapman-Enskog expansion has a wider range of application than expected (i.e. is able to describe kinetic boundary layers, which seems rather strange) or is completely meaningless for domains with boundaries, and one has to reorder the terms of the expansion according to a different criterion (in the case of shear flow problems a possible reordering has been suggested by Trilling [56]). This reordering can be also obtained "a posteriori" by re-expanding the Chapman-Enskog equations in powers of the mean free path, while keeping the Navier-Stokes equations as the leading term [55], [57].

In order to illustrate this discussion, let us consider the case of plane Couette flow at low Mach numbers; if one applies the Chapman-Enskog procedure to the linearized Boltzmann equation and truncates the series, he finds that the relevant component of the mass velocity, $u(x)$, satisfies the following differential equation:

$$\frac{d^2u}{dx^2} + Q_1 l^2 \frac{d^4u}{dx^4} + \cdots Q_N l^{2N} \frac{d^{2N+2}u}{dx^{2N+2}} = 0, \tag{6.1}$$

where l is the mean free path and $Q_1, \ldots, Q_N$ numerical coefficients (depending upon the law of interaction and the temperature of the gas, which is constant in the linearized description). If one solves equation (6.1), he finds

$$u(x) = Ax + B + \sum_{n=1}^{2N} C_n \exp\,[k_n(x/l)] \tag{6.2}$$

where A, B, C_n are arbitrary coefficients while the k_n are obtained by solving an algebraic equation of Nth degree with coefficients $Q_N, \ldots, Q_1, 1$ (if r_p ($p = 1, \ldots, N$) are the roots of this equation, then $k_{2p-1} = (r_p)^{1/2}$, $k_{2p} = -(r_p)^{1/2}$).

It is clear that the exponential terms describe space transients which are important in the kinetic layers: therefore, they must be thrown out if the Chapman-Enskog expansion is believed to be wrong there! The same result is achieved if we expand $u(x)$ into a series of powers of l^2,

$$u(x) = u_0(x) + l^2u_1(x) + \cdots + l^{2q}u_q(x) + \cdots, \tag{6.3}$$

and substitute into equation (6.1). In fact we find

$$\frac{d^2u_0}{dx^2} = 0,$$

$$\frac{d^2u_1}{dx^2} + Q_1\frac{d^4u_0}{dx^4} = 0, \tag{6.4}$$

$$\frac{d^2u_2}{dx^2} + Q_1\frac{d^4u_1}{dx^4} + Q_2\frac{d^6u_0}{dx^6} = 0.$$

The solution of this chain of equations gives

$$u_q = A_q x + B_q, \qquad (q = 0, 1, \ldots), \tag{6.5}$$

i.e.

$$u(x) = Ax + B \qquad \left(A = \sum_{q=0}^{\infty} A_q l^q;\ B = \sum_{q=0}^{\infty} B_q l^q\right) \tag{6.6}$$

as was to be shown. The constants A and B can now be determined by means of the inner-outer procedure which furnishes one boundary condition for each boundary. As a matter of fact, once the expansion has been reordered in such a way as to be amenable to a treatment by the inner-outer expansion procedure, the treatment is standard **[55]**; however, since the boundary curvature effects are usually of the same order of magnitude as the space gradients contributions **[47]**, one must be careful in using the inner-outer expansion procedure for curved boundaries.

A final question which is worth asking is whether the corrections connected with higher order terms in the Chapman-Enskog series are likely to be important in any regime or not. Although there are even examples of truncated Chapman-Enskog series furnishing exact solutions outside the kinetic layers **[55]**, **[58]**, it is difficult to give an answer to the above question; in fact, the question is meaningless unless asked within the frame of specific boundary value problems, since the boundary data (and the shape of the boundary) fix the order of magnitude of the various terms. Besides this, one has to take into account **[55]**, **[58]** that some overall quantities which are likely to be compared with experimental data, as the flow rate in Poiseuille flow, involve an integration over the flow region; now, the kinetic layers give to this integral a contribution of the same order of magnitude as the higher order terms in the Chapman-Enskog expansion.

Taking into account all these features considerably complicates the problem but is required in order to consider the question in a meaningful frame; once again the method of elementary solutions **[55]**, **[58]**, **[17]** seems to be capable of offering a systematic tool for approaching the problem and thus narrowing, if possible, the gap between the nearly-free and slip flow regimes.

7. **External problems.** Since a basic theory exists for the linearized Boltzmann equation in a bounded domain (§3) and many typical problems of flows and heat transfer in channels have been solved by means of a linearized treatment (§4), it seems natural to ask how much of the program successfully developed for the case of internal problems can be extended to cover the case of external problems, such as, e.g. flow past a solid body.

It is clear from the beginning that the extension of the above mentioned program to the case of external problems cannot be carried out without modifications. On one hand we have the fact that even the simplest problem of the family, i.e. flow past a sphere, involves variations of density, temperature, and two components of the mass velocity, and dependence upon two space variables; it is true that the dependence on the angular variable can be easily found by analytical means, but, even if we adopt the BGK model, the set of integral equations for the four above mentioned macroscopic quantities requires a sensibly larger amount of manipulations and computation than, e.g. the corresponding set for cylindrical Couette flow. Although much numerical computation can be avoided by using variational techniques, as discussed in §5, the amount of manipulations to be done can be discouraging. On the other hand, the basic theory of existence, uniqueness, convergence of model solutions, reviewed in §3, cannot be extended to unbounded domains without some basic modifications unknown at present (although the use of the V and E spaces defined in §3 seems to be promising).

What is more important, however, is that we cannot rely upon a blind linearization of the Boltzmann equation, even if the usual conditions for linearization (e.g. a very small Mach number) are met. As a matter of fact, linearization is, in general, not uniformly valid, and one can prove **[59]** that the linearized Boltzmann equation cannot be used in a naive sense: in other words, a kinetic treatment of the flow past solid bodies does not avoid the so-called Stokes and Whitehead paradoxes, well known from the Stokes expansion in continuum fluid-dynamics **[51]**. We know, of course, that these singular perturbation problems must be treated by more refined techniques (inner-outer expansions); it is therefore only natural that a naive approach to a general theory of the linearized Boltzmann equation for external flows fails, especially in the two-dimensional case. Also, the solution of such interesting problems as flow past an infinite cylinder of a flat plate at small Mach numbers cannot be properly attacked by means of a linearized treatment till we realize that the linearized Boltzmann equation cannot claim validity outside a surface which is large but finite (for small but finite M). One can prove a simple lemma of basic importance for discussing external flows: if h satisfies the steady linearized Boltzmann equation (equation (3.3) with $s = 0$) (the basic Maxwellian is assumed to have zero mass velocity with respect to the inner surface Σ_B, where boundary conditions of the general class discussed in §2 (with $h_0 = 0$) apply) and also satisfies

$$\lim_{\Sigma\to\infty} \int_\Sigma dS \int \boldsymbol{v}\cdot\boldsymbol{n}[h(\boldsymbol{v})]^2 F(\boldsymbol{v})\, d\boldsymbol{v} = 0, \tag{7.1}$$

where Σ is a closed surface, then h is a constant independent of $\boldsymbol{x}$ and $\boldsymbol{v}$.

If all the assumptions made in this lemma were to be satisfied in actual flows, the only possible solution would correspond to a state of rest for the gas surrounding the body (or bodies) bounded by the surface Σ_B; if we want to avoid this conclusion, we have to give up equation (7.1) (whose physical meaning is that space infinity does not conceal a sink of entropy). In the case of two dimensions, however, equation (7.1) can be shown to follow from boundedness of the mass velocity at space infinity plus an asymptotic analysis of the solutions of the Boltzmann equation **[59]**; therefore the Stokes paradox is not avoided in kinetic theory, i.e. boundedness at infinity plus satisfactory boundary conditions at the body surface imply that the only possible state of two-dimensional flow is the state of rest. In fact, the only modification introduced by a kinetic treatment is that we have now a kinetic layer a mean free path thick in the vicinity of the body (this layer can be very large but is always finite); then we have a large layer extending to distances of order l/M (l being the mean free path and M the Mach number), where the linearization assumption still holds and the flow can be essentially described in terms of a Stokes solution; for $r \gg l/M$, however, the flow is strongly influenced by the conditions at infinity, cannot be described by the Boltzmann equation linearized about a state of rest, and can be essentially described in terms of the Oseen solution.

The same picture, of course, holds for three-dimensional flows, but, insofar as we do not intend to iterate to find terms in M^2, we can rely upon the linearized Boltzmann equation in this case; the only consequence is that we cannot build an L^2-theory of existence, uniqueness, etc. by simply imitating the procedure valid for bounded domains, because we can only ask boundedness, not square integrability.

If the linearization is only the first step of an iteration procedure, then we must take into account the nonuniform validity of the expansion even in three-dimensions and construct proper inner-outer expansions [**59**].

8. **Concluding remarks.** The boundary value problem for the linearized Boltzmann equation, as presented in the above survey, seems to have reached a satisfactory status: this does not mean that all the problems have been solved and all the questions answered, but that satisfactory foundations have been laid down for constructing a complete basic theory and tools have been found for solving particular problems to a satisfactory degree of accuracy. With regard to the fundamentals, the most interesting questions concern the influence of the cutoffs on the theory and the possibility of passing to the limit of vanishing cutoff (of either the radial or the angular type): the only known result is that the origin is an isolated point eigenvalue for potentials which are harder than Maxwell's, even if no cutoff is introduced [**9**]. This result, though basic for the problem of approach to equilibrium, seems an insufficient basis for a complete treatment of the boundary value problem.

Other interesting open problems, which have been reviewed in the above sections, concern the meaning (if any) of the Chapman-Enskog expansion in connection with boundary value problems, the practical accuracy of sufficiently simple models, the influence of boundary conditions, the basic theory for boundary value problems in unbounded domains. There are, of course, other topics which have been purposely left out from the present treatment, in spite of their interest; most of them arise in connection with nonlinear problems, an area still virtually untouched in spite of many interesting efforts and results.

Acknowledgement. This survey was prepared when the author was visiting the Massachusetts Institute of Technology and his researches were supported by the National Science Foundation, Grant No. GP-6084. Much of the basic material, however, had been developed at Applicazioni e Ricerche Scientifiche, Milano (Italy), under contract AF 61 (052)-881 with the Air Force Office of Scientific Research through the European Office of Aerospace Research, OAR, United States Air Force.

REFERENCES

1. C. Cercignani, *Elementary solutions of the linearized gas-dynamics Boltzmann equation and their application to the slip-flow problem*, Ann. Phys. **20** (1962), 219.

2. H. Grad, "Asymptotic theory of the Boltzmann equation, II," in *Rarefied gas dynamics*, Vol. I, edited by J. A. Laurmann, Academic Press, New York, 1963, p. 21.

3. ———*Theory of the Boltzmann equation*, AFOSR 64-1377, MF 40, Courant Inst. Math. Sci., New York Univ., New York 1964.

4. C. Cercignani, *Unsteady solutions of kinetic model with velocity dependent collision frequency*, Ann. Phys. **40** (1966), 454.

5. P. L. Bhatnagar, E. P. Gross, and M. Krook, *A model for collision processes in gases*, Phys. Rev. **94** (1964), 511.

6. P. Welander, *On the temperature jump in a rarefied gas*, Ark. Fys. **7** (1954), 506–553.

7. E. P. Gross and E. A. Jackson, *Kinetic models and the linearized Boltzmann equation*, Phys. Fluids **2** (1959), 432.

8. L. Sirovich, *Kinetic modeling of gas mixtures*, Phys. Fluids **5** (1962), 908.

9. C. Cercignani, *On Boltzmann equation with cutoff potentials*, Phys. Fluids **10** (1967), 2097.

10. J. C. Maxwell, "On stresses in rarefied gases arising from inequalities of temperature. Appendix" in *The scientific papers of James Clerk Maxwell*, Dover, New York, 1965, p. 703.

11. E. P. Gross, E. A. Jackson and S. Ziering, *Boundary value problems in kinetic theory of gases*, Ann. Phys. **1** (1957), 141.

12. C. Cercignani, *Existence, uniqueness and convergence of the solutions of models in kinetic theory*, J. Math. Phys. (to appear).

13. ———*Existence and uniqueness in the large for boundary value problems in kinetic theory*, J. Math. Phys. **8** (1967), 1953.

14. Y. Pao, *Boundary value problems for the linearized and weakly nonlinear Boltzmann equation*, J. Math. Phys. (to appear).

15. D. R. Willis, *Comparison of kinetic theory analyses of linearized Couette flow*, Phys. Fluids **5** (1962), 127.

16. Y. Pao, Private communication.

17. C. Cercignani, *Elementary solutions of linearized kinetic models and boundary value problems in the kinetic theory of gases*, Brown Univ. Rep., 1965.

18. S. Albertoni, C. Cercignani and L. Gotusso, *Numerical evaluation of the slip coefficient*, Phys. Fluids **6** (1963), 993.

19. L. H. Holway, Jr., *Sound propagation in a rarefied gas calculated from kinetic theory by means of a statistical model*, Raytheon Co. Rep. T-578 (1964).

20. C. Cercignani and G. Tironi, *Some applications of a linearized kinetic model with correct Prandtl number*, Nuovo Cimento **43** (1966), 64.

21. C. Cercignani, *The method of elementary solutions for kinetic models with velocity-dependent collision frequency*, Ann. Phys. **40** (1966), 469.

22. P. Foresti, *Tecniche variazionali e metodi analitici per un recente modello dell'equazione di Boltzmann*, Ph.D. thesis, Univ. of Milano, 1967.

23. S. K. Loyalka and J. H. Ferziger, *Model dependence of the slip coefficient*, Phys. Fluids **10** (1967), 1833.

24. C. Cercignani, *The Kramers problems for a not completely diffusing wall*, J. Math. Anal. Appl. **10** (1965), 568.

25. C. Cercignani and F. Sernagiotto, *The method of elementary solutions for time-dependent problems in linearized kinetic theory*, Ann. Phys. **30** (1964), 154.

26. ———"Rayleigh's problem at low Mach numbers according to kinetic theory" in *Rarefied Gas dynamics*, Vol. I, edited by J. H. de Leeuw, Academic Press, New York, 1965, p. 332.

27.———"Some results about a kinetic model with velocity-dependent collision frequency" in *Rarefied gas dynamics*, Vol. I, edited by C. L. Brundin, Academic Press, New York, 1967, p. 381.

28. H. Weitzner, "Steady-state oscillations in a gas" in *Rarefied gas dynamics*, Vol. I, edited by J. H. de Leeuw, Academic Press, New York, 1965, p. 1.

29. R. J. Mason, "Forced sound propagation in gases of arbitrary density" in *Rarefied gas dynamics*, Vol. I, edited by J. H. de Leeuw, Academic Press, New York, 1965, p. 48.

30. A. S. Ostrowski and D. J. Kleitman, *Steady-state oscillations in gases*, Nuovo Cimento **44B** (1966), 49.

31. R. J. Mason, "Forced sound propagation induced by a fractionally accommodating piston" in *Rarefied gas dynamics*, Vol. I, edited by C. L. Brundin, Academic Press, New York, 1967, p. 395.

32. P. Bassanini, C. Cercignani and C. D. Pagani, *Comparison of kinetic theory analyses of linearized heat transfer between parallel plates*, Internat. J. Heat and Mass Transfer **10** (1967), 447.

33. C. Cercignani and A. Daneri, *Flow of rarefied gas between two parallel plates*, J. Appl. Phys. **34** (1963), 3509.

34. A. B. Huang and D. P. Giddens, "The discrete ordinate method for the linearized boundary value problems in kinetic theory of gases" in *Rarefied gas dynamics*, Vol. I, edited by C. L. Brundin, Academic Press, New York, 1967, p. 481.

35. C. Cercignani, "Plane Poiseuille flow and Knudsen minimum effect" in *Rarefied gas dynamics*, Vol. II, edited by J. A. Laurmann, Academic Press, New York, 1963, p. 92.

36. ———*Plane Couette flow according to the method of elementary solutions*, J. Math. Anal. Appl. **11** (1965), 93.

37. ———*Plane Poiseuille flow according to the method of elementary solutions*, J. Math. Anal. Appl. **12** (1965), 254.

38. D. R. Willis, "Heat transfer in a rarefied gas between parallel plates at large temperature ratios" in *Rarefied gas dynamics*, Vol. I, edited by J. A. Laurmann, Academic Press, New York, 1963, p. 209.

39. C. Cercignani and G. Tironi, "Nonlinear heat transfer between two parallel plates according to a model with correct Prandtl number" in *Rarefied gas dynamics*, Vol. I, edited by C. L. Brundin, Academic Press, New York, 1967, p. 441.

40. C. Cercignani and F. Sernagiotto, *Cylindrical Poiseuille flow of a rarefied gas*, Phys. Fluids **9** (1966), 40.

41. ———*Cylindrical Couette flow of a rarefied gas*, Phys. Fluids **10** (1967), 1200.

42. P. Bassanini, C. Cercignani and F. Sernagiotto, *Flow of a rarefied gas in a tube of annular section*, Phys. Fluids **9** (1966), 1174.

43. P. Bassanini, C. Cercignani and P. Schwendimann, "The problem of a cylinder rotating in a rarefied gas" in *Rarefied gas dynamics*, Vol. I, edited by C. L. Brundin, Academic Press, New York, 1967, p. 505.

44. D. Anderson, *On the steady Krook kinetic equation*. Part 2, J. Plasma Phys. **1** (1967), 255.

45. C. Cercignani and C. D. Pagani, "Variational approach to rarefied flows in cylindrical and spherical geometry" in *Rarefied gas dynamics*, edited by C. L. Brundin, Vol. I, Academic Press, New York, 1967, p. 555.

46. ———*Variational approach to rarefied gas dynamics*, Phys. Fluids **9** (1966), 1167.

47. J. H. Ferziger, *Flow of a rarefied gas through a cylindrical tube*, Phys. Fluids **10** (1967), 1448.

48. C. Cercignani, "Alcune osservazioni sulla teoria del trasporto" in *Fisica del reattore*, C.N.R., Roma, 1966, p. 633.

49. S. Chapman and T. G. Cowling, *Mathematical theory of nonuniform gases*, Cambridge Univ. Press, Cambridge, England, 1952.

50. H. Grad, "Principles of the kinetic theory of gases" in *Handbuch der Physik*, Vol. XII, Springer, Berlin, 1958.

51. M. V. Dyke, *Perturbation methods in fluid mechanics*, Academic Press, New York, 1964.

52. H. Grad, *Asymptotic theory of the Boltzmann equation*. II, Phys. Fluids **6** (1963), 147.

53. L. Sirovich, *Formal and asymptotic solutions in kinetic theory*, Phys. Fluids **6** (1963), 218.

54. J. E. McCune, T. F. Morse and G. Sandri "On the relaxation of gases toward continuum flow" in *Rarefied gas dynamics*, Vol. I, edited by J. A. Laurmann, Academic Press, New York, 1963, p. 102.

55. C. Cercignani, *Higher order slip according to the linearized Boltzmann equation*, Univ. of California Rep. NAS-64-18 (1964).

56. L. Trilling, *Asymptotic solution of the Boltzmann-Krook equation for the Rayleigh shear flow problem*, Phys. Fluids **7** (1964), 1681.

57. R. Schamberg, *The fundamental differential equations and the boundary conditions for high speed slip flow and their application to several specific problems*, Ph.D. thesis, California Inst. of Tech., 1947.

58. C. Cercignani, *Flow of rarefied gases supported by density and temperature gradients*, Univ. of California Rep. NAS-64-16 (1964).

59. ———*The Stokes paradox in kinetic theory*, Phys. Fluids **11** (1968), 303

Massachusetts Institute of Technology

Singular and nonuniform limits of solutions of the Boltzmann equation[1]

Harold Grad

Abstract. A survey is made of various singular and nonuniform limits of solutions of the Boltzmann equation. Included are the limits of large and small mean free path, fast and slow molecules, initial, boundary, and shock layers, and singular features of the collision integral. Several new results are given including an analysis of the Hilbert expansion in steady flow, of the kinetic boundary layer including a complete formal theory of slip coefficients, and of the profile of a very strong shock.

Table of Contents

[1] The work presented here was supported by the Air Force Office of Scientific Research under Grant number AF-AFOSR-815-67.

1. Introduction. The study of singularities and nonuniform limits appears to be a major preoccupation of applied mathematicians. Aside from the intrinsic interest of such problems, there is good reason for this concentration. A good existence theorem will include as a corollary continuous or smooth dependence of the solution on parameters. If a parameter which occurs in the problem has a natural range of variation, then one can expect the behavior at the end-points to be singular; otherwise the solution could be continued. Thus a detailed description of the singular endpoints, together with continuous dependence in between, should give a uniformly valid qualitative picture of the entire problem.

There is also a quantitative profit that can be gained from knowledge of the singular endpoints. It is frequently possible to "subtract off" the singular part and then obtain quite accurate answers for the remainder using relatively unsophisticated or *ad hoc* procedures. In kinetic theory, this combination of relatively recent precise investigations of many special limiting cases with the older moment and polynomial approximations offers the likelihood of quite accurate practical calculations even if convergence is not yet established.

In a previous talk at this Symposium,[2] existence theorems covering a certain area of kinetic theory were given together with a rigorous analysis of the behavior at one singular limit (small mean free path). We now survey all (or at least many) of the singular limits to which solutions of the Boltzmann equation fall prey. Since the available existence theory does not cover this extended territory, the present treatment will be formal rather than rigorous.

For coherence in the oral presentation, the emphasis was placed on a broad survey rather than on the new material. We reverse this emphasis in the written text since there are other sources available for the details of many of the nonuniform limits which are here coordinated.[3] The new material consists of a unified formal treatment of the small mean free path limit in steady boundary value problems (§§7–10) and an analysis of the profile of a very strong shock (§11).

The boundary value problem separates into an analysis of the interior and an evaluation of asymptotic boundary conditions. For the interior analysis, an asymptotic procedure is given which is superior to the direct Hilbert expansion or the conventional Chapman-Enskog expansion; (but in special cases this reduces to a degenerate form of the Hilbert expansion). This analysis of the interior of the domain is supplemented by a formally complete description of the slip flow boundary conditions in a general domain. Numerical evaluation of the slip coefficients requires solution of certain special one-dimensional problems generalizing the Kramers problem.

The strong (strictly speaking, infinite strength) shock profile is reduced to a relatively nonsingular problem by introducing the correct scaling in physical space and in velocity space after subtracting off a suitable singularity. This treatment can be qualitatively interpreted as a more rational refinement of the *ad hoc*

[2] New York City, April 22, 1964, Grad, [**23**].
[3] References will be given as appropriate.

Mott-Smith bimodal approximation. In the correct scaling, it appears that a very simple interpolation can be made between the infinitely strong and weak shock profiles.

A survey of various singular limiting cases is presented in §§2–6, followed by more detailed consideration of steady slip flow in §§7–10 and of the shock layer in §11. We conclude in §12 with some opinions on future trends in kinetic theory.

We should like to acknowledge many useful discussions with Dr. Yoshio Sone in connection with the material in §10.

2. Nonuniform limits. The argument of the Boltzmann equation is a *distribution function*, $f(\boldsymbol{v}, \boldsymbol{x}, t)$. This represents the density at a given instant in the phase space of a molecule; $\boldsymbol{x}$ = position and $\boldsymbol{v}$ = velocity. We find nonuniform limits at each extreme of every argument of f. There are nonuniform limits as $x \to 0$ (boundary layer) and as $x \to \infty$; also as $t \to 0$ (initial layer) and as $t \to \infty$ (approach to equilibrium); also for slow molecules, $v \to 0$, and fast molecules, $v \to \infty$. If dimensionless variables are introduced for $\boldsymbol{v}$, $\boldsymbol{x}$, and t ($\boldsymbol{v}$ relative to a representative mean thermal speed, $\boldsymbol{x}$ and t relative to some given scale),[4] then a single dimensionless parameter, ε, appears in the Boltzmann equation. This is the ratio of mean free path to the given scale length or, equivalently, of the mean collision time to the scale time. Both limits, $\varepsilon \to 0$ (continuum) and $\varepsilon \to \infty$ (free flow) are singular.

Since macroscopic fluid dynamics occurs as a limiting case, $\varepsilon \to 0$, in the theory of the Boltzmann equation, all singular and nonuniform limits of classical fluid dynamics are automatically included. These encompass an assortment of boundary layers, wakes, shock layers, various linearizations, behavior at infinity, small and large Mach number and Reynolds number, and many more. The singular *approach* to fluid dynamics as $\varepsilon \to 0$ is, of course, supplemental to the strictly fluid nonuniformities.

An interesting point is the fact that the parameter ε which gives rise to fluid dynamics in the limit $\varepsilon \to 0$ occurs within fluid dynamics proper as a *finite* parameter, $\varepsilon = \mathrm{M}/\mathrm{Re}$ (Knudsen number equals Mach over Reynolds number). From fluid dynamics alone we could not suspect that finite (i.e. not small) values of M/Re are intrinsically unreliable. The limit $\varepsilon \to 0$ is singular in both fluid dynamics and kinetic theory, with the two sets of complications superposed. For example, a fluid dynamic boundary layer expansion in ε ($\mathrm{Re} \to \infty$ at finite M) requires supplementation by kinetic effects, usually at second order in ε.

The fact that fluid dynamics emerges from the Boltzmann equation in a certain limit is intimately tied to the laws of conservation of mass, momentum, and energy. The conservation laws result from certain elementary properties of the collision integral, viz. that the nullspace is nontrivial (and finite dimensional). This situation can be compared to the case of neutron transport where there is a similar equation,

[4] We prevent the appearance of a Mach number in the equation by taking the v scale equal to the ratio of the x and t scales.

also called the Boltzmann equation.[5] In the neutron case, conservation does not hold. As a result solutions will, in general, either grow exponentially or decay exponentially on the time scale of the mean collision time. To identify the borderline case between growth and decay (e.g. in its dependence on geometry) is one of the basic problems in neutron transport. In a sense, the Boltzmann equation for a gas (or *Boltzmann's equation*) represents a set of measure zero from the point of view of neutron transport. But it is exactly this borderline quality which gives a gas its unique and distinctive features (including fluid flow) and makes a significant portion of the mathematical analysis much more delicate.

Our primary concern is with nonuniformities which are peculiar to kinetic theory (one of which is the singular connection between the fluid and kinetic descriptions, cf. §5). Many nonuniformities arise from the specific form of the kernel of the collision term of the Boltzmann equation which represents the nature of the intermolecular force. It is important to realize that "the" Boltzmann equation is not a specific equation but a generic term for a class of equations (similar in extent to the term *integral equation*). There are various types of collision kernels, some more singular than others. In particular, one feature that can be extracted from the collision kernel is the velocity-dependent collision frequency $\nu(v)$ and the related velocity-dependent mean free path $\lambda(v) = v/\nu(v)$. Much of the singular behavior for small and large v can be related to the corresponding limiting behavior of $\nu(v)$ and $\lambda(v)$, cf. §§3 and 4.

One class of collision kernels about which almost nothing is known (in a basic sense the most important class) is that for a noncutoff intermolecular potential.[6] In this case no simple collision frequency $\nu(v)$ or mean free path $\lambda(v)$ can be defined, and the nature of the collision operator (e.g. whether it is an integral operator, singular integral operator, differential operator, etc.) is not known.[7] For a certain type of cutoff (in angle of deflection), a relatively complete mathematical theory has been developed **[20]**. For another cutoff (intermolecular distance) a few results have been obtained **[8]**. One would like to know, on physical grounds, that "many" properties of solutions of the Boltzmann equation are insensitive to the type of cutoff and are, indeed, similar to those for noncutoff potentials. The one case in which this is known to be true is in the formula for the calculation of transport coefficients **[3]**. What is probably the case more generally is that the precise nature of certain mathematical singularities is sensitively dependent on the nature of the cutoff (or on the exact decay of a noncutoff potential), even though some gross properties of the gas flow will not be so sensitive.

We have mentioned that all the singular limits of classical fluid dynamics are contained within kinetic theory. This is not quite so; some of the bad behavior may be wiped out. In particular, nonlinear and linearized results are related quite differently in the fluid and kinetic versions. The streaming term of the Boltzmann

---

[5] Cf. N. Corngold, *Some recent results in the theory of the transport of thermal neutrons*, these Proceedings, pp. 79–91.

[6] For a result in a special case, see **[42]**, **[61]**.

[7] A preliminary approach to this problem is described in **[10]**.

equation is strictly linear; nonlinearity enters only through the quadratic collision term. But fluid equations are obtained from the Boltzmann equation by a procedure which operates only on the linear streaming terms, explicitly ignoring the collision term; the relevant moments of the collision term vanish because of conservation. The complete set of fluid conservation equations is therefore linear (in the proper variables) until it is "simplified" by use of the strongly nonlinear Maxwellian distribution function. The nonlinearity of the fluid equations (and the associated equations of state) is imposed by the nonlinear Boltzmann collision term, but only in a rather devious manner.

To illustrate this divergence between the two theories, consider the flow at finite Mach number past a small sphere or cylinder (small compared to a mean free path). The fluid solution is definitely nonlinear regardless of the size of the body. The kinetic solution also yields a strong disturbance near the body. But this strong disturbance is governed by the strictly linear streaming terms of the Boltzmann equation; the collision term is negligible until about a mean free path away, at which distance the disturbance has become small. The limiting flow is given as a finite amplitude solution of a linear equation (free flow). The leading term of the departure from free flow is nonlinear in three dimensions but is linear in two dimensions.

A similar example is given by high frequency sound, e.g. as generated by an oscillating wall.[8] The conventional fluid linearization, which takes derivatives as well as fluid perturbations to be small, becomes increasingly poor as the frequency and wave number become large. But in the kinetic analysis, a disturbance with frequency which is large compared to the collision frequency is dominated by the streaming terms, and the nonlinear collision term offers only a minor correction to the behavior of even a finite amplitude wave. For a fixed small amplitude of the disturbance, linearization of the Boltzmann equation will be most unreliable at some intermediate frequency.

A similar conclusion is reached for the initial value problem. For the Boltzmann equation, the natural time scale is a mean collision time. One would expect the discrepancy between a linear approximation and the correct nonlinear solution to grow with time on this time scale. But if the disturbance is very long compared to the mean free path, a much longer (macroscopic) time scale emerges. These expectations are verified by quantitative estimates in linear and nonlinear existence theorems **[23]**. Thus the worst case with regard to linearization is, again, the intermediate region with wave length comparable to (or probably somewhat larger than) the mean free path.

The Navier-Stokes level of the Chapman-Enskog expansion amounts to a linearization in velocity space (small deviation from local Maxwellian) in a problem which may be strongly nonlinear in physical space. But even the linearization in velocity space is not uniformly valid for large velocity (as evidenced by the

[8] For the solution with a relaxation model see **[31]** and **[55]**; with the Boltzmann equation at high frequency see **[24]**.

distribution function becoming negative). Even in a problem in which the fluid disturbance is small, i.e., linear in physical space, the correct treatment of the high velocity tail is always nonlinear; the tail of the distribution, although small, is not small relative to a Maxwellian **[36]**.

Nonlinearity can even introduce a dissipative mechanism where there is none in the linear theory. In the limit of small mean free path, ε, one aspect of dissipation occurs rapidly on a time scale ε, during which the Boltzmann H-function becomes approximately equal to the thermodynamic entropy S; this is then followed by a much slower viscous dissipation on a time scale $1/\varepsilon$. The linear problem is nondissipative on the intermediate time scale $O(1)$. But in a nonlinear problem, shock waves provide additional dissipation (which involves H and S differently) on the intermediate time scale.

The moment and polynomial approximations which were originally introduced to break away from the fluid regime **[13]** are essentially *ad hoc*. They usually supply fairly good approximations in a low order approximation, but no rapid improvements are found even with considerably extra labor **[38]**. But, if an explicit singularity or nonuniformity which prevents rapid convergence is recognized, the polynomial expansion can be altered to suit, and, although still *ad hoc*, the level of accuracy may be greatly improved even at low order. The most elementary example is the inclusion of two-sidedness in the distribution by using half range expansions **[25]**, **[28]**.

Another example is given by the fundamental solution in steady flow. For a relaxation model of the Boltzmann equation the Fourier transform of the solution is explicit **[16]**. For the actual Boltzmann equation we are not so fortunate. However, the singularity near the source *is* explicit to a certain number of terms in an expansion. Subtracting this off leaves a residue which is bounded and can be expected to be well-approximated by a simple moment expansion (Euler, Navier-Stokes, or Thirteen Moment). In addition, the leading term at infinity is governed by fluid dynamics and is therefore given correctly by a moment approximation. This scheme has been carried out **[62]**; unfortunately the singularity at the origin, although "explicit," involves very complex high-dimensional integrals.

A more successful application of this technique is to the strong shock. We recall that for a Navier-Stokes profile, proper scaling (viz. introducing the downstream mean free path as a reference length) gives a shock thickness which does not vary by more than a factor two or three and a profile which is not very far from universal as the strength varies from zero to infinity **[14]**. We now find that for the Boltzmann equation, after subtracting off an explicit singularity in velocity space, there is a definite limiting profile in v as well as x for the distribution function at infinite strength. This limiting distribution function can be approximated by relatively crude methods (e.g., by a local Maxwellian, which gives no result at all if the singularity is not removed). One conclusion is that the shock profile can be qualitatively approximated over the entire range from zero to infinite strength by simple interpolation (§11).

This very brief summary of singular limits is continued in the next two sections with a slightly more elaborate discussion of the nonuniformities which are introduced by the collision kernel and at the large mean free path limit.

3. The spectrum—fast and slow molecules. With the exception of the finite nullspace of the collision operator (which expresses conservation and leads to fluid dynamics), probably the single most important qualitative feature of the linear Boltzmann equation is the almost universal occurrence of a continuous spectrum. There are two sources of a continuum, the streaming operator, $\boldsymbol{v} \cdot \nabla$, and the collision operator L.

The streaming operator alone, as in

$$\partial f/\partial t + \boldsymbol{v} \cdot \partial f/\partial \boldsymbol{x} = 0, \tag{3.1}$$

has the entire imaginary axis as a continuum. This leads to the decay in time of an arbitrary initial function $f(\boldsymbol{v}, \boldsymbol{x}, 0)$ **[17]** and the decay in space of the signal induced by an oscillating wall in a collisionless gas **[27]**. It is also the dominant effect at high frequency even in the presence of collisions.

The collision operator by itself,

$$\partial f/\partial t + L(f) = 0 \tag{3.2}$$

has a continuum covering part of the real axis for most molecular force laws which have been investigated up to the present time **[20]**.

Since the operator L is symmetric and $\boldsymbol{v} \cdot \nabla$ is antisymmetric, and both operators are unbounded, the spectrum of the sum, $\boldsymbol{v} \cdot \nabla + L$, is not evident. There are a large number of interesting perturbation problems with a few proved results and many plausible conjectures. It is relatively easy to make reliable plausible estimates of the continuous spectrum. This yields valuable and detailed qualitative information about solutions of the Boltzmann equation. In particular, many singular features are immediately visible from a study of the continuous spectrum. The continuous spectrum is as easily found for the actual Boltzmann Equation as for a model and was, in fact, first investigated in the former case.

For the initial value problem (3.2), the continuum is the set of values taken by the collision frequency $\nu(v)$. For a class of potentials called *hard* we have $\nu(0) = 1$ increasing to $\nu(\infty) = \infty$, for a *soft* potential, $\nu(0) = 1$ decreasing to $\nu(\infty) = 0$. In a gas of hard molecules the high velocity tail of an initial distribution $f(\boldsymbol{v}, 0)$ will instantly equilibrate, this effect gradually penetrating towards the center. On the other hand, a soft gas will reach equilibrium in the center of the distribution in a time $O(1)$, and this will spread more and more slowly to the tail.

A more interesting illustration is given by a boundary value problem,

$$v_1 \frac{\partial f}{\partial x_1} + L(f) = 0. \tag{3.3}$$

The continuous spectrum (plausible conjecture) is in this case determined as the set of values taken by the reciprocal of the free path, $1/\lambda_1(v) = \nu(v)/v_1$. In most cases the continuum covers the entire real axis. Because $\lambda_1 = 0$ for $v_1 = 0$, there is nonuniform convergence in velocity space for the distribution function near a wall. The difference between $f(\boldsymbol{v}, \boldsymbol{x})$ and the emitted wall distribution $f^+(\boldsymbol{v}, 0)$ $(v_1 > 0)$ does not approach zero in a velocity boundary layer of thickness $v_1 = O(x)$. As a consequence, certain moments (e.g. temperature, tangential velocity) have infinite slopes at the wall.[9]

At the opposite extreme we have $\lambda_1 \to \infty$ for $v_1 \to \infty$ (with the single exception of hard spheres). In this case, fast particles will exert an influence at long distances. It has been found, for example, that the weak shock profile has a nonexponential decay at infinity, apparently contradicting the Navier-Stokes result **[30]**. The reason is the large penetration of fast particles past the bulk of the shock transition. There is an inversion of limits in that the Navier-Stokes profile is correct for a relatively greater portion of the profile as the shock strength decreases. The effect is small because there are exponentially few fast particles. Whether this effect is important in a strong shock has not yet been determined.

In the presence of a wall oscillating at a fixed frequency,

$$i\omega f + v_1\, \partial f/\partial x_1 + L(f) = 0 \tag{3.4}$$

the spectrum [i.e. of $\partial f/\partial x_1 = \mathscr{L}(f)$] moves into the complex plane. Many qualitative features of the solutions can be predicted just from a study of the continuous spectrum, e.g. the existence for large ω of subboundary layers in the approach of f to the wall distribution f^+, and the disappearance of a sound wave (i.e. of all discrete normal modes) for sufficiently high ω. (See footnote 9.)

The situation is qualitatively similar (e.g. with regard to the complex continuous spectrum swallowing up the normal modes at small wavelength) for an initial value problem at fixed wavelength,

$$\partial f/\partial t + iv_1 k_1 f + L(f) = 0. \tag{3.5}$$

A more detailed description of these spectra is discussed in **[21]**, **[22]**, **[24]**.

4. Large mean free path—The ubiquitous logarithm. In the limit of large mean free path we write $\varepsilon = \mathrm{K}$ (Knudsen number). There are two distinct reasons why $\mathrm{K} \to \infty$ does not approach uniformly to the free flow limit. One reason is that $\lambda(v) \to 0$ for small v so that there is always a class of slow particles with small free path no matter how large K may be. A second reason is that in an exterior or otherwise unbounded domain there will always be distant collisions.

There is a simple mathematical lemma which, although mathematically trivial, is responsible for many important phenomena which are physically profound and subtle, including both of the above mentioned sources of nonuniformity at large K. The divergent integral $\int dr/r$ can be given a finite value by introducing a convergence

---

[9] For a relaxation model see **[43]**, **[44]**; for the Boltzmann equation see **[24]**.

factor

$$I(\varepsilon) = \int_1^\infty \phi(\varepsilon r)\,\frac{dr}{r} \tag{4.1}$$

(the divergence near $r = 0$, which we dismiss with the finite limit $r = 1$, can be studied in exactly the same way). We assume that $\phi(0) = 1$ and $\phi(s)$ is smooth near $s = 0$, also that ϕ is integrable at infinity. Writing

$$I = \int_\varepsilon^\infty \phi(s)\,\frac{ds}{s} = \int_\varepsilon^1 (\phi - 1)\,\frac{ds}{s} + \int_\varepsilon^1 \frac{ds}{s} + \int_1^\infty \phi\,\frac{ds}{s} \tag{4.2}$$

we see that the first and third integrals are finite as $\varepsilon \to 0$. Consequently,

$$I(\varepsilon) = \log 1/\varepsilon + O(1). \tag{4.3}$$

We observe that the dominant term $\log 1/\varepsilon$ is completely independent of the nature of the convergence factor, depending only on the length scale ε; e.g. its value would be the same if ε were replaced by 2ε. Also the dominant contribution to the integral comes from values of r which are large compared to the lower limit, $r = 1$, but are small compared to $r = 1/\varepsilon$ [where ϕ deviates from the value $\phi(0) = 1$]. The error term, $O(1)$ in (4.3), does depend on the entire function $\phi(s)$. In other words, the leading term of an expansion of $I(\varepsilon)$ is local and extremely insensitive, whereas higher order terms are global and more complicated.

The special choice

$$\begin{aligned} \phi(s) &= 1, \qquad 0 < s < a \\ \phi(s) &= 0, \qquad s > a \end{aligned} \tag{4.4}$$

which amounts to cutting off the simple integral $\int dr/r$ at $r = a/\varepsilon$, also gives a leading term which is independent of the cutoff parameter a.

An integral of the form (4.1) arises commonly when a problem presents two length scales, as in a boundary layer analysis. More generally,

$$J(\varepsilon) = \int_1^\infty \phi(\varepsilon r)\,\frac{dr}{r^n} \tag{4.5}$$

can be expanded as

$$J = \frac{\phi(0)}{n-1} + \frac{\phi'(0)}{(n-1)(n-2)}\,\varepsilon + \cdots + \frac{\phi^{(n-1)}(0)}{(n-1)!}\,\varepsilon^{n-1}[\log 1/\varepsilon + O(1)], \tag{4.6}$$

a polynomial in ε ending with the appearance of a logarithm. All the listed coefficients are local [depending on derivatives of ϕ at $r = 0$]; higher order terms in the expansion depend globally on ϕ.

Similar expansions can be made for

$$\mathrm{K}(\varepsilon) = \int_1^\infty \phi(\varepsilon r)\psi(r)\,dr \tag{4.7}$$

in terms of as many moments $\int r^n\psi(r)\,dr$ as are finite and an equivalent number of derivatives of ϕ at the origin. The boundary layer interpretation of this result states that the dominant part of the expansion of $K(\varepsilon)$ (up to and including the logarithm) can be evaluated locally, within the boundary layer, without the necessity of matching to the exterior region.

A common tool in investigating solutions of the Boltzmann equation is to write it as an integral equation. In the integrand will appear a factor

$$\exp\,[-\nu(v)x/v_1] = \exp\,[-x/\lambda_1(v)]. \tag{4.8}$$

Near a wall, $x \sim 0$, we can expand the exponent in powers of x except for small v_1 where $\lambda_1 \sim 0$. Integrals with respect to v of such an expression will yield a leading term $x \log 1/x$ (or higher order terms $x + x^2 + \cdots + x^r \log 1/x$ for moments which have powers of v_1 as a factor).

For large mean free path, the relevant factor appears as

$$\exp\,[-x/K\lambda_1(v)]. \tag{4.9}$$

With x fixed and nonzero, this "convergence factor" is regular in $1/K$ except where $\lambda_1 \sim 0$. Again we obtain $(1/K)\log K$ as the leading term for an integral over this expression. But this is correct only in one dimension. At a point definitely removed from a wall, the dimension of the manifold on which the appropriately defined $\lambda(v)$ vanishes depends on the dimensionality of the problem. In one dimension (e.g. two parallel flat plates), $\lambda = 0$ when $v_1 = 0$ and (v_2, v_3) are arbitrary. In two dimensions (e.g. two concentric cylinders), $\lambda = 0$ when $v_r = v_\theta = 0$ and v_z is arbitrary. In three dimensions (e.g. two concentric spheres, or any bounded interior domain), $\lambda = 0$ only when the vector $\mathbf{v} = 0$. Therefore, in two dimensions there will be a term $1/K$ followed by $(\log K)/K^2$, and in three dimensions two terms, $1/K + 1/K^2$, followed by $(\log K)/K^3$. Furthermore, the evaluation of these coefficients is local, depending essentially on the free flow distribution, whereas any higher order terms pose difficulties of matching in velocity space.

In an external domain we have, in addition to the slow particles treated above, the effect of distant particles. The same integral (4.9) appears as a convergence factor, now integrated with respect to x from infinity instead of from a wall. In two dimensions (e.g. flow past a cylinder), the function which is multiplied by this exponential convergence factor decays as $1/x$ (this represents the free flow disturbance which is proportional to the angle subtended by the object at a distance x). Thus the leading term is $\log K$. An additional factor $1/K$ is explicit in the integral which is evaluated; thus the modification of the free flow drag is $(\log K)/K$. Similarly, in three dimensions where the influence function decays as $1/x^2$, we obtain $1/K + (\log K)/K^2$ as the leading terms [after free flow which is scaled to be $O(1)$].

In an exterior domain where both the slow and distant particle nonuniformities are present (arising from integration over v and x respectively), it is the distant particle effect which dominates, always entering at lower order. Alternatively, the logarithm enters at a lower order in a two-dimensional exterior problem, say drag past a cylinder, than in a two-dimensional interior problem, such as drag between two concentric cylinders.

We have been intentionally vague in not stating above exactly what integral is being evaluated. The descriptions given are basic in the sense that there exists some interesting macroscopic quantity with the logarithm appearing at the given order. Other observables, involving integrals over functions which vanish at small velocity or involving some special symmetry, may exhibit a logarithm only later on in the expansion. For example, in linearized one-dimensional heat flow, although the temperature gradient at the midplane is $(\log K)/K$ **[58]**, the ratio of total heat flow to the difference in wall temperature (effective conductivity) is $1 + 1/K$ (because the heat flow is a moment of f with a factor v_1). But in a *non-linear* problem with a finite temperature difference, the effective conductivity (normalized at fixed total mass in the system) will behave like $1 + (\log K)/K$ because of a loss of symmetry about the midplane.

The integrals which arise in this subject will, in general, involve combinations of boundary layers $x \sim 0$, $v \sim 0$ and nonuniformities $x \sim \infty$, $K \sim \infty$. Although the basic concepts and mathematical analysis of these linked nonuniformities are elementary, the bookkeeping can be quite complicated, involving the number of dimensions, the degree of symmetry, and the specific observable. A complete list has not been written down, although many specific examples have been studied, e.g. **[58]**, **[7]**, **[48]**, **[43]**, **[44]**, **[37]**, **[24]**, **[62]**.

It is interesting to note that the same mathematical lemma, and approximately the same physical situation, occurs in the *derivation* of a kinetic equation. To lowest order in the appropriate expansion parameter through which the Boltzmann equation arises, there is no formal difficulty. To higher order, including dense gas corrections, one finds a nonuniformity arising from slow particles and another from distant particles[10] **[15]**, **[11]**, and many others.

Essentially the same mathematical situation also arises in evaluating transport coefficients in a plasma. A formal divergence, arising from the long range Coulomb potential, is eliminated on introducing the concept of a shielded potential **[47]**. Without performing any of the lengthy computations, it is immediately evident that the dominant term of the electrical conductivity (i.e. the Coulomb logarithm) must arise with identical numerical coefficients whether we employ the Fokker-Planck equation or the Boltzmann equation, and whether we use a spherical Debye shielding cloud, or a more sophisticated distorted shielding cloud, or whether we simply cut off the divergence at the Debye length **[18]**. The only fact that we must know is the scale length of the shielding, without regard to the details of the shielding process. But to carry the computation beyond the dominant term is formidable and requires a correct model for the shielding.

5. Small mean free path—Introduction. The limit of small mean free path probably represents the most extreme of all nonuniformities to be found within

[10] This was first pointed out in 1958 **[15]** and has been intensively studied more recently, e.g. **[11]** and many others.

the scope of the Boltzmann equation, but it is also the best understood and documented. In this limit, the gas description is reduced from a distribution function $f(\boldsymbol{v}, \boldsymbol{x}, t)$ to the macroscopic fluid coordinates $\rho(\boldsymbol{x}, t)$, $T(\boldsymbol{x}, t)$, $\boldsymbol{u}(\boldsymbol{x}, t)$.

The Boltzmann equation can be written in skeleton form as

$$\partial f/\partial t + \boldsymbol{v} \cdot \partial f/\partial \boldsymbol{x} = Q(f, f) \tag{5.1}$$

where Q is a quadratic operator on f which vanishes when applied to a local Maxwellian

$$f_M = \frac{\rho}{(2\pi RT)^{3/2}} \exp\left[-\frac{(\boldsymbol{v}-\boldsymbol{u})^2}{2RT}\right]. \tag{5.2}$$

By *local* we mean that ρ, T, and $\boldsymbol{u}$ may be arbitrary functions of $\boldsymbol{x}$ and t. Under appropriate scaling, the Boltzmann equation becomes

$$\frac{\partial f}{\partial t} + \boldsymbol{v} \cdot \frac{\partial f}{\partial \boldsymbol{x}} = (1/\varepsilon)Q(f, f) \tag{5.3}$$

or in a linearized version

$$\frac{\partial f}{\partial t} + \boldsymbol{v} \cdot \frac{\partial f}{\partial \boldsymbol{x}} + \frac{1}{\varepsilon} L(f) = 0. \tag{5.4}$$

The nonlinear form can be obtained by the substitution $f \to f/\varepsilon$ in (5.1) and the linear version by the substitution $f \to (1/\varepsilon)f_M + f$ where f_M is here an absolute Maxwellian with constant parameters ρ, T, $\boldsymbol{u}$. The significance of the substitution $f \to f/\varepsilon$ is that for small ε the density is large and therefore the mean free path is small. In the linearization, ε represents the mean free path of the unperturbed constant state, f_M/ε.

It is evident that the limit $\varepsilon \to 0$ is singular since all derivatives are lost. For small ε we should expect to have $Q(f, f) \sim 0$ and $f \sim f_M$ except in regions where $\partial/\partial t$ or $\partial/\partial \boldsymbol{x}$ become unbounded. To be more definite, take a fixed domain with appropriate fixed boundary and initial conditions on f, independent of the parameter ε. The qualitative behavior for small ε of the solution $f(\boldsymbol{v}, \boldsymbol{x}, t; \varepsilon)$ is described by a conjecture **[15]** which states that f approximates f_M, and the fluid parameters in f_M obey fluid equations, except where this is evidently impossible. Specifically, there are three locations where f cannot approximate f_M closely:

(1) in an initial layer, since the initial value $f(v, x, 0)$ is arbitrary;

(2) in a boundary layer where the "half" distribution of molecules leaving the wall is, in principle, arbitrary;

(3) in a shock layer where one cannot conclude that $Q(f, f) \sim 0$ since $\partial/\partial t$ and $\partial/\partial \boldsymbol{x}$ are of order $1/\varepsilon$ (the shock thickness scales as ε).

The conjecture states that, except in these three layers of nonuniformity, the Hilbert or Chapman-Enskog formal expansions in ε are asymptotic to the actual solution $f(\boldsymbol{v}, \boldsymbol{x}, t; \varepsilon)$. This conjecture has been promoted to a proved theorem in linear and weakly nonlinear formulations of the initial value problem **[19]**, **[20]**, **[23]**.

Accepting the conjecture more generally, the question immediately arises of matching the asymptotic expansions across the three types of boundary layers within which they are not valid. The fluid equations (which can take several forms) yield solutions only when they are completed with initial data, boundary data, and possibly shock jump conditions. The relation between the given Boltzmann boundary data (half distribution) and the proper asymptotic fluid boundary data requires analysis of a boundary layer, and similarly for the initial and shock layers. A crude analysis of the boundary layer was given by Maxwell and led to the concept of *boundary slip* (of tangential velocity and temperature). In the same terminology we can talk of *initial slip* and *shock slip*. The initial slip represents the amount by which the true initial values of the fluid state (given as moments of the assigned initial distribution function) must be altered so that, when used as initial values in fluid equations, they give rise to the correct future fluid state as it would be found by solving the Boltzmann equation. Similarly, the shock slip represents the correction to the classical Hugoniot condition which is required for a fluid calculation to yield the same result as solution of the Boltzmann equation. A related concept of shock slip was later found within fluid dynamics proper as a modification of the inviscid Hugoniot conditions to take into account the interaction between a shock and a boundary layer [**39**]. Although viscous fluid equations can presumably give a smooth shock transition, to approximate the kinetic solution correctly they must be allowed to have discontinuities.

The small mean free path limit has been split into two problems. The first, in the interior (away from initial, boundary, and shock layers), is solved by a Hilbert or Chapman-Enskog expansion (except, as we shall see later, when they require modification). The second consists in the analysis of three kinetic boundary layers in order to complete the fluid formulation with initial, boundary, and jump conditions. Note that the type of boundary data that is required for inviscid fluid equations is quite different from what is required for viscous equations. The kinetic boundary layer analysis will therefore differ depending on which fluid model is selected.

6. The initial value problem. The solution to the problem of initial slip is essentially complete [**19**], [**20**], [**23**]. For the nonlinear problem, the solution is formal; an algorithm is given for the computation of the asymptotic initial fluid state in terms of the given initial distribution function. For the linear and weakly nonlinear problems, the solution $f(\boldsymbol{v}, \boldsymbol{x}, t; \varepsilon)$ has been shown to exist, and the formal expansions in ε have been proved to be asymptotic to this solution.

To be more precise, let $f(\boldsymbol{v}, \boldsymbol{x}, t; \varepsilon)$ be the solution of the Boltzmann equation which takes the initial value $f(\boldsymbol{v}, \boldsymbol{x}, 0)$. From $f(\boldsymbol{v}, \boldsymbol{x}, 0)$, we extract its moments $\rho(\boldsymbol{x}, 0)$, $\boldsymbol{u}(\boldsymbol{x}, 0)$, $T(\boldsymbol{x}, 0)$ and solve the inviscid Euler equations with these initial values. Let $f_E(\boldsymbol{v}, \boldsymbol{x}, t)$ be the local Maxwellian formed from these Euler solutions $\rho_E(\boldsymbol{x}, t)$, $\boldsymbol{u}_E(\boldsymbol{x}, t)$, $T_E(\boldsymbol{x}, t)$. Then

$$\lim_{\varepsilon \to 0} f(\boldsymbol{v}, \boldsymbol{x}, t; \varepsilon) = f_E(\boldsymbol{v}, \boldsymbol{x}, t) \tag{6.1}$$

for any fixed $t > 0$ and in a variety of norms in $(\boldsymbol{x}, \boldsymbol{v})$ (including L_2, maximum, and others). In the limit $\varepsilon \to 0$ the Boltzmann equation disappears; but its solutions have a well-defined limit. The moments ρ, T, $\boldsymbol{u}$ are continuous at $t = 0$ uniformly in ε, but f is not; nor are $\partial \boldsymbol{u}/\partial t$ and $\partial T/\partial t$.

The Navier-Stokes equations must be supplied with corrected initial data, $\boldsymbol{u}_N(\boldsymbol{x}, 0); \varepsilon = \boldsymbol{u}_0(\boldsymbol{x}, 0) + \varepsilon \boldsymbol{u}_1(\boldsymbol{x}, 0)$, etc. Here $\boldsymbol{u}_0(\boldsymbol{x}, 0)$ is the actual initial velocity [a moment of $f(\boldsymbol{v}, \boldsymbol{x}, 0)$], and the correction $\boldsymbol{u}_1(\boldsymbol{x}, 0)$ is obtained from $f(\boldsymbol{v}, \boldsymbol{x}, 0)$ by a definite analytic procedure. The solution $\rho_N(\boldsymbol{x}, t; \varepsilon)$, $T_N(\boldsymbol{x}, t; \varepsilon)$, $\boldsymbol{u}_N(\boldsymbol{x}, t; \varepsilon)$ of the Navier-Stokes equations (in which ε multiplies the viscosity and heat conductivity) is used to construct a Chapman-Enskog distribution function $f_N = f_0 + \varepsilon f_1$ where f_0 is locally Maxwellian with ρ_N, T_N, $\boldsymbol{u}_N$ as parameters, and f_1 depends on gradients of $\boldsymbol{u}_N$ and T_N. It is then shown (in various norms) that

$$\text{(6.2)} \qquad \lim_{\varepsilon \to 0} \frac{1}{\varepsilon} |f - f_N| = 0,$$

this time uniformly in $t > t_0$.

The quantities ρ_1, T_1, $\boldsymbol{u}_1$, represent the initial slip, i.e. the discrepancy between the initial fluid state and the initial value which is required for the Navier-Stokes solution to approximate the Boltzmann solution correctly. The first order initial slip vanishes if $f(\boldsymbol{v}, \boldsymbol{x}, 0)$ is a local Maxwellian. The initial slip vanishes to higher order if the initial distribution is a Chapman-Enskog function; but this requires the initial function $f(\boldsymbol{v}, \boldsymbol{x}, 0; \varepsilon)$ to depend explicitly on ε.

Proofs of these statements make use of strong existence theorems with estimates of f and of space derivatives of f that are uniform in the singular parameter ε.

The statement that a Boltzmann solution is approximated by fluid equations outside the regions of nonuniformity is somewhat flexible. The Hilbert theory (linear or nonlinear) gives a formal expansion of f in powers of ε governed by the inviscid Euler equations to lowest order and by successive inhomogeneous Euler equations to higher order. More precisely, the higher order equations are the inhomogeneous variational equations of the Euler system (linear with variable coefficients, having the same characteristics as the Euler equations). The Chapman-Enskog theory also gives the Euler equations to lowest order; but this is followed by the Navier-Stokes equations, the Burnett equations, and a sequence of partial differential equations in ρ, T, and $\boldsymbol{u}$ of successively higher order in space derivatives. In other words, the Chapman-Enskog theory gives a formal expansion of the macroscopic *equations* which govern the fluid's evolution, whereas the Hilbert theory gives a formal expansion of the *solution*.

Both expansions, Hilbert and Chapman-Enskog, are asymptotic. But the ranges of validity are not the same. Each expansion, when truncated, satisfies the Boltzmann equation with a remainder,

$$\text{(6.3)} \qquad \frac{\partial f_n}{\partial t} + \boldsymbol{v} \cdot \frac{\partial f_n}{\partial \boldsymbol{x}} + \frac{1}{\varepsilon} L(f_n) = \varepsilon^n R_n(\boldsymbol{v}, \boldsymbol{x}, t; \varepsilon).$$

The Hilbert remainder, which is obtained by solving inhomogeneous Euler equations, grows algebraically as t^n. The Navier-Stokes remainder decays as $t \to \infty$. This is why the Hilbert expansion is asymptotic for only limited time, $t_0 < t < t_1$, whereas the Navier-Stokes result is uniformly valid for $t > t_0$. The higher order Chapman-Enskog approximations, Burnett, etc. are only formal at the present time in the absence of any estimates of the boundedness, growth, or decay of their solutions.

The proof that a given formal expansion is asymptotic to a solution of the Boltzmann equation is quite oblivious to the recipe which gave rise to the expansion. All that needs to be verified is that the expansion satisfies the Boltzmann equation term by term with a remainder $\varepsilon^n R_n(v, x, t; \varepsilon)$ where R is suitably bounded. We may construct such a series by *ad hoc* means as well as by expansion. For example, from the Hilbert or the Chapman-Enskog formulas we can easily construct a unique sequence starting with Euler, Navier-Stokes, and continuing with *inhomogeneous* Navier-Stokes equations. It is trivial to verify that this sequence is asymptotically equivalent to either the Hilbert or Chapman-Enskog series. This formal series satisfies the Boltzmann equation with an error term R_n which decays for large t at any order $n > 1$. This series is therefore superior to the Hilbert series in range of validity, and it is superior to the Chapman-Enskog series since the theory of the associated partial differential equations is known. Although this procedure is arbitrary for the initial value problem, we shall find that it arises naturally in slow steady flow (§8).

The solution of the initial slip problem can be given a simple abstract formulation in the case of the linear Boltzmann equation (5.4). In terms of a suitable normalization, the local Maxwellian takes the form

(6.4)
$$\omega = \frac{1}{(2\pi)^{3/2}} \exp\left(-\tfrac{1}{2}v^2\right)$$

and the nullspace of the collision operator L,

(6.5)
$$L(\psi_\alpha) = 0, \qquad \alpha = 0, 1, 2, 3, 4$$

is given by

(6.6)
$$\psi_0 = \omega^{1/2},$$
$$\psi_i = v_i \omega^{1/2} \qquad i = 1, 2, 3,$$
$$\psi_4 = (v^2 - 3)(\omega/6)^{1/2}.$$

The ψ_α are orthonormal in the L_2 norm,

(6.7)
$$(f, g) = \int fg \, dv.$$

The (linearized) fluid state is given by the projection of f into the nullspace of L,

(6.8)
$$\hat{f} = \sum \rho_\alpha \psi_\alpha$$

where

$$(6.9)\qquad \begin{aligned} (\psi_0, f) &= \rho_0 = \rho, \\ (\psi_i, f) &= \rho_i = u_i \qquad i = 1, 2, 3, \\ (\psi_4, f) &= \rho_4 = (\tfrac{3}{2})^{1/2} T. \end{aligned}$$

We introduce

$$(6.10)\qquad f = \hat{f} + \tilde{f}$$

where $\tilde{f}$ is the projection into the orthogonal complement to the nullspace.

The basis of the Hilbert theory is the hypothesis that $f(\boldsymbol{v}, \boldsymbol{x}, t; \varepsilon)$ be expressible as a formal power series in ε. The essential conclusion is that to every given fluid state, $\rho_\alpha(\boldsymbol{x})$, is assigned a unique distribution function **[26]**

$$(6.11)\qquad f_H(\boldsymbol{v}, \boldsymbol{x}; \varepsilon) = f_H^0 + \varepsilon f_H^1 + \cdots.$$

An equivalent statement is that to a given projection $\hat{f}(\boldsymbol{v}, \boldsymbol{x})$, there is assigned a unique distribution f_H; in particular, the leading term f_H^0 is exactly the given fluid state $\hat{f}$, and all succeeding terms are in the orthogonal complement. The manifold of special functions f_H is formally invariant under solution of the Boltzmann equation. In other words, a special function f_H taken as an initial value for the Boltzmann equation will give rise to a solution which remains within the space of functions f_H. A complementary result has also been proved **[19]**. The hypothesis that $f(\boldsymbol{v}, \boldsymbol{x}, \tau; \varepsilon)$, (where $\tau = t/\varepsilon$) is a power series in ε and that f decays to zero implies that to a given projection $\tilde{f}(\boldsymbol{v}, \boldsymbol{x})$ is assigned a unique distribution function

$$(6.12)\qquad f_G = f_G^0 + \varepsilon f_G^1 + \cdots,$$

where f_G^0 is just the given projection $\tilde{f}$, and all succeeding terms lie in the nullspace (i.e. they are strictly fluid components). This procedure also defines an asymptotically invariant space of solutions of the Boltzmann equation.

These two results can be combined to obtain the unique resolution

$$(6.13)\qquad f(\boldsymbol{v}, \boldsymbol{x}) = f_H(\boldsymbol{v}, \boldsymbol{x}; \varepsilon) + f_G(\boldsymbol{v}, \boldsymbol{x}; \varepsilon)$$

of an arbitrary distribution f, which we choose to be independent of ε. To simplify the notation we write h for f_H and g for f_G. The orthogonal decomposition $f = \tilde{f} + \hat{f}$ of the given function $f(\boldsymbol{v}, \boldsymbol{x})$ is used to obtain the leading terms

$$(6.14)\qquad h_0^0 = \hat{f}, \qquad g_0^0 = \tilde{f}$$

which generate the two series

$$(6.15)\qquad \begin{aligned} h_0(\boldsymbol{v}, \boldsymbol{x}; \varepsilon) &= h_0^0 + \varepsilon h_0^1 + \cdots, \\ g_0(\boldsymbol{v}, \boldsymbol{x}; \varepsilon) &= g_0^0 + \varepsilon g_0^1 + \cdots. \end{aligned}$$

The difference

$$(6.16)\qquad f - h_0 - g_0 = \varepsilon f_1(\boldsymbol{v}, \boldsymbol{x}; \varepsilon)$$

is similarly decomposed, $f_1 = \tilde{f}_1 + \hat{f}_1$ to generate the two series h_1 and g_1, etc.

In this way we obtain (6.13) as the series

$$f_H = f_0 + \varepsilon h_1 + \cdots,$$
(6.17)
$$f_G = g_0 + \varepsilon g_1 + \cdots.$$

Let $f(\boldsymbol{v}, \boldsymbol{x}, t; \varepsilon)$ be the solution of the Boltzmann equation with $f(\boldsymbol{v}, \boldsymbol{x})$ as initial value. Taking as an initial value for the Boltzmann equation the Hilbert component f_H of the resolution (6.13) of $f(\boldsymbol{v}, \boldsymbol{x})$ yields a solution which is asymptotically equal to $f(\boldsymbol{v}, \boldsymbol{x}, t; \varepsilon)$ after a time $O(\varepsilon)$. The complementary solution, taking f_G as initial value, describes a transient. Although the time evolution of f_H is governed by fluid equations, the entire initial function $f(\boldsymbol{v}, \boldsymbol{x})$ is reflected in the fluid initial data, $f_H(\boldsymbol{v}, \boldsymbol{x}; \varepsilon)$.

If the collision operator L has discrete eigenvalues other than zero, then the "transient" component f_G can be further decomposed into fluidlike components which satisfy partial differential equations (similar to Euler, Navier-Stokes, etc.), but which decay exponentially in time.

If $f(\boldsymbol{v}, \boldsymbol{x})$ has n derivatives with respect to $\boldsymbol{x}$, the separation into f_H and f_G can be carried out only to order ε^{n-1}. The problem of discontinuous or more general nonsmooth initial data remains an open question.[11] For Maxwell molecules, the expansion f_H is explicitly convergent, as a polynomial in ε, when $f(\boldsymbol{v}, \boldsymbol{x})$ is a polynomial in $\boldsymbol{x}$ [**19**]. One can expect convergence of the series for f_H if f is sufficiently smooth (analyticity is not enough). A conjecture made in [**19**], based on Maxwellian molecules, is that convergence requires f to be an entire function of $\boldsymbol{x}$ of order one-half. In the special case of hard spheres, a single Fourier coefficient, $f = \phi(\boldsymbol{v})e^{ikx}$, has been shown to give rise to a convergent series for f_H when k is sufficiently small [**32**]. It is not clear whether an entire function of order one (like e^{ikx}) is sufficient for convergence in general, or whether there is an essential difference between hard spheres and Maxwellian molecules.

7. The boundary value problem—General discussion. The boundary layer presents a more difficult problem than the initial layer for at least two reasons. First of all, the basic proof of existence is more difficult. There have been great strides recently in existence theory for the boundary value problem,[12] but these results are not yet adequate to the demands of the small mean free path limit. Neither the required uniform bounds as $\varepsilon \to 0$ nor estimates of derivatives are yet available. This is not simply a technicality, since space derivatives do become infinite near a boundary.

---

[11] For a model of the Boltzmann equation, Y. Sone [*Phys. Fluids*, to appear] has obtained the asymptotic fluid behavior for a special class of discontinuous initial data.

[12] C. Cercignani [*J. Math. Phys.*, to appear] in a general bounded domain for the linear equation and Y. P. Pao [*J. Math. Phys.*, to appear] for the weakly nonlinear equation in a bounded one-dimensional domain.

The second difficulty is that the limit $\varepsilon \to 0$ is ambiguous even in fluid dynamics. We recall that the viscosity and heat conductivity coefficients are proportional to ε. In a Stokes linearization about a free stream $U = 0$, the flow is essentially independent of the value of the viscosity coefficient; cf. slow flow about a sphere or Poisseuille flow with a parabolic velocity profile. On the other hand, in an Oseen linearization about a finite velocity U, the limit $\mu \to 0$ is singular and gives an inviscid limiting flow except within a boundary layer and wake. This boundary layer, of thickness $O(\varepsilon^{1/2})$, is distinct from the kinetic boundary layer which is $O(\varepsilon)$.

To obtain a definite result in the kinetic problem, we must at least distinguish between the various macroscopic cases. One distinction is the value of the Reynolds number. In the Stokes linearization we have set $\mathrm{Re} = 0$ by setting the free stream velocity equal to zero before $\mu \to 0$. In the Oseen linearization Re is finite, but it becomes infinite in the subsequent limit $\mu \to 0$. We shall find it convenient to interpolate between these cases by taking an Oseen linearization after which we allow U to approach zero with ε. In this limit Re remains finite.

The interior of the fluid will be described by a formal expansion which is equivalent to Hilbert's in the initial value problem. By supplementing this expansion with conjectured existence theorems, we shall also be able to imitate the rigorous analysis of the initial layer by a plausible analysis of the boundary layer. This will give the asymptotic boundary conditions which are needed to complete the formal fluid expansion in the interior.

In the case of the Stokes linearization and the more general finite Re linearization just described, we find that the formal Hilbert expansion automatically gives Navier-Stokes followed by successive inhomogeneous Navier-Stokes equations, rather than Euler and inhomogeneous Euler equations as in the initial value problem. The reason is that the steady flow Euler equations are too degenerate. They are not a determined system of differential equations which yield solutions merely by supplementation with boundary conditions. In common with most asymptotic expansions, a given term in the standard Hilbert expansion is not uniquely determined until we turn to the next higher order from which we extract compatibility conditions. In the case of steady flow, the additional degeneracy requires us to extract information from *two* succeeding levels to complete the determination of a given order. Thus the expansion of a solution of the Boltzmann equation in powers of ε is essentially different in steady flow and time-dependent flow.

In the case $\mathrm{Re} \to \infty$, a restriction to the viscous boundary layer rather than the entire interior allows a similar result (inhomogeneous Navier-Stokes equations) to be obtained **[52]**.

For the general case, nonlinear and nonsteady, we can find an asymptotic fluid description by an *ad hoc* procedure similar to the one described above (§6) in which we proposed inhomogeneous Navier-Stokes equations for the initial value problem. The important point with regard to the boundary layer is that the boundary conditions which it must supply are, to any order in ε, just what is appropriate to a Navier-Stokes system.

8. Steady Hilbert expansion for finite Reynolds number. We take the linear Boltzmann equation

$$\frac{\partial f}{\partial t} + \boldsymbol{v} \cdot \frac{\partial f}{\partial \boldsymbol{x}} + \frac{1}{\varepsilon} L(f) = 0 \tag{8.1}$$

in the normalized and dimensionless form described in equations (6.4)–(6.9). A Stokes linearization in steady flow is simply

$$\boldsymbol{v} \cdot \frac{\partial f}{\partial \boldsymbol{x}} + \frac{1}{\varepsilon} L(f) = 0 \tag{8.2}$$

and an Oseen linearization about the free stream velocity εU yields

$$\varepsilon \boldsymbol{U} \cdot \frac{\partial f}{\partial \boldsymbol{x}} + \boldsymbol{v} \cdot \frac{\partial f}{\partial \boldsymbol{x}} + \frac{1}{\varepsilon} L(f) = 0 \tag{8.3}$$

(the molecular velocity $\boldsymbol{v}$ is relative to the stream at infinity, as is appropriate in taking moments of f). We note the linearized equation of state

$$p = \rho + T \tag{8.4}$$

and the conservation equations [appropriate to (8.1)]

$$\begin{gathered} \frac{\partial \rho}{\partial t} + \nabla \cdot \boldsymbol{u} = 0, \\ \frac{\partial u_i}{\partial t} + \frac{\partial p}{\partial x_i} + \frac{\partial p_{ij}}{\partial x_j} = 0, \\ \frac{\partial}{\partial t}\left(\frac{3}{2} T\right) + \nabla \cdot \boldsymbol{u} + \nabla \cdot \boldsymbol{q} = 0. \end{gathered} \tag{8.5}$$

The dimensionless stress deviator and heat flow vector are the projections

$$\begin{aligned} p_{ij} &= (v_i v_j - \tfrac{1}{3} v^2 \delta_{ij}, f), \\ q_i &= (\tfrac{1}{2}[v^2 - 5] v_i, f). \end{aligned} \tag{8.6}$$

The Euler equations are obtained by setting $p_{ij} = q_i = 0$ in (8.5), and the Navier-Stokes equations result from

$$\begin{aligned} p_{ij} &= -\mu \left\{ \frac{\partial u_i}{\partial x_j} + \frac{\partial u_j}{\partial x_i} - \frac{2}{3} \delta_{ij} \nabla \cdot \boldsymbol{u} \right\}, \\ q_i &= -\lambda \frac{\partial T}{\partial x_i}. \end{aligned} \tag{8.7}$$

The full Chapman-Enskog expansion takes the form [19]

$$\frac{\partial \rho}{\partial t} + \nabla \cdot \boldsymbol{u} = 0,$$

$$\frac{\partial \boldsymbol{u}}{\partial t} + \nabla p = \sum_0^\infty \varepsilon^{2n+1} \Delta^n[\mu_n \Delta \boldsymbol{u} + \alpha_n \nabla(\nabla \cdot \boldsymbol{u})]$$

$$+ \sum_1^\infty \varepsilon^{2n} \Delta^n[\beta_n \nabla T + \gamma_n \nabla \rho], \tag{8.8}$$

$$\frac{\partial}{\partial t}\left(\frac{3}{2} T\right) + \nabla \cdot \boldsymbol{u} = \sum_0^\infty \varepsilon^{2n+1} \Delta^{n+1}[\lambda_n T + \delta_n \rho]$$

$$+ \sum_1^\infty \varepsilon^{2n} \zeta_n \Delta^n(\nabla \cdot \boldsymbol{u}),$$

where the coefficients μ_n, λ_n, α_n, β_n, γ_n, δ_n, ζ_n depend on the molecular model. In particular, $\mu = \varepsilon\mu_0$ is the (dimensionless) viscosity coefficient and $\lambda = \varepsilon\lambda_0$ is the heat conductivity ($\alpha_0 = \mu_0/3$ and $\delta_0 = 0$). The steady flow Stokes linearization is obtained by setting $\partial/\partial t = 0$, and the Oseen linearization is obtained by replacing $\partial/\partial t$ by $\boldsymbol{U} \cdot \nabla$.

In a steady Stokes flow the Euler equations are

$$\nabla \cdot \boldsymbol{u} = 0, \qquad \nabla p = 0 \tag{8.9}$$

and the Navier-Stokes equations are

$$\nabla \cdot \boldsymbol{u} = 0,$$

$$\nabla p + \mu \operatorname{curl} \operatorname{curl} \boldsymbol{u} = 0, \tag{8.10}$$

$$\Delta T = 0.$$

The conventional Hilbert expansion

$$f = \sum \varepsilon^n f_n \tag{8.11}$$

substituted into (8.2) yields the sequence of compatibility relations:

$$\nabla \cdot \boldsymbol{u}_0 = 0, \qquad \nabla p_0 = 0 \tag{8.12}$$

at zero order and

$$\nabla \cdot \boldsymbol{u}_n = 0,$$

$$\nabla p_n + \mu_0 \operatorname{curl} \operatorname{curl} \boldsymbol{u}_{n-1} = 0, \tag{8.13}$$

$$\Delta T_{n-1} = 0,$$

for $n = 1, 2, \ldots$. We recall that $\mu_0 = \mu/\varepsilon$ is a scaled viscosity coefficient which is of order one. Regrouping the variables,

$$
\begin{aligned}
&p_0,\\
(8.14)\qquad &\boldsymbol{u}_0, p_1, T_0\\
&\boldsymbol{u}_n, p_{n+1}, T_n \qquad n = 1, 2, \ldots,
\end{aligned}
$$

gives $p_0 =$ constant to lowest order and a well-posed Navier-Stokes system for each set $(\boldsymbol{u}_n, p_{n+1}, T_n)$.

The Stokes linearization gives a degenerate result in that the inhomogeneous terms vanish at any order[13] [this is easily seen by induction in (8.8)]. Since the same differential system occurs at each order, the sum of any finite number of terms in the Hilbert expansion (8.11) is exactly a Navier-Stokes flow; ε enters only through the boundary conditions which must be found outside this theory and which we expect to be a formal power series in ε.

We turn to the Oseen linearization for which typical problems are the flow around a stationary object or the flow through a channel. Taking $\varepsilon \boldsymbol{U}$ as the unperturbed flow velocity, the formal Hilbert expansion (8.11) with the same grouping of variables (8.14) gives

$$(8.15)\qquad p_0 = \text{constant}$$

followed by

$$
\begin{aligned}
&\nabla \cdot \boldsymbol{u}_0 = 0,\\
(8.16)\qquad &\boldsymbol{U} \cdot \nabla \boldsymbol{u}_0 + \nabla p_1 + \mu_0 \operatorname{curl} \operatorname{curl} \boldsymbol{u}_0 = 0,\\
&(5/2)\boldsymbol{U} \cdot \nabla T_0 - \lambda_0 \Delta T_0 = 0,
\end{aligned}
$$

then

$$
\begin{aligned}
&\nabla \cdot \boldsymbol{u}_1 = \boldsymbol{U} \cdot \nabla T_0,\\
(8.17)\qquad &\boldsymbol{U} \cdot \nabla \boldsymbol{u}_1 + \nabla p_2 + \mu_0 \operatorname{curl} \operatorname{curl} \boldsymbol{u}_1 = \sigma(\boldsymbol{U} \cdot \nabla) \nabla T_0,\\
&(5/2)\boldsymbol{U} \cdot \nabla T_1 - \lambda_0 \Delta T_1 = \boldsymbol{U} \cdot \nabla p_1,
\end{aligned}
$$

$[\sigma = (1/3)\mu_0 + (5/2)(\beta_1 - \gamma_1)/\lambda_0]$ and further inhomogeneous forms of the same differential system. The system of equations (8.16) is the standard incompressible Oseen linearization of the Navier-Stokes equations (with scaled viscosity and heat conductivity).

In the general case, $\mathrm{Re} \neq 0$, kinetic theory enters more substantially than in the special case $\mathrm{Re} = 0$, where only the boundary conditions are modified. The reason, qualitatively, is that the real characteristics $\boldsymbol{U} \cdot \nabla$ of the Oseen linearization carry the boundary values into the flow.

9. Large Reynolds number. With a finite value for the unperturbed speed U, the Reynolds number becomes large as $1/\varepsilon$, and there can be a double boundary layer. There is no advantage to linearization in this generality, so we may just

[13] This degeneracy was first pointed out in the special case of one-dimensional heat flow **[54]**.

as well consider the full nonlinear flow. A formal Hilbert expansion yields Euler equations and inhomogeneous Euler equations as in the initial value problem. This system is mathematically well posed provided that all streamlines intersect a boundary (or go to infinity). If we adopt the direct Hilbert expansion, we must look for asymptotic boundary conditions which are appropriate to the Euler equations, the boundary conditions to be extracted from an independent analysis of the *combined* viscous and kinetic boundary layers. A boundary condition on u_n (reflecting the displaced boundary) will be a power series in $\varepsilon^{1/2}$. The first order term is taken from a classical viscous boundary layer analysis. The second order boundary term $O(\varepsilon)$ is a combination of a kinetic contribution and a second order viscous effect, and it is to be used in an inhomogeneous Euler equation. In particular, a second order fluid dynamic boundary layer analysis is not correct without the kinetic modification.

Another procedure is to give up the attempt to examine the whole flow and concentrate instead on the viscous boundary layer alone. The stretched variables in conventional boundary layer analysis involve ε (since $\mu \sim \varepsilon$). This reordering of terms, when combined with the Hilbert expansion, again gives Navier-Stokes (in a boundary layer approximation) and inhomogeneous Navier-Stokes equations [**52**]. Any such higher order boundary layer analysis must be matched to an appropriate fluid formulation in the interior, e.g. inhomogeneous Euler or Navier-Stokes.

The most general procedure, applicable uniformly to the entire fluid domain, is to follow the *ad hoc* rule described previously. The complete (nonlinear and nonsteady) Chapman-Enskog expansion, involving differential equations of increasing order, is asymptotically equivalent to a sequence of Navier-Stokes and inhomogeneous Navier-Stokes equations with inhomogeneous terms of increasing order. Analysis of the kinetic boundary layer should give boundary conditions for the Navier-Stokes equations as a formal power series. The number of boundary conditions does not change with the order of approximation. Since ε occurs nontrivially in the Navier-Stokes equations, their solutions will not be simple power series in ε. Nevertheless this procedure does produce a sequence of fluid approximations which is asymptotic to the solution of the Boltzmann equation to any order ε^n in a region which includes the viscous boundary layer and excludes only the kinetic boundary layer.

It is important to note that there is an essentially arbitrary choice of fluid procedures available for approximating to the interior fluid flow, excluding the kinetic boundary layer. The number and type of boundary conditions that must be supplied by the kinetic boundary layer analysis depends on the particular fluid choice that is made (Euler, Navier-Stokes, Burnett, etc.). In other words, the kinetic boundary layer analysis cannot be made universally but only in conjunction with a particular treatment of the interior. The common worry that Burnett and higher order Chapman-Enskog equations presumably require an increasing number of boundary conditions poses no real difficulty. There is no compelling reason to choose the Chapman-Enskog procedure which is only one of a number of valid formal series. The choice we recommend, a sequence of Navier-Stokes systems with inhomogeneous terms of increasing order and similar slip boundary conditions, has the overwhelming advantage of including the viscous boundary layer (as distinguished from the Hilbert expansion) and of utilizing a familiar system of differential equations (as opposed to the Chapman-Enskog expansion). For reasons that will appear in the next section, the inhomogeneous terms in the differential equations and the finite slip boundary conditions enter nontrivially at the same order. Therefore, a second order boundary layer investigation with slip is incomplete without modifying the Navier-Stokes equations.

10. The boundary layer—Slip flow. The initial layer is studied by introducing a stretched time scale, $\tau = t/\varepsilon$, in terms of which the Boltzmann equation

$$\partial f/\partial \tau + L(f) = -\varepsilon \boldsymbol{v} \cdot \partial f/\partial \boldsymbol{x} \tag{10.1}$$

exhibits weak coupling of spatial gradients with a purely time-dependent problem. Similarly, the boundary layer can be studied by introducing $y = x_1/\varepsilon$ as the scaled distance from the wall and $\boldsymbol{x}' = (x_2, x_3)$ as unscaled coordinates within a parallel surface $x_1 =$ constant to obtain an approximately one-dimensional Boltzmann equation

$$v_1(\partial f/\partial y) + L(f) = O(\varepsilon). \tag{10.2}$$

This estimate [ignoring the term $O(\varepsilon)$] is sufficient to calculate first order slip coefficients. We can easily refine this estimate

$$v_1(\partial f/\partial y) + L(f) = -\varepsilon \boldsymbol{v}' \cdot \partial f/\partial \boldsymbol{x}' + O(\varepsilon^2). \tag{10.3}$$

In order to simplify the calculations (in particular keeping all tensor moments of f unchanged), we do not transform $\boldsymbol{v}$, keeping the original rectangular coordinates in velocity space.

The slip boundary conditions arise from matching solutions of the boundary layer equation (10.3) with the Hilbert expansion in the interior. To do this we shall have to postulate certain existence theorems, both for the full Boltzmann equation in a general interior domain D, and for the one-dimensional equation in a half space. The two types of existence theorems are quite distinct. For example, $u_n = 0$ can be imposed on the entire boundary of D, but conservation of mass implies $u_n = 0$ at one end (or at infinity) if it holds at the opposite end of a one-dimensional problem. Also, suitable regularity and boundedness conditions at infinity are needed for the unbounded domain.

The basic boundary condition involves specifying the half distribution function f^-, $\boldsymbol{v} \cdot \boldsymbol{n} < 0$, where $\boldsymbol{n}$ is the outward normal to D. To be able to guarantee mass conservation ($u_n = 0$) at the boundary, we specify f^- modulo a constant multiple of ψ_0 [cf. (6.6)]

$$f^- = \phi^- + c_0 \psi_0. \tag{10.4}$$

The parameter c_0 (which is a point function over the boundary of the domain) is to be adjusted to satisfy $u_n = 0$. A special case of this boundary condition is diffuse reflection for which ϕ^- is explicit,

$$f^- = c_0\psi_0 + \boldsymbol{u}'_w \cdot \boldsymbol{\psi}' + \left(\frac{3}{2}\right)^{1/2} T_w\psi_4. \tag{10.5}$$

Here $\boldsymbol{u}'_w$ and T_w are the tangential wall velocity and the wall temperature, and $\boldsymbol{\psi}'$ is the tangential component of the vector ψ_i (6.6). The parameters $\boldsymbol{u}'_w$ and T_w are given functions on the boundary, and c_0 is to be determined such that $u_n = 0$. This is the linearized version of the statement that the emitted distribution is a Maxwellian at the local wall velocity and temperature.

We now list the conjectured theorems for an interior domain D:

THEOREM 1. *A unique steady solution exists subject to f^- given on the boundary of D.*

THEOREM 2. *A steady solution exists with $u_n = 0$ subject to* (10.4) *where ϕ^- is given and c_0 is left open; the solution is unique if the total mass, $\int_D \rho\, dx$ (or $\int_D p\, dx$) is specified.*

COROLLARY. *A unique solution exists subject to the diffuse boundary condition* (10.5) *with specified wall velocity and temperature, provided that the total mass of the system is specified.*

The results quoted in these theorems have been proved in a certain weighted L_2 norm.[14]

In a one-dimensional steady flow there are five constants of the motion. In a nonlinear version these are

$$\begin{gathered} \rho u_x, \\ \rho u_x^2 + p + p_{xx}, \\ \rho u_x u_y + p_{xy}, \\ \rho u_x u_z + p_{xz}, \\ \rho u_x(e + \tfrac{1}{2}u^2) + u_x(p + p_{xx}) + q_x, \end{gathered} \tag{10.6}$$

and in the linearized version (about zero free stream) they can be taken as

$$\begin{gathered} u_x, \\ p + p_{xx}, \\ p_{xy}, p_{xz}, \\ q_x. \end{gathered} \tag{10.7}$$

The basic conjectured existence theorem is

THEOREM 3. *A unique steady solution exists in a half space $x > 0$ if at $x = 0$, f^- is given modulo $c_0\psi_0$* (10.4), *and if, in addition, the five constants* (10.7) *are specified.*

[14] C. Cercignani [*J. Math. Phys.*, to appear].

Far from the imposed boundary condition at $x = 0$, we expect a Navier-Stokes flow with linear temperature gradient, $dT/dx = \text{const.}$, and velocity gradient, $du'/dx = \text{const.}$, $\boldsymbol{u}' = (u_y, u_z)$. Setting three of the constants equal to zero,

$$p_{xy} = p_{xz} = q_x = 0 \tag{10.8}$$

guarantees that the flow is bounded at infinity. The flow at infinity will then be a constant Maxwellian characterized by five parameters. To produce a flow which decays to zero at infinity, we must impose five additional conditions. There are two constants (10.7) left which we can set equal to zero, $u_x = p = 0$. This leaves three finite constants, $\boldsymbol{u}'$ and T, at infinity. Subtracting these from the given boundary function f^-, we are led to

THEOREM 4. *There exists a unique steady solution in the half space, vanishing at infinity and satisfying the boundary condition*

$$f^- = \phi^- + c_0\psi_0 - \boldsymbol{u}' \cdot \boldsymbol{\psi}' - (\tfrac{3}{2})^{1/2} T\psi_4 \tag{10.9}$$

where ϕ^- *is arbitrarily given and* $(c_0, \boldsymbol{u}', T)$ *are adjustable parameters.*

The interpretation of Theorem 4 is that to any given emitted distribution ϕ^- are assigned unique values of the slip velocity $\boldsymbol{u}'$ and temperature T. The parameter count (which would give *four* boundary conditions at infinity for the elementary boundary condition, $f^- =$ given) can be easily discovered as a plausible consequence of the basic Theorem 1 applied to a finite one-dimensional domain, say $0 < x < x_1$, on letting $x_1 \to \infty$.[15] The application of Theorem 2 to this domain is somewhat different. Two adjustable constants c_0 are available at the two ends even though $u_x = 0$ at one end implies $u_x = 0$ at the other. The counting is correct, however, since the second constant c_0 is needed to adjust the total mass, $\int \rho\, dx$.

We shall also refer to the

COROLLARY. *Theorem 4 holds if the Boltzmann equation is replaced by an inhomogeneous equation*

$$v_1(\partial f/\partial y) + L(f) = \psi \tag{10.10}$$

where ψ *decays rapidly as* $y \to \infty$. *In this case the moments* (10.7) *are not constant but are taken to be zero at* $y = \infty$.

The slip coefficients, $\boldsymbol{u}'$ and T in (10.9), will depend linearly on the given boundary function ϕ^- and the inhomogeneous term ψ.

Let us now consider a general domain D with smooth boundary, taken with the diffuse reflection boundary condition (10.5), $\boldsymbol{u}'_w$ and T_w being specified on the boundary; (the boundary may be interior or exterior and may have disjoint components, as, for example, in a flow between two cylinders or spheres). The Boltzmann equation is to be solved in the original scaling,

$$\boldsymbol{v} \cdot \partial f/\partial \boldsymbol{x} + (1/\varepsilon)L(f) = 0. \tag{10.11}$$

[15] In a special case in which the parameter count is simpler, this theorem has been proved for a model of the Boltzmann equation; Y. P. Pao [*Arch. Rational Mech. Anal.*, to appear].

As in the initial layer analysis, we iterate back and forth between the interior and boundary layer. To start, we solve the Navier-Stokes equations in the given domain D subject to $u_n = 0$ and the given boundary conditions for $\boldsymbol{u}'_w$ and T_w. Using this solution we construct a Chapman-Enskog function $f_0 + \varepsilon f_1$; f_0 is a local Maxwellian with the Navier-Stokes parameters, and f_1 is a more complicated function of $\boldsymbol{v}$ (obtained by solving integral equations) involving gradients of the Navier-Stokes velocity and temperature. The Maxwellian function f_0 satisfies the kinetic boundary condition (10.5) exactly (and determines a specific value for the parameter c_0). The difference between the exact solution to the Boltzmann equation and the Navier-Stokes approximation,

$$f - (f_0 + \varepsilon f_1) = \varepsilon g, \tag{10.12}$$

satisfies an inhomogeneous Boltzmann equation (related to the Burnett approximation) and is subject to a boundary condition induced by the Navier-Stokes residual f_1,

$$g^- = c_1\psi_0 - f_1^-. \tag{10.13}$$

Turning to the one-dimensional boundary layer equation (which to this order is simply the homogeneous one-dimensional Boltzmann equation), we find a solution g_0 (Theorem 4) which decays at infinity and satisfies the boundary condition

$$g_0^- = -f_1^- + c_1\psi_0 - \boldsymbol{u}'_1 \cdot \boldsymbol{\psi}' - \left(\frac{3}{2}\right)^{1/2} T_1\psi_4. \tag{10.14}$$

Having determined the slip coefficients $\boldsymbol{u}'_1$ and T_1 (also c_1), we then compute a Navier-Stokes function g_1 in D which takes the boundary values $\boldsymbol{u}'_1$, T_1 (and $u_n = 0$). The sum, $g_0 + g_1 = g$ satisfies the boundary condition (10.13) (including the determination of c_1) and leaves a residue which is $O(\varepsilon^2)$.

The first-order slip coefficients $\boldsymbol{u}'_1$ and T_1 are computed as the solution of a one-dimensional problem (Theorem 4) with a boundary condition f_1^- which is taken from the first-order Chapman-Enskog function. This function, f_1, is a linear combination of velocity and temperature gradients, $\partial u_i/\partial x_j$ and $\partial T/\partial x_i$ (or p_{ij} and q_i), taken from the zero-order Navier-Stokes solution. A scalar T_1 and tangential vector $\boldsymbol{u}'_1$ can be combined with the normal n_i as linear functions of p_{ij} and q_i only in the combinations

$$\begin{aligned} T_1 &= \alpha\tau_n + \beta q_n, \\ \boldsymbol{u}'_1 &= \gamma\boldsymbol{\tau}' + \delta\boldsymbol{q}', \end{aligned} \tag{10.15}$$

where $\tau_i = p_{ij}n_j$ is the vector force on the wall, and $\tau_n = \boldsymbol{\tau} \cdot \boldsymbol{n}$, $\boldsymbol{\tau}' = \boldsymbol{\tau} - \boldsymbol{n}\tau_n$ are its normal and tangential components. The coefficient β is the conventional temperature slip due to normal heat flow, and γ describes the slip velocity $\boldsymbol{u}'_1$ induced by the tangential stress $\boldsymbol{\tau}'$. The slip $\boldsymbol{u}'_1$ caused by $\boldsymbol{q}'$ is called thermal creep.

The term $\alpha\tau_n$ is ostensibly of higher order since τ_n always vanishes at the boundary in any Navier-Stokes steady flow with no-slip boundary conditions.[16]

[16] This was pointed out to the author by Y. Sone.

But it would be retained in a time-dependent boundary value problem, or in any problem (such as with the thirteen moment equations) in which no strict ordering with respect to a parameter is followed.

The formulation for second and higher order slip boundary conditions is quite similar. As in the analysis of the initial layer, we note that both the interior and boundary layer expansions individually satisfy the Boltzmann equation without interference. The two expansions are coupled only through boundary conditions. The interior (Hilbert) expansion can be written

$$\begin{aligned} f_H = f_0^0 &+ \varepsilon f_1^0 + \varepsilon^2 f_2^0 + \cdots \\ &+ \varepsilon f_0^1 + \varepsilon^2 f_1^1 + \cdots \\ &+ \varepsilon^2 f_0^2 + \cdots \end{aligned} \tag{10.16}$$

where f_0^0, f_0^1, f_0^2 are locally Maxwellian; f_1^0 and f_2^0 are the Navier-Stokes and Burnett residues induced by f_0^0; f_1^1 is the Navier-Stokes residue induced by f_0^1. All the terms on a given line in (10.16) are uniquely determined once the leading (Maxwellian) term is specified. The Maxwellian f_0^0 is a solution of the Navier-Stokes equations subject to the given boundary conditions, $\boldsymbol{u}'_w$ and T_w. The higher order Maxwellians f_0^1 and f_0^2 satisfy the Navier-Stokes equations, but with slip boundary conditions $\boldsymbol{u}'_1$, T_1 and $\boldsymbol{u}'_2$, T_2 which are to be found from the boundary layer analysis.

The determination of f_0^1 (from the first order boundary conditions, $\boldsymbol{u}'_1$ and T_1) is described above. The second order boundary conditions for f_0^2 are found (Theorem 4, Corollary) by solving the inhomogeneous one-dimensional boundary layer equation

$$v_1(\partial g^1/\partial y) + L(g^1) = -\boldsymbol{v}' \cdot \partial g^0/\partial \boldsymbol{x}', \tag{10.17}$$

where g^0 is the already known first order boundary layer solution [called g_0 in (10.14)], and g^1 is to be found subject to the boundary condition

$$(g^1)^- = -(f_2^0 + f_1^1)^- + c\psi_0 - \boldsymbol{u}'_2 \cdot \boldsymbol{\psi}' - (\tfrac{3}{2})^{1/2} T_2 \psi_4. \tag{10.18}$$

The Navier-Stokes term $(f_1^1)^-$ in the boundary condition leads to exactly the same slip coefficients $(\alpha, \beta, \gamma, \delta)$ as in (10.15). The Burnett function f_2^0 involves second derivatives of $\boldsymbol{u}^0$ and T^0 (or first derivatives of q_i^0 and p_{ij}^0) and gives rise to seven slip coefficients:

$$\begin{aligned} T_2 &= ak_n + b\sigma_n + c(\boldsymbol{n} \cdot \nabla p), \\ \boldsymbol{u}'_2 &= d\boldsymbol{k}' + e\boldsymbol{\sigma}' + f\nabla' p + g\boldsymbol{\theta}' \end{aligned} \tag{10.19}$$

where

$$\begin{aligned} \sigma_i &= n_r n_s(\partial p_{ir}/\partial x_s), \\ \theta_i &= n_r n_s(\partial p_{rs}/\partial x_i) \qquad (\sigma_n = \theta_n), \\ k_i &= n_r(\partial q_i/\partial x_r). \end{aligned} \tag{10.20}$$

The terms in ∇p are alternative forms equivalent to $\partial p_{ij}/\partial x_j$. The function g^0 (generated from f_1^0 as a boundary condition) takes the form

$$g^0 = p_{ij}(x_2, x_3)v_i v_j \phi(v_1, v_2^2 + v_3^2, y) + q_i(x_2, x_3)v_i \psi(v_1, v_2^2 + v_3^2, y). \tag{10.21}$$

In evaluating $\boldsymbol{v}' \cdot \partial g^0/\partial \boldsymbol{x}'$ for (10.17) we have contributions from $\boldsymbol{v}' \cdot \partial p_{ij}/\partial \boldsymbol{x}'$ and $\boldsymbol{v}' \cdot \partial q_i/\partial \boldsymbol{x}'$ which are easily seen to be included within the format (10.19) (in fact, only five of the possible seven slip coefficients arise from this source, since $\boldsymbol{\sigma}$ and ∇p occur only through the combination $\boldsymbol{\sigma} + \nabla p$). But there are also contributions from $\boldsymbol{v}' \cdot \partial\phi/\partial \boldsymbol{x}'$ and $\boldsymbol{v}' \cdot \partial\psi/\partial \boldsymbol{x}'$ which are derivatives holding $\boldsymbol{v}$ fixed in a given rectangular coordinate system, whereas the components $v_1 = \boldsymbol{v} \cdot \boldsymbol{n}$ and $v_2^2 + v_3^2 = |\boldsymbol{v} \times \boldsymbol{n}|^2$ vary with the normal $\boldsymbol{n}$ to the boundary. A simple computation gives the result

$$\boldsymbol{v}' \cdot \partial\phi/\partial \boldsymbol{x}' = (\phi_1 - 2v_n\phi_2)(v_i \kappa_{ij} v_j) \tag{10.22}$$

and similarly for ψ, where ϕ_1 and ϕ_2 are the derivatives of ϕ with respect to its arguments v_1 and $v_2^2 + v_3^2$ respectively and

$$\kappa_{ij} = \kappa' a_i a_j + \kappa'' b_i b_j \tag{10.23}$$

is a surface tensor which describes the curvature of the boundary of the domain; κ' and κ'' are the principal curvatures along the principal directions a_i and b_i; curvature is positive toward the interior of the domain. From the products $p_{ij}\kappa_{rs}$ and $q_i\kappa_{rs}$ in $\boldsymbol{v}' \cdot \partial g^0/\partial \boldsymbol{x}'$ we obtain seven further slip coefficients,

$$\begin{aligned} T_2 &= \alpha_2 \tau_n \bar{\kappa} + \beta_2 q_n \bar{\kappa} + \lambda_2 p_{rs}\kappa_{rs}, \\ \boldsymbol{u}_2' &= \gamma_2 \boldsymbol{\tau}' \bar{\kappa} + \delta_2 \boldsymbol{q}' \bar{\kappa} + \mu_2 \boldsymbol{\kappa} \cdot \boldsymbol{\tau} + \nu_2 \boldsymbol{\kappa} \cdot \boldsymbol{q}, \end{aligned} \tag{10.24}$$

where

$$\bar{\kappa} = \tfrac{1}{2}\kappa_{rr} = \tfrac{1}{2}(\kappa' + \kappa'') \tag{10.25}$$

is the mean curvature (note that $\boldsymbol{\kappa} \cdot \boldsymbol{\tau} = \kappa_{ij}\tau_j$ and $\boldsymbol{\kappa} \cdot \boldsymbol{q}$ are necessarily tangential to the boundary). The four coefficients $\alpha_2, \beta_2, \gamma_2, \delta_2$ are similar in effect to $\alpha, \beta, \gamma, \delta$, but with an extra factor $\bar{\kappa}$.

Since the second order boundary layer equation (10.17) is inhomogeneous, the moments (10.7) which are zero at $y = \infty$ are finite at $y = 0$ (cf. Corollary to Theorem 4). The normal velocity, u_n, is therefore different from zero [the adjustable constant c in (10.18) is required to make $u_n = 0$ at $y = \infty$]. We have an effective displacement of the boundary given by the Navier-Stokes slip condition

$$u_n = \hat{a}(k_n + 2\bar{\kappa}q_n) + \hat{b}(\sigma_n + 2\bar{\kappa}\tau_n + \partial p/\partial x_n - \kappa_{rs}p_{rs}). \tag{10.26}$$

The presence of only two slip coefficients is a consequence of $\nabla \cdot \boldsymbol{u} = 0$ which allows u_n to be evaluated in terms of the first order mean slip velocity $\hat{\mathbf{u}}' = \int_0^\infty \boldsymbol{u}'(y)\,dy$ [obtained from $\int_0^\infty g^0(y)\,dy$ in (10.21)] using $\oint \hat{\mathbf{u}}' \cdot d\boldsymbol{x}$ for paths on the boundary. In computing drag, heat transfer at the boundary, etc., the jump in the values of other moments (10.17) across the boundary layer would be needed.

The slip coefficients α_2 and λ_2 in (10.24) and the terms $2\bar{\kappa}\tau_n - \kappa_{rs}p_{rs}$ in (10.26) can be considered to be of higher order just as was pointed out for the first order coefficient α; also the coefficients μ_2 in (10.24) and g in (10.19) coalesce since $\theta' + 2\boldsymbol{\tau} \cdot \boldsymbol{\varkappa} = 0$ when $\tau_n = 0$. (See footnote 16.) But α must be kept as a second order coefficient.

In addition to the explicit appearance of boundary curvature in (10.24), it is also implicit in the Navier-Stokes expression for p_{ij} in terms of $\partial u_i/\partial x_j$ and also in the global relation between T_w and the boundary values of q_n and k_n through the solution of an elliptic equation.

A rough estimate can be made of the first order slip coefficients following Maxwell; for diffuse reflection (cf. **[15]**)

$$\alpha = -\tfrac{1}{4}, \qquad \beta = (\pi/8)^{1/2} \sim 0.627$$
$$\gamma = (\pi/2)^{1/2} \sim 1.235, \qquad \delta = -\tfrac{1}{5}. \tag{10.27}$$

More accurate values have been computed for β and γ using half range polynomial expansions; for Maxwellian molecules **[63]**

$$\beta \sim 0.766 \qquad \gamma \sim 1.48 \tag{10.28}$$

For a relaxation model of the Boltzmann equation essentially precise values have been calculated for β **[56]**, **[45]**, for γ **[59]**, **[1]**, and for δ **[46]**,

$$\beta \sim 0.737 \qquad \gamma \sim 1.437 \qquad \delta \sim -0.306 \tag{10.29}$$

and for five of the sixteen second order coefficients; $e - f$ **[9]**, $(2b - c)$ and $\hat{b}$ **[49]**, $(\gamma_2 + 2e - 4f)$ and δ_2 [Sone and Yamamato, private communication],

$$2b - c \sim -0.1980\lambda, \qquad e - f \sim 0.960\lambda,$$
$$\gamma_2 + 2e - 4f \sim 2.55\lambda, \qquad \delta_2 \sim 0.189\lambda, \tag{10.30}$$
$$\hat{b} \sim 0.2927\lambda.$$

The normalization (10.15) in terms of stress and heat flow instead of velocity and temperature gradients makes the values of the first order slip coefficients independent of the essentially arbitrary parameter which enters into a relaxation model. But the multiplicative factor for the second order coefficients depends on this parameter, i.e. on the definition of a mean free path [which in (10.30) is $\lambda = (8RT/\pi)^{1/2}/\nu$].

We have found by consistent formal expansion that the Navier-Stokes system with no-slip boundary conditions is followed, at next order, by an inhomogeneous Navier-Stokes system (comparable to Burnett) subject to the slip boundary conditions (10.15). In particular, slip boundary conditions are formally inconsistent with use of the ordinary Navier-Stokes equations. This is in contrast to the initial value problem where the first occurrence of the *homogeneous* Navier-Stokes equations is accompanied by nontrivial slip initial conditions. The reason for the discrepancy is that the standard diffuse reflection boundary condition states that f^- is a local Maxwellian. If the initial function $f(\boldsymbol{v}, \boldsymbol{x}, 0)$ is taken to be locally Maxwellian, the initial slip also recedes one notch, and it appears for the first time at the Burnett (or inhomogeneous Navier-Stokes) level.

We have presented the boundary layer theory for a Stokes linearization of the fluid equations. It is applicable after trivial modifications [viz. restoring $\partial p_{ij}/\partial x_j$ for $\partial p/\partial x_i$ in (10.19)] for an Oseen linearization and, to first order, even for the nonlinear steady Navier-Stokes equations. In the nonlinear case, the expansion becomes one in $\varepsilon^{1/2}$ rather than ε. The linear first order slip theory is valid since the kinetic layer is thinner than the viscous layer by a factor $O(\varepsilon^{1/2})$ (we require that $M\varepsilon^{1/2}$ be small). The entire analysis is for smooth boundaries; the behavior near a corner, leading edge, etc., requires separate investigation. In a time-dependent problem, the first order slip boundary condition (10.15) is unchanged (after leaving the initial layer). The second order theory is modified by the occurrence of an additional inhomogeneous term, $-\partial g^0/\partial t$, in (10.17).

Implicit in the analysis made above is a conjectured existence theorem for the steady Navier-Stokes equations in an interior domain D given the vector velocity and temperature on the boundary and the total mass in D.

11. Infinitely strong shock profile. We recall that for a weak shock the Navier-Stokes profile is explicit and universal **[50]**, **[14]**,

$$\phi(x) = \tanh \varepsilon x/\lambda^*. \tag{11.1}$$

Here ϕ represents any of the flow variables, density, pressure, temperature, or velocity, suitably normalized to vary between -1 and $+1$; ε is a shock strength parameter, $0 < \varepsilon < 1$, e.g.

$$\varepsilon = (p_1 - p_0)/(p_1 + p_0) \tag{11.2}$$

(the subscripts 0 and 1 refer to the front-upstream and back-downstream respectively); λ^* is a specific mean free path. Expressed in terms of the viscosity coefficient

$$\lambda^* = (35/12)a_0(\mu/m) \tag{11.3}$$

where $m = \rho u$ is the mass flow constant and a_0 is a numerical constant which is exactly unity for Maxwellian molecules ($\mu \sim T$) and differs from unity by less than five percent for all potentials between Maxwellian and hard spheres ($\mu \sim T^{1/2}$).

The weak shock profile can also be found from the Boltzmann equation by a formal expansion in ε.[17] Scaling $\varepsilon x = y$ yields a Boltzmann equation

$$v(\partial f/\partial y) = (1/\varepsilon)Q(f, f) \tag{11.4}$$

in which the parameter ε appears exactly as in the Hilbert theory. But, as in the steady flow linearization (§8), the fluid compatibility relations are degenerate, and

[17] P. N. Hu, private communication.

we do not obtain the standard Hilbert or Chapman-Enskog sequence of fluid equations. The steady flow Euler equations in one dimension can be completely integrated. The compatibility relations are therefore essentially algebraic rather than differential equations. The lowest order compatibility requirement (corresponding to a local Maxwellian f_0) yields a constant state. To next order, the variational Euler equations are a linear homogeneous algebraic system; the condition for a nontrivial solution (vanishing determinant) is that the zero order constant state be sonic. We recognize this as a bifurcation equation; in searching for a weak shock, we are looking for a nonunique solution which branches off from the trivial constant state. To next order, the classical G.I. Taylor solution (11.1) is found. This expansion in ε is not strictly comparable to Navier-Stokes, Burnett, etc. since the compatibility relations are not a sequence of fluid dynamic equations. But a similar singular perturbation applied to the Navier-Stokes equations will yield the same leading term; applied to the thirteen moment equations it yields two correct terms, etc. (See footnote 17.)

For finite shock strength the Navier-Stokes profile can be solved numerically by integration of a simple plane direction field **[12]** or by expansion in ε **[14]** or even explicitly for the special value $a_0 = 20/21$ **[2]**, **[51]**. A paradoxical early result was that the shock thickness for strong shocks could become arbitrarily small or arbitrarily large compared to the mean free path, depending on the temperature dependence of the viscosity coefficient. This difficulty was resolved **[14]** by the simple expedient of introducing the downstream mean free path, λ_1, as the reference length instead of the upstream mean free path λ_0. Not only is the limiting thickness finite in terms of λ_1, but there is a definite limiting profile for each of the flow variables. The importance of this choice of scaling lies in the fact that λ_1/λ_0 is unbounded in a strong shock (except in the special case of hard spheres).

We now suggest that for the Boltzmann equation there is a finite limiting profile at infinite strength for the distribution function $f(v, x)$ when the correct scaling is introduced in velocity space, viz. at the downstream sound-speed, $c_1 = (\frac{5}{3}RT_1)^{1/2}$. With this scaling, the upstream Maxwellian reduces to a δ-function ($c_0 \sim 0$) centered at $u_0/c_1 = 4/5^{1/2}$; cf. Figure 1, where the upstream Maxwellian is indicated by constant f contours centered at $u_1/c_1 = 1/5^{1/2}$. The crucial point is that the entire upstream distribution is at a definite thermal velocity when measured on the downstream scale. Even though the mean free path is strongly velocity dependent, the mean free path of an upstream particle when considered to impinge upon the downstream distribution is, within a factor of order unity, equal to the mean free path for self relaxation of the upstream distribution.

Based on this qualitative remark, we introduce the *Ansatz*

(11.5) $$\bar{f} = f_\delta + f$$

where

(11.6) $$f_\delta = \rho_\delta(x)\, \delta(\boldsymbol{v} - \boldsymbol{u}_0)$$

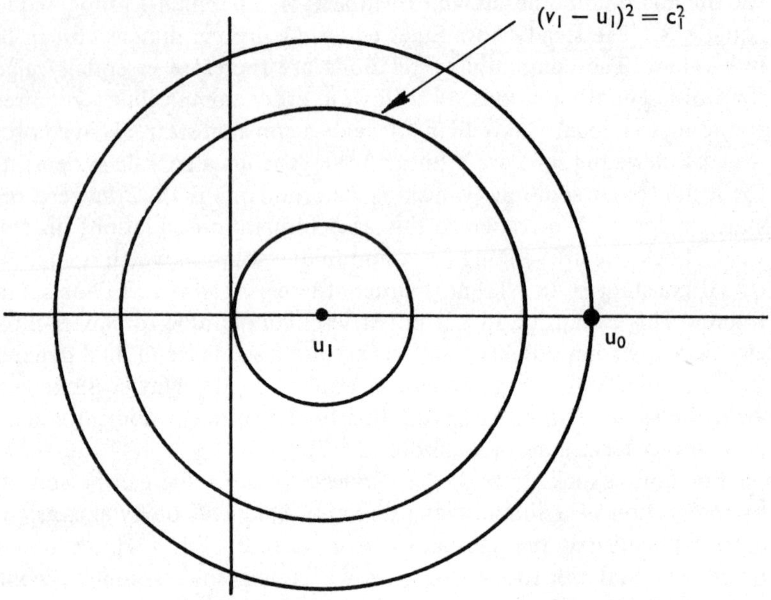

FIGURE 1.

and f is taken to be nonsingular on the downstream scale (it may be unbounded near $v = u_0$, but it is, nevertheless, an integrable function and not a distribution).

For any cutoff potential we write

$$Q(\bar{f},\bar{f}) = Q_1(\bar{f},\bar{f}) - \bar{\nu}\bar{f} \tag{11.7}$$

where

$$Q_1(\bar{f},\bar{f}) = \int \bar{f}'\bar{f}'\, V\, d\omega dv_1, \tag{11.8}$$

$$\bar{\nu}(x) = \int \bar{f}_1\, V\, d\omega dv_1 = \nu + \nu_\delta \tag{11.9}$$

in a standard notation. In the expansion of the collision term $Q(f_\delta + f, f_\delta + f)$ we identify $Q(f_\delta, f_\delta)$ and νf_δ as δ-functions (a collision between two molecules in f_δ remains within f_δ) and all other terms as legitimate functions on the scale f. Consequently, the Boltzmann equation splits into

$$u_0(\partial f_\delta/\partial x) + \nu_0 f_\delta = Q(f_\delta, f_\delta) \tag{11.10}$$

and

$$v(\partial f/\partial x) + \nu_\delta f - 2Q_1(f_\delta, f) = Q(f,f) \tag{11.11}$$

where, in $\nu(v, x)$ we set $v = u_0$,

$$\nu_0(x) = \nu(u_0, x). \tag{11.12}$$

The mass conservation moment of (11.10) is

$$u_0(d\rho_\delta/dx) + \nu_0\rho_\delta = 0. \tag{11.13}$$

We have

$$\rho_\delta(x) = \rho_0 \exp\left(-\frac{1}{u_0}\int_{-\infty}^{x} \nu_0\, dx\right) \tag{11.14}$$

and therefore f_δ (11.6) is given explicitly in terms of the "moment" ν_0 of f. Substituting this expression for f_δ into (11.11), we obtain a nonlinear integro-differential equation in f alone. This Boltzmann-like equation, provided that it has a solution satisfying the boundary conditions

$$\begin{aligned} f(-\infty) &= 0, \\ f(+\infty) = f_1 &= \frac{\rho_1}{(2\pi RT_1)^{3/2}} \exp\left[-\frac{(v-u_1)^2}{2RT_1}\right] \end{aligned} \tag{11.15}$$

gives *the* solution to the infinite strength shock profile. No shock strength parameter remains, and the downstream parameters ρ_1, u_1, RT_1, are easily scaled out of the problem. The only parameters that remain are those implicit in the molecular force law. The equation (11.11) for f is, in appearance, at least as complicated as the original Boltzmann equation. But it has the great advantage of being nonsingular, or, at least, less singular than the Boltzmann equation. In particular, this will allow us to use naive approximations for f which would be worthless for the evaluation of $\bar{f}$.

The simplest approximation to any distribution f is the locally Maxwellian; in one dimension this involves three moments. For the total distribution, $\bar{f}$, this is too crude to give a profile at any strength; it leads at most to the jump conditions and a discontinuous shock front. Applied to f, this approximation becomes valuable.

Taking the three moments $(1, v_1, \frac{1}{2}v^2)$ of the full Boltzmann equation, without approximation, yields conservation of mass, momentum, and energy,

$$\begin{aligned} \int v_1\bar{f}dv &= m, \\ \int v_1^2\bar{f}dv &= P, \\ \int \tfrac{1}{2}v_1v^2\bar{f}dv &= Q, \end{aligned} \tag{11.16}$$

where m, P, Q are constants. For f_δ these moments (x-dependent) are

$$\begin{aligned} \int v_1 f_\delta dv &= \rho_\delta u_0, \\ \int v_1^2 f_\delta dv &= \rho_\delta u_0^2, \\ \int \tfrac{1}{2}v_1v^2 f_\delta dv &= \tfrac{1}{2}\rho_\delta u_0^3. \end{aligned} \tag{11.17}$$

For f, using the locally Maxwellian approximation, we have

$$\int v_1 f dv = \rho u,$$

$$\int v_1^2 f dv = \rho u^2 + p, \tag{11.18}$$

$$\int \tfrac{1}{2} v_1 v^2 f dv = \tfrac{1}{2}\rho u^3 + \tfrac{5}{2} p u,$$

where $\rho(x)$, $u(x)$, $p(x)$ are the standard moments of f. The constants m, P, Q can be evaluated at $x = -\infty$ where $\bar{f} = f_\delta$,

$$m = \rho_0 u_0, \qquad P = \rho_0 u_0^2, \qquad Q = \tfrac{1}{2}\rho_0 u_0^3. \tag{11.19}$$

Adding (11.17) to (11.18), we obtain

$$\begin{aligned} \rho_\delta u_0 + \rho u &= \rho_0 u_0, \\ \rho_\delta u_0^2 + \rho u^2 + p &= \rho_0 u_0^2, \\ \rho_\delta u_0^3 + \rho u^3 + 5pu &= \rho_0 u_0^3. \end{aligned} \tag{11.20}$$

Eliminating p and ρ from these equations leaves an equation in u,

$$(\rho_\delta - \rho_0) u_0 (4u^2 - 5uu_0 + u_0^2) = 0. \tag{11.21}$$

If ρ_δ is not identically equal to ρ_0, then

$$4u^2 - 5uu_0 + u_0^2 = 0. \tag{11.22}$$

This algebraic equation has one solution $u(x) = u_0$, which we dismiss as being incompatible with the boundary condition at plus infinity, and another solution

$$u(x) = \tfrac{1}{4} u_0 = u_1. \tag{11.23}$$

Returning to (11.20), we also find

$$\rho = 4(\rho_0 - \rho_\delta), \qquad p = \tfrac{3}{4}(\rho_0 - \rho_\delta) u_0^2. \tag{11.24}$$

In particular,

$$RT = p/\rho = (3/16) u_0^2 = RT_1. \tag{11.25}$$

Both u and T are constant throughout the flow, leaving $\rho(x)$ as the only nontrivial moment. This is the entire information to be gleaned from the moment equations, but there remains equation (11.13) for ρ_δ. The moment ν_0 of f is a function of temperature and is proportional to ρ, $\nu_0 = \alpha(T)\rho$. Since $T = T_1$ is constant, we can write

$$\nu_0 = \alpha_1 \rho \tag{11.26}$$

where α_1 is a constant, and, between (11.13) and (11.24) obtain

$$d\rho_\delta / dx = -(4\alpha_1 / u_0)(\rho_0 - \rho_\delta)\rho_\delta. \tag{11.27}$$

This has the solution

$$\rho_\delta = \rho_0 e^{-y}/(1 + e^{-y}) \tag{11.28}$$

and

$$\rho = \rho_1/(1 + e^{-y}), \tag{11.29}$$

$(\rho_1 = 4\rho_0)$, also

$$\bar{\rho} = \rho + \rho_\delta = \frac{\rho_1 + \rho_0 e^{-y}}{1 + e^{-y}} = \rho_0[5/2 + (3/2)\tanh y/2] \tag{11.30}$$

where

$$y = x/\hat{\lambda} \tag{11.31}$$

and

$$\hat{\lambda} = u_0/\alpha_1\rho_1 = u_0/\hat{\nu}. \tag{11.32}$$

The quantity $\hat{\nu} = \alpha_1\rho_1$ is the collision frequency of a molecule at speed u_0 in a Maxwellian f_1; or, it is the collision frequency of a molecule at speed $u_0 - u_1 = 3u_1 = (3RT_1)^{1/2}$ in a rest frame Maxwellian at density ρ_1 and temperature T_1.

The net velocity $\bar{u}$ is easily obtained from $\bar{\rho}\bar{u} = \rho_0 u_0$,

$$\bar{u} = u_0(1 + e^{-y})/(4 + e^{-y}). \tag{11.33}$$

The total moments $\bar{p}$ and $R\bar{T}$ are obtained by integrating over $\bar{f} = f + f_\delta$,

$$\begin{aligned} \bar{p} &= p_1(1 + \tfrac{1}{2}e^{-y})/[(1 + e^{-y})(1 + \tfrac{1}{4}e^{-y})], \\ \bar{T} &= T_1(1 + \tfrac{1}{2}e^{-y})/(1 + \tfrac{1}{4}e^{-y})^2. \end{aligned} \tag{11.34}$$

The $\bar{T}$ profile is flatter at the downstream end than any other variable; the departure from T_1 is e^{-2y} instead of e^{-y}. This is particularly interesting because a very slight overshoot of $\bar{T}$ is sometimes found at this end [**40**]. Our approximation is too crude to decide this question. But the higher order departure of $T_1 - \bar{T}$ indicates that this effect may be quite sensitive to the exact approximation used. Separate parallel and transverse temperatures are sometimes defined; they are

$$\begin{aligned} T_{\parallel} &= T_1(1 + e^{-y})/(1 + \tfrac{1}{4}e^{-y})^2, \\ T_{\perp} &= T_1/(1 + \tfrac{1}{4}e^{-y}). \end{aligned} \tag{11.35}$$

The greatest anisotropy is at the upstream end where $T_{\parallel}/T_{\perp} \sim 4$.

The fact that this very crude approximation yields any answer at all indicates that a moderate improvement, e.g. equivalent to thirteen moments (or a modification such as that of [**4**]) should yield quite accurate results. Investigation of the shock equation (11.11) for the behavior at $x \to \pm\infty$ and $v \to \infty$ might suggest better approximations. Also, if f is unbounded at $v = u_0$ [this appears likely from the term $Q_1(f, f_\delta)$ in (11.11)] a bimodal approximation to f would seem to be called for. For f this would involve a sum of three terms, two peaked at u_0 with spreads of T_0 and T_1 and one at u_1; this is essentially trimodal.

Our three-moment approximation is closely related to the Mott-Smith bimodal assumption. But the specific relation (11.14) for ρ replaces the more or less arbitrary moment equation which is needed to supplement the Mott-Smith prescription; also, the fact that u and T are constant arises naturally rather than by fiat.

Elaborations of the Mott-Smith approximation (the most accurate is probably [**40**]) are found to be consistent with our statement that the correct length scale for a strong shock is the downstream mean free path. It is likely that any of these *ad hoc* approximations with enough adjustable parameters is fairly accurate. But an equivalent amount of effort is likely to produce much more accurate results if applied to the infinite strength equation (11.11). This has yet to be done.

The separation (11.5) into a "molecular beam" f_δ plus remainder has been introduced, independently, by I. M. Scholnick and D. L. Turcotte (private communication). The remainder, f, is approximated by using the Navier-Stokes equations. This use of the Navier-Stokes equations can only be considered as an *ad hoc* procedure, possibly more accurate than our local Maxwellian, but less easily assessed than, say, a thirteen moment approximation to f. In the present context the Navier-Stokes approximation for second and third moments of f has no intrinsic significance, as it does in a small mean free path expansion.

For the determination of f, it is sufficient to know that f_δ is concentrated near $v = u_0$. But as a separate problem, we can investigate the structure of the distribution f_δ on the small scale $v - u_0 \sim (RT_0)^{1/2}$. For hard potentials (harder than a Maxwellian fifth power), self-collisions between f_δ and itself occur at a much slower rate than f_δ with f. On the distance scale that f_δ decays to zero, we can set $(Qf_\delta, f_\delta) = 0$. Equation (11.10) can be solved

$$f_\delta = f_0 \exp\left(-\frac{1}{u_0}\int_{-\infty}^{x} \nu_0\, dx\right), \tag{11.36}$$

and we find that f_δ remains Maxwellian while it decays.

The fact that the approximate profile (11.30) for an infinite strength shock is a hyperbolic tangent just like the exact weak shock profile, allows us to make a very simple interpolation for the shock thickness that should be accurate to better than a factor two for all strengths. The Navier-Stokes weak shock formula (11.1) extrapolates to an infinite strength shock thickness ($\varepsilon = 1$) of just λ^*; (we, of course, use the downstream temperature for $\mu(T)$ to evaluate λ^*). The equivalent thickness from the present theory is $2\hat{\lambda}$ (11.30). For hard spheres a simple calculation shows that $\hat{\lambda} \sim 0.98\lambda^*$. In other words, the infinite strength shock thickness estimated from (11.11) by a local Maxwellian approximation is just twice the weak Navier-Stokes thickness, λ^*, extrapolated to infinite strength.[18]

---

[18] The Navier-Stokes thickness for hard spheres calculated exactly for an infinite strength shock is some 40% smaller than the weak shock extrapolation—see Figure 14 in [**14**]. Empirically, it is more accurate (as well as simpler) to extrapolate the weak shock formula than to calculate the correct Navier-Stokes profile.

The hyperbolic tangent profile is exact for a weak shock but can only be expected to be a rough approximation for strong shocks. However, the very strong asymmetry that is found in strong shock profiles using a relaxation model for the Boltzmann equation **[29]** seems to be specific to the model itself. The creation term of the Boltzmann equation, $\int f'f'_1$, is replaced by a Maxwellian source at the local temperature, which approximates $T_0 \sim 0$ for the downstream part of the profile. This is quite different from our scaling which states that molecules are created with a velocity spread comparable to T_1 for the whole profile. More precisely, the correct upstream limiting values of the moments u and T of the residual distribution f are not known (they turn out to be u_1 and T_1 in our locally Maxwellian approximation to f). Only if it were to turn out that $u \to u_0$ and $T \to 0$ for the actual solution at minus infinity would our *Ansatz* (11.5) be qualitatively comparable with a model equation.

12. Conclusion. Taking fluid dynamics as our model, we can look at the theory of the Boltzmann equation as a subject which is several orders of magnitude more extensive, containing an endless variety of boundary layers and nonuniform limits. Enough evidence has been accumulated to indicate that a large part of the subject is accessible to standard methods, particularly asymptotics and boundary layer analysis. But some aspects, possibly including noncutoff potentials, nonselfadjoint spectra, nonlinear existence in the large, may require significantly new approaches.

In trying to project from the past, we can distinguish three qualitatively different eras: transport coefficients, *ad hoc* polynomial and moment methods, and the blossoming of more precise mathematical investigations.

Until 1949 the subject of kinetic theory was essentially coextensive with the Chapman-Enskog theory. The *raison d'être* of the Boltzmann equation was to compute transport coefficients in order to complete the classical equations of continuum mechanics. The phrase "solve the Boltzmann equation" is synonymous, in this era, with the operation of solving the Hilbert integral equation. The concept of the Boltzmann equation as an entity in itself, to be solved subject to initial and boundary conditions was missing (except for Carleman, whose singular relation to the subject was only revealed posthumously **[6]**).

The break with this preoccupation with transport coefficients was made by the introduction of the concept of interpolating different levels of description of a gas **[13]** between the (at that time) almost inaccessible Boltzmann description via a distribution function and the macroscopic description via fluid moments through the Chapman-Enskog hierarchy. Polynomial and moment approximations gave an *ad hoc* scheme for this interpolation.

In a few specific problems (viz. linearized and one-dimensional) high order polynomial expansions were carried out **[53]**, **[54]**, **[35]**, **[38]**. More accurate (but more special and more complicated) methods of parameter fitting were the introduction of half range polynomial expansions **[25]**, **[28]** and bimodal approximations of varying degrees of complexity **[34]**, **[40]**. Quite general schemes have been introduced which perturb by polynomial expansion a suitable weight function chosen to fit the problem [**27**], [**4**]. All these techniques are essentially *ad hoc*, and the ratio of accuracy to effort (convergence is unknown) is a sensitive function of the cleverness of the performer.

Mathematical precision has developed more recently in two directions: in general estimates of convergence (i.e. in existence theory) and in precise estimates of the behavior of solutions in special problems and singular limits.

The text of this paper serves as a summary of the increased precision which has developed in special problems and singular limits.

The early existence theory was confined to spatially homogeneous problems [**5**], [**57**], [**33**] or gave estimates which became intolerable after one or two collision times [**6**], [**15**]; or one or two mean free paths [**60**]. More recently a strong existence theory has been developed for the linearized initial value problem [**20**]. Weaker but very promising results have been obtained in boundary value problems [Cercignani, to appear], and a start has been made toward a nonlinear theory, [**23**] also [Pao, to appear].

The judicious application of *ad hoc* parameter fitting directed by the growing supply of precise information about singularities offers the most promising hope for accurate practical calculations. We have already pointed out the application to the fundamental solution and the strong shock wave. There are endless variations, such as to extend the precise result for high frequency sound [**24**] to lower frequencies where it may overlap the normal mode analysis from the other direction [**53**], [**38**], [**41**]; or to modify a half range expansion to incorporate the velocity boundary layer at $v = 0$.

Extrapolating towards the future, we submit the following as a suggestion with regard to the most promising directions and most urgent needs:

(1) stronger existence theorems for boundary value problems to try to bring the level up to what is known for the initial value problem;

(2) further exploitation of the abundance of known singular and nonuniform limits;

(3) development of numerical techniques to be able to make use of the many explicit answers (in near free flow) which are given as multidimensional integrals, and to serve as numerical experiments for comparison with theory;

(4) study of infinite range potentials (which is the real goal of all kinetic investigations), either to discover properties of the collision operator or to solve specific elementary flows.

It is safe to say that kinetic theory for monatomic rarefied gases has reached a stage where the obstructions are mathematical techniques rather than physical principles.

References

1. S. Albertoni, C. Cercignani, and L. Gotusso, Phys. Fluids **6** (1963), 993.

2. R. Becker, Z. Physik **8** (1922), 321.

3. D. Burnett, Proc. London Math. Soc., (2) **40** (1936), 382.

4. D. S. Butler and W. M. Anderson, "Shock structure calculations by an orthogonal expansion method," *Rarefied gas dynamics, fifth symposium*, vol. 1, Academic Press, New York, 1967.

5. T. Carleman, Acta Math. **60** (1933), 91.
6. ———, *Problèmes mathématiques dans la théorie cinétique des gaz*, Almqvist and Wiksells, Uppsala, 1957.
7. C. Cercignani, Ann. Phys. **20** (1962), 219.
8. ———, Phys. Fluids **10** (1967), 2097.
9. C. Cercignani and F. Sernagiotto, Phys. Fluids **9** (1966), 40.
10. L. Finkelstein, Phys. Fluids **8** (1965), 431.
11. E. A. Frieman and R. Goldman, J. Math. Phys. **7** (1966), 2153.
12. D. Gilbarg and D. Paolucci, J. Rational Mech. Anal. **2** (1953), 617.
13. H. Grad, Comm. Pure Appl. Math. **2** (1949), 331.
14. ———, Comm. Pure Appl. Math. **5** (1952), 257.
15. ———, "Principles of the kinetic theory of gases" in *Handbuch der Physik*, vol. XII, Springer, Berlin, 1958.
16. ———, "Equations of flow in a rarefied atmosphere" in *Proceedings of the conference on aerodynamics of the upper atmosphere*, Rand Corp., 1959.
17. ———, Comm. Pure Appl. Math. **14** (1961), 323.
18. ———, "Modern kinetic theory of plasmas" in *Proceedings of fifth international conference on ionization phenomena in gases*, (Munich, 1961), vol. 2, Amsterdam, North-Holland, 1962.
19. ———, Phys. Fluids **6** (1963), 147.
20. ———, "Asymptotic theory of the Boltzmann equation. II," *Rarefied gas dynamics, third symposium*, vol. 1, Academic Press, New York, 1963.
21. ———, "Accuracy and limits of applicability of solutions of equations of transport; Dilute monatomic gases," *Proceedings of international seminar on transport properties of gases*, (Brown University), 1964.
22. ———, "Theory of the Boltzmann equation," in *Proceedings of the eleventh international congress of applied mechanics*, (Munich), August, 1964.
23. ———, "Asymptotic equivalence of the Navier-Stokes and nonlinear Boltzmann equations" in *Proc. Sympos. Appl. Math.*, vol. XVII, Amer. Math. Soc., Providence, R.I., 1965.
24. ———, J. Soc. Indust. Appl. Math. **14** (1966), 935.
25. E. P. Gross and S. Ziering, Phys. Fluids **1** (1958), 215.
26. D. Hilbert, *Grundzüge einer allgemeinen Theorie der linearen Integralgleichungen*, Teubner, Vienna, 1912; also Chelsea, New York, 1953.
27. D. Kahn and D. Mintzer, Phys. Fluids **8** (1965), 1090.
28. L. Lees, *A kinetic theory description of rarefied gases*, GALCIT Memo. 51, California Institute of Technology, 1959.
29. H. W. Liepmann, R. Narasimha, and M. T. Chanine, Phys. Fluids **5** (1962), 1313.
30. G. Ya. Lyubarski, JETP **40** (1961), 1050.
31. R. J. Mason, "Forced sound propagation in gases of arbitrary density" in *Rarefied gas dynamics, fourth symposium*, vol. 1, Academic Press, New York, 1965.
32. J. A. McLennan, Phys. Fluids **8** (1965), 1580.
33. D. Morgenstern, Proc. Nat. Acad. Sci. U.S.A. **40** (1954), 719.
34. H. M. Mott-Smith, Phys. Rev. **82** (1951), 885.
35. ———, *A new approach to the kinetic theory of gases*, Lincoln Laboratory, MIT, 1954.
36. R. Narasimha, "The structure of the distribution function in gas kinetic flows" in *Rarefied gas dynamics, fifth symposium*, vol. 1, Academic Press, New York, 1967.
37. Y. P. Pao, "A uniformly valid asymptotic theory of rarefied flows under nearly free molecular conditions. II" in *Rarefied gas dynamics, fourth symposium*, vol. 1, Academic Press, New York, 1965.
38. C. L. Pekeris, Z. Alterman, L. Finkelstein, and K. Frankowski, Phys. Fluids **5** (1962), 1608.
39. R. F. Probstein and Y. S. Pan, "Shock structure and the leading edge problem" in *Rarefied gas dynamics, third symposium*, vol. 2, Academic Press, New York, 1963.
40. H. Salwen, G. E. Grosch, and S. Ziering, Phys. Fluids **7** (1964), 180.
41. L. Sirovich and J. K. Thurber, Phys. Fluids **6** (1963), 218.
42. J. J. Smolderen, "Near free molecular behavior of gases with infinite collision cross section," in *Rarefied gas dynamics, fourth symposium*, vol. 1, Academic Press, New York, 1965.
43. Y. Sone, Phys. Fluids **7** (1964), 470.
44. ———, J. Phys. Soc. Japan **19** (1964), 1463.
45. ———, J. Phys. Soc. Japan **21** (1966), 1620.
46. ———, J. Phys. Soc. Japan **21** (1966), 1836.
47. L. Spitzer, *Physics of fully ionized gases*, Interscience, New York, 1956.

48. C. H. Su, Phys. Fluids **7** (1964), 1867.
49. K. Tamada, J. Phys. Soc. Japan **22** (1967), 1284.
50. G. I. Taylor, Proc. Roy. Soc. Ser. A. **84** (1910), 371.
51. L. H. Thomas, J. Chem. Phys. **12** (1944), 449.
52. L. Trilling, Phys. Fluids **7** (1964), 1681.
53. C. S. Wang Chang and G. E. Uhlenbeck, *On the propagation of sound in monatomic gases* Engr. Research Inst., Univ. of Michigan, 1952.
54. ———, *The heat transport between two parallel plates as functions of the Knudsen number* Engr. Res. Inst., Univ. of Michigan, 1953.
55. H. Weitzner, "Steady state oscillations of a gas" in *Rarefied gas dynamics, fourth symposium*, vol. 1, Academic Press, New York, 1965.
56. P. Welander, Ark. Fys. **7** (1954), 507.
57. E. Wild, Proc. Cambridge Philos. Soc. **47** (1951), 602.
58. D. R. Willis, *A study of some nearly free molecular flow problems*, Aeronautical Engineering Laboratory, Report 440, Princeton University, 1958.
59. ———, *Linear Couette flow at arbitrary Knudson number*, KTH Aero. TN-52, Royal Inst. of Tech., Stockholm, 1960.
60. ———, "The effect of the molecular model on solutions to linearized couette flow with large Knudsen number" in *Rarefied gas dynamics, second symposium*, Academic Press, New York, 1961.
61. ———, *The effect of the molecular model on rarefied gas flows*, Rand Memo. RM-4638-PR, 1965.
62. E. Y. Yu, Phys. Fluids **10** (1967), 2466.
63. S. Ziering, Phys. Fluids **3** (1960), 503.

COURANT INSTITUTE OF MATHEMATICAL SCIENCES
NEW YORK UNIVERSITY.

Recent results in plasma kinetic theory[1]

Edward A. Frieman

I. **Introduction.** In a previous paper [**1**] the author attempted a summary of the state of development of some of the major facets of plasma kinetic theory up to 1965 and pointed out some of the unsolved problems and current speculations of workers in the field that existed at that time. The present review essentially consists of an updating of the previous summary with special emphasis on new results and more sophisticated ways of looking at old results. In §II we describe equilibrium, in §III nonequilibrium stable plasmas, in §IV nonequilibrium unstable plasmas, and in §V some new attempts at nonanalytic expansions.

II. **Basic equations and equilibrium.** The starting point for our considerations is the formal Liouville equation

$$(\partial/\partial t)f_{N,\infty} + \{f_{N,\infty}, H\} = 0 \tag{1}$$

governing the distribution function of all N particles and the infinity of radiation oscillators describing the radiation field. We define the usual reduced distribution functions

$$f_s = V^s \int dx^{N-s} \prod_{\lambda} dq_\lambda \, dp_\lambda f_{N,\infty} \tag{2}$$

normalized with the volume, V; and the ensemble average $\langle\psi\rangle$ of a phase function ψ,

$$\langle\psi\rangle = \int d\Gamma \, \psi f_{N,\infty}. \tag{3}$$

[1] This work was performed under the auspices of the Air Force Office of Scientific Research, Contract No. AF49(638)-1555.

The technique of Klimontovich [2] and Dupree [3] utilizes the particular phase function

(4) $$\rho = (1/n) \sum_{i=1}^{N} \delta(\boldsymbol{x} - \boldsymbol{x}_i)\delta(\boldsymbol{v} - \boldsymbol{v}_i)$$

normalized to the average density n. For simplicity we consider a system of electrons, radiation oscillators, and a smeared out ion background.

Upon using (2), (3), and (4) we find the chain of definitions

(5) $$\langle \rho \rangle = f_1,$$

(6) $$\langle \rho(1)\rho(2) \rangle = \Delta(1, 2)f_1(1) + f_2(1, 2),$$

(7) $$\langle \rho(1)\rho(2)\rho(3) \rangle = \Delta(1, 2)\, \Delta(1, 3)f_1(1) + \Delta(1, 2)f_2(1, 3) + \Delta(2, 3)f_2(1, 2) + \Delta(1, 3)f_2(2, 3) + f_3(1, 2, 3),$$

where

(8) $$\Delta(1, 2) = (1/n)\, \delta(\boldsymbol{x}_1 - \boldsymbol{x}_2)\, \delta(\boldsymbol{v}_1 - \boldsymbol{v}_2).$$

The equation of evolution for ρ is

(9) $$\left\{\partial/\partial t + \boldsymbol{v} \cdot \nabla_x - (e/m)(\boldsymbol{E} + \boldsymbol{v} \times \boldsymbol{B}/c) \cdot \nabla_v\right\}\rho = 0$$

to which must be appended the Maxwell equations

(10) $$\nabla \times \boldsymbol{E} = -(1/c)(\partial \boldsymbol{B}/\partial t),$$

(11) $$\nabla \times \boldsymbol{B} = -\frac{4\pi}{c}\,\text{en} \int . \boldsymbol{v}\rho + \frac{1}{c}\frac{\partial \boldsymbol{E}}{\partial t},$$

and the initial conditions

(12) $$\nabla \cdot \boldsymbol{E} = -4\pi\,\text{en} \int \rho \; d\boldsymbol{v} + 4\pi\,\text{en}.$$

(13) $$\nabla \cdot \boldsymbol{B} = 0.$$

These equations demonstrate a useful feature of the Klimontovich-Dupree technique in that the radiation field is taken into account in a compact manner without the necessity of explicitly introducing the radiation oscillators. A drawback, however, is that the statistics of the radiation field are not completely treated by the present formulation.

We now ensemble average (9) and obtain

(14) $$\frac{\partial}{\partial t}\langle \rho \rangle + \boldsymbol{v} \cdot \nabla_x \langle \rho \rangle - \frac{e}{m}\langle (\boldsymbol{E} + \boldsymbol{v} \times \boldsymbol{B}/c) \cdot \nabla_v \rho \rangle = 0.$$

In order to calculate the third term of (14) we must multiply (9) by the appropriate quantities and then ensemble average. It is clear that in this manner we build up a hierarchy of equations that is completely equivalent to the usual BBGKY hierarchy but includes the effects of radiation.

The usefulness of the technique is further enhanced by the decomposition

$$\rho = \langle\rho\rangle + \delta\rho, \tag{15}$$

$$\boldsymbol{E} = \langle\boldsymbol{E}\rangle + \delta\boldsymbol{E}, \tag{16}$$

$$\boldsymbol{B} = \langle\boldsymbol{B}\rangle + \delta\boldsymbol{B}, \tag{17}$$

which explicitly displays the fluctuations. We then find

(18)

$$\frac{\partial}{\partial t}\langle\rho\rangle + \boldsymbol{v}\cdot\nabla_x\langle\rho\rangle - \frac{e}{m}(\langle\boldsymbol{E}\rangle + \boldsymbol{v}\times\langle\boldsymbol{B}\rangle/c)\cdot\nabla_v\langle\rho\rangle = \frac{e}{m}\langle(\delta\boldsymbol{E} + \boldsymbol{v}\times\delta\boldsymbol{B}/c)\cdot\nabla_v\delta\rho\rangle$$

and

$$\begin{aligned}\left(\frac{\partial}{\partial t} + \boldsymbol{v}\cdot\nabla_x - \frac{e}{m}(\langle\boldsymbol{E}\rangle + \boldsymbol{v}\times\langle\boldsymbol{B}\rangle/c)\cdot\nabla_v\right)\delta\rho - \frac{e}{m}(\delta\boldsymbol{E} + \boldsymbol{v}\times\delta\boldsymbol{B}/c)\cdot\nabla_v\,\delta\rho \\ = \frac{e}{m}(\delta\boldsymbol{E} + \boldsymbol{v}\times\delta\boldsymbol{B}/c)\cdot\nabla_v\langle\rho\rangle - \frac{e}{m}\langle(\delta\boldsymbol{E} + \boldsymbol{v}\times\delta\boldsymbol{B}/c)\cdot\nabla_v\,\delta\rho\rangle.\end{aligned} \tag{19}$$

It is clear that a hierarchy of equations results if we attempt to solve (19) to yield the collisional and wave effects appearing in the term on the right-hand side of (18).

The usual cluster expansion in which

$$f_2(1,2) = f_1(1)f_1(2) + g(1,2), \tag{20}$$

$$f_3(1,2,3) = f_1(1)f_1(2)f_1(3) + \sum_{\{1,2,3\}} f_1(1)g(2,3) + h(1,2,3), \tag{21}$$

defines the correlation functions g, h, etc. It then follows that

$$\langle\delta\rho(1)\,\delta\rho(2)\rangle = \Delta(1,2)f_1(1) + g(1,2), \tag{22}$$

$$\begin{aligned}\langle\delta\rho(1)\,\delta\rho(2)\,\delta\rho(3)\rangle = f_1(1)\,\Delta(1,2)\,\Delta(1,3) + g(1,2)\,\Delta(1,3) \\ + g(1,3)\,\Delta(1,2) + g(2,3)\,\Delta(1,2) + h(1,2,3),\end{aligned} \tag{23}$$

which relates the ensemble average of products of fluctuations to the correlation functions. It should be noted that in writing (22) and (23) the usual $N, V\to\infty$ thermodynamic limit has been used.

Using these relations, (18) and (19) yield the BBGKY hierarchy which, for a spatially homogeneous system, is

$$\frac{\partial f_1^{(1)}}{\partial t} = \frac{n}{m}\int dx_2\,dv_2\,\nabla_{x_1}\phi(12)\cdot\nabla_{v_1}g(12), \tag{24}$$

(25)

$$\left\{\frac{\partial}{\partial t}+\sum_{1,2} \boldsymbol{v}_1 \cdot \nabla_{x_1} - \frac{1}{m} \nabla_{x_1}\phi(12) \cdot (\nabla_{v_1} - \nabla_{v_2})\right\} g(12)$$

$$= \sum_{1,2} \frac{1}{m} \nabla_{x_1}\phi(12) \cdot \nabla_{v_1} f_1(1) f_1(2) + \sum_{1,2} \frac{n}{m} \int d\boldsymbol{x}_3\, d\boldsymbol{v}_3\, \nabla_{x_1}\phi(13) \cdot \nabla_{v_1}[f_1(1) g(23) + h(123)],$$

(26)

$$\frac{\partial h}{\partial t} + \sum_{123} \left\{ \boldsymbol{v}_1 \cdot \nabla_{x_1} h - \frac{1}{m} \nabla_{x_1}\phi(12) \cdot \nabla_{v_1}[h(123) + f_1(1) g(23)] \right\}$$

$$= \sum_{123} \frac{n}{m} \int d\boldsymbol{x}_4\, d\boldsymbol{v}_4\, \nabla_{x_1}\phi(14) \cdot \nabla_{v_1}[g(12)g(34) + g(13)g(24) + f_1(1)h(234) + k(1234)].$$

In writing (24)–(26) we have further specialized to the case in which the radiation field is negligible.

We now proceed to consider thermal equilibrium. We know that all the time derivative terms must vanish, but we then lose the ability to follow the evolution of the system into the equilibrium state. Thus we must append additional information to get a unique solution. We do this by assuming that f_N is a canonical distribution

$$f_N = Ae^{-H/\theta}, \tag{27}$$

$$\theta = kT. \tag{28}$$

It then follows that all the velocity distributions are Maxwellian. We then find that the spatial distributions (for which we use the same notation) are governed by

(29)

$$\nabla_{x_1} g(12) + \frac{1}{\theta} \nabla_{x_1}\phi(12)\,[1 + g(12)] + \frac{n}{\theta} \int d\boldsymbol{x}_3\, d\boldsymbol{v}_3\, \nabla_{x_1}\phi(13)\,[g(23) + h(123)] = 0,$$

$$\begin{aligned} \nabla_{x_1} h &+ \frac{n}{\theta} \int d\boldsymbol{x}_4\, d\boldsymbol{v}_4\, \nabla_{x_1}\phi(14)\, h(234) \\ &+ \left\{ g(12)\left[\frac{1}{\theta} \nabla_{x_1}\phi(13) + \frac{n}{\theta} \int d\boldsymbol{x}_4\, d\boldsymbol{v}_4\, \nabla_{x_1}\phi(14)\, g(34)\right] \right. \\ &\left. + \frac{1}{\theta} \nabla_{x_1}\phi(13)\,[h(123) + g(23)] + (2 \to 3) \right\} = 0. \end{aligned} \tag{30}$$

Introducing the Debye length λ_D as a length unit and the dimensionless potential

$$\phi = 1/x \tag{31}$$

we find that the natural expansion parameter ϵ defined by

$$\epsilon = 1/4\pi n \lambda_D^3 \tag{32}$$

appears throughout the problem. It is well known that for almost all classical plasmas ϵ is much smaller than unity. Equations (29) and (30) then become

$$\nabla_{x_1} g(12) + \epsilon \nabla_{x_1} \phi\,[1 + g(12)] + \frac{1}{4\pi}\int dx_3\, dv_3\, \nabla_{x_1}\phi(13)\,[g(23) + h(123)] = 0 \tag{33}$$

and

$$\begin{aligned}\nabla_{x_1} h + \frac{1}{4\pi}\int dx_4\, dv_4\, \nabla_{x_1}\phi(14)\, h(234) \\ + \Big\{ g(12)\Big[\epsilon\, \nabla_{x_1}\phi(12) + \frac{1}{4\pi}\int dx_4\, dv_4\, \nabla_{x_1}\phi(14)\, g(34)\Big] \\ + \epsilon\, \nabla_{x_1}\phi(13)\,[h(123) + g(23)] + (2 \to 3)\Big\} = 0.\end{aligned} \tag{34}$$

It is natural then to look for solutions of (33) and (34) as power series in ϵ. These solutions have been obtained by many authors. A detailed summary of this work, including a complete list of references, is given by Tappert [**4**]. The result, correct to two orders in ϵ, is

$$\begin{aligned} g(x) = -(\epsilon/x)e^{-x} + (\epsilon^2/2x^2)e^{-2x} + (\epsilon^2/2x)e^{-x}\{\tfrac{1}{4}(x-3)[\ln 3 - E(x)] \\ -\tfrac{1}{4}(x+3)e^{2x}E(3x) - \tfrac{1}{3}(1 - e^{-x})\} + O(\epsilon^3).\end{aligned} \tag{35}$$

For $x \ll 1$, (35) becomes

$$g(x) \cong -(\epsilon/x)e^{-x} + \epsilon^2[e^{-2x}/2x^2 + \ln x + (\ln 3 + \gamma - \tfrac{1}{2}) + O(x)] \tag{36}$$

while for $x \gg 1$, (35) yields

$$g(x) \cong -\frac{\epsilon}{x}e^{-x} + \epsilon^2 e^{-x}\Big[\frac{\ln 3}{8} - \frac{1}{x}\Big(\frac{1}{6} + \frac{3\ln 3}{8}\Big) + O(e^{-x})\Big]. \tag{37}$$

We thus see that the power series expansion diverges for both small and large values of x. We call the region in which the particle separation is of order λ_D, the central region.

In the inner region then a nonanalytic expansion in ϵ must be sought. The solution is

$$g_{\text{in}} = -1 + \exp\Big(-\frac{\epsilon}{x}e^{-x}\Big) + \cdots \tag{38}$$

As x becomes large in this solution, we recover (36).

For the outer region, a different asymptotic analysis must be performed. A new and complete discussion of this is given in [**4**]. The result is

$$g_{\text{out}} = \frac{\epsilon}{\lambda}[1 + O(\epsilon)]\exp\Big(-\Big(1 + \frac{\epsilon \ln 3}{8}\Big)x\Big). \tag{39}$$

We note that as x becomes small in (39) we recover (37).

Thus we see that a complete solution has been found. Physically, the plasma behaves like a Boltzmann gas in the inner region where $x \sim e^2/kT$. In the central region where $x \sim \lambda_D$, the classical Debye shielding result obtains and for $x \sim \lambda_D/\epsilon$, a correction to the Debye length is found.

These results lead us to believe that in the nonequilibrium, time-dependent kinetic problem, at least three different asymptotic expansions must be used in the analysis.

III. **The nonequilibrium stable plasma.** We briefly recall, for illustrative purposes, the standard weak coupling result **[5]**, **[6]** which results from (24)–(26). We introduce an expansion parameter λ defined by

(40) $$\lambda = \langle\phi\rangle/mv_{av}^2$$

where $\langle\phi\rangle$ is a typical potential energy of interaction and mv_{av}^2 is of the order of a characteristic kinetic energy. We then expand (24)–(26) in powers of λ using a multiple time scale procedure and assume

(41) $$g = O(\lambda).$$

(42) $$h = O(\lambda^2).$$

We find

(43) $$\partial f_1^{(0)}/\partial t = 0$$

and

(44) $$\partial f_1^{(1)}/\partial t + \partial f_1^{(0)}/\partial \lambda t = 0.$$

Equation (44) yields

(45) $$\partial f_1^{(1)}/\partial t = \partial f_1^{(0)}/\partial \lambda t = 0$$

upon removal of secular terms in the expansion.

To next order there results

(46) $$\frac{\partial f_1^{(2)}}{\partial t} + \frac{\partial f_1^{(1)}}{\partial \lambda t} + \frac{\partial f_1^{(0)}}{\partial \lambda^2 t} = \frac{n}{m}\int d\boldsymbol{x}_2\, d\boldsymbol{v}_2 \frac{\partial \phi(12)}{\partial \boldsymbol{x}_1}\cdot\frac{\partial g^{(1)}}{\partial \boldsymbol{v}_1}$$

and

(47) $$\left(\frac{\partial}{\partial t} + (\boldsymbol{v}_1 - \boldsymbol{v}_2)\cdot\nabla_x\right) g^{(1)} = \frac{1}{m}\nabla_{x_1}\phi(12)\cdot(\nabla_{v_1} - \nabla_{v_2}) f_1^{(0)}(1) f_1^{(0)}(2).$$

Upon integrating (46) with respect to t we find

(48) $$\frac{\partial f_1^{(1)}}{\partial \lambda t} + \frac{\partial f_1^{(0)}}{\partial \lambda^2 t} = \frac{n}{m}\int d\boldsymbol{x}_2\, d\boldsymbol{v}_2\, \nabla_{x_1}\phi(12)\cdot\nabla_{v_1} g^{(1)} \quad (t\to\infty, 12)$$

and

(49) $$\frac{\partial f_1^{(2)}}{\partial t} = \frac{n}{m}\int d\boldsymbol{x}_2\, d\boldsymbol{v}_2\, \nabla_{x_1}\phi(12)\cdot\nabla_{v_1}(g^{(1)}(t) - g^{(1)}(\infty))$$

upon removal of secular terms. We further integrate (48) with respect to λt and get

$$\partial f_1^{(1)}/\partial \lambda t = 0 \tag{50}$$

in addition to

$$\frac{\partial f_1^{(0)}}{\partial \lambda^2 t} = \frac{n}{m} \int dx_2\, dv_2\, \nabla_{x_1}\phi(12) \cdot \nabla_{v_1} g^{(1)} \qquad (\infty, \lambda^2 t, 12). \tag{51}$$

In order for these results to be formally valid both (49) and (51) must yield acceptable behavior. Upon solving (47) and substituting the result in (51), the Fokker-Planck equation is found. The H-theorem then guarantees that $f_1^{(0)}$ approaches a Maxwellian. The solution of (49) depends on the choice of $g^{(1)}$ initially. For sufficiently smooth initial conditions, this equation will also predict a decay.

There are a number of important points that emerge from such a demonstration. First, we see how the Bogoliubov synchronization hypothesis is validated in at least a formal manner. Second, we establish a lowest order irreversible kinetic equation. Third, there appears to be a formal way of carrying out this procedure to higher order in the relevant expansion parameter. Fourth, the procedure appears to lead to divergences in every case in which it has been tried **[6]**.

In the central region, for the plasma case, the analogue of (51) is the well-known Balescu [7]-Lenard **[8]**-Guernsey **[9]** equation. It diverges for impact parameters in the inner region, as one might expect on physical grounds. It also appears that the analogue of (49) is ill-behaved in that the right-hand side of the equation does not decay fast enough. The first point is corrected by reordering the equations in a manner appropriate to the inner region with the result that a Boltzmann type of behavior then appears **[10]**. Repairing the analogue of (49) is, however, a more arduous task. It was pointed out earlier **[1]** that the probable cause for this difficulty lies in the fact that the Landau damping rate becomes exponentially small in the outer region (which contributes heavily to the time-dependence of the process) and that one should take into account the collisional damping of the plasma oscillations. This has now been done by Rostoker and Matsuda **[11]**, Kaufmann **[12]**, Tappert **[4]**, and Rogister and Oberman **[13]**. The results show that the slow damping is in fact replaced by the faster collisional damping, and it thus appears that the classical Fokker-Planck equation used by Spitzer **[14]** with a lower and upper cutoff is indeed correct and is the lowest order asymptotic result in an expansion in ϵ.

It is interesting to note that in the Boltzmann case similar phenomena appear **[15]**, **[16]**, **[17]** and the divergences can again be removed by introducing new physics on the scale of the free path.

Recent work by Rostoker, Aamodt, and Eldridge **[18]** shows that a very complicated kinetic equation results when radiation fields are included in the plasma case and the plasma is immersed in a static magnetic field. In addition, the analogue of (49) yields acceptable behavior due to the strong damping brought about by synchrotron radiation. However, when the static magnetic field is set to zero, the traverse electron plasma oscillations exhibit no Landau damping since the phase velocity of the waves is greater than that of light. Nevertheless, an analysis by Hsuan [**19**] indicates that the Simon-Harris [**20**] kinetic equation is correct since the equation analogous to (49) does yield acceptable behavior in this case. The reason is that the dispersion of the waves provides a sufficiently strong damping mechanism to allow the Bogoliubov synchronization process to take place. It is well known that no radiation appears to lowest order. A higher order calculation by Dupree [**21**] indicates that to next order bremsstrahlung appears as expected on physical grounds.

IV. **Unstable plasmas.** In the unstable plasma domain there is a body of theory that has accumulated, vaguely termed "quasi-linear." The basic notion involves the treatment of weakly turbulent plasmas by balancing the growth of a weakly unstable set of plasma waves against the stabilizing effect of weak nonlinearities and various collisional processes. A weakly turbulent state is defined by the wave energy being much larger than the thermal equilibrium fluctuation wave energy but much smaller than the total thermal energy in the plasma. It is a state that does not seem to cover a wide variety of situations in hydrodynamics but does seem experimentally to occur quite often in plasmas.

To treat this problem formally we go back to the hierarchy for a spatially homogeneous plasma and look for an ordering in ϵ, in which γ, the growth rate, is formally related to ϵ. It is found that if

$$\gamma \sim \epsilon^{1/2}, \tag{52}$$

$$g \sim \epsilon^{1/2}, \tag{53}$$

$$h \sim \epsilon, \tag{54}$$

are chosen, a complete set of kinetic equations that jointly describe the variation of E_k, the electric field wave spectrum and f_1 can be found. A complete discussion with references to earlier work is given by Davidson [**22**].

The major new physical effects that appear are

1. three-wave mode coupling,

$$\omega(\boldsymbol{k}) - \omega(\boldsymbol{1}) - \omega(\boldsymbol{k} - \boldsymbol{1}) \sim 0; \tag{55}$$

2. wave-particle resonant diffusion,

$$\omega(\boldsymbol{k}) - \boldsymbol{k} \cdot \boldsymbol{v} \sim 0; \tag{56}$$

3. higher order Landau damping,

$$\omega(\boldsymbol{k}) - \omega(\boldsymbol{1}) - (\boldsymbol{k} - \boldsymbol{1}) \cdot \boldsymbol{v} \sim 0; \tag{57}$$

4 particle-particle interactions.

The theory described here essentially operates on three time scales, a fast scale

$$\tau_0 \sim 1/\omega p, \tag{58}$$

a slower scale

$$\tau_1 \sim 1/\gamma, \tag{59}$$

and the slowest scale

(60) $$\tau_2 \sim 1/\nu$$

where ν is a typical collision frequency.

In summary, one can say that the present quasilinear theory yields reasonable results. However, there are two physical processes, which can be important and which the present theory neglects. First, as E_k becomes of order of the thermal equilibrium fluctuation level and as f approaches a stable distribution, γ should become smaller, and for long wavelengths collisional damping should again become dominant. Such a reordered theory has been carried out by Rogister and Oberman **[13]** and represents a generalization of the theory given in **[22]**.

The second major new piece of physics is the effect of trapping and other processes that are nonexpandable in ϵ. It is well known that for a charged particle moving approximately at the phase velocity, ω/k, of an electrostatic wave of amplitude E_k, the trapping becomes important after a time

(61) $$\tau_{\text{trap}} \sim (m/eE_k k)^{1/2}$$

and thus involves the square root of the amplitude. A detailed nonlinear theory for a single wave has been given by O'Neil **[23]**. However, if we deal with a wave packet of width Δk, the transit time across the packet of a particle traveling at the phase velocity is given by

(62) $$\tau_{\text{transit}} \sim 1/\Delta k(V_{\text{phase}} - V_{\text{group}}).$$

The theory of **[22]** assumes

(63) $$\tau_{\text{transit}} \ll \tau_{\text{trap}}$$

so that we expect a particle transits across the wave packet before it has a chance to be trapped. The effect of this is that various phase mixing terms coming from free streaming operators $e^{-i\mathbf{k}\cdot\mathbf{v}t}$ appear and yield nonanalytic $t^{-3/2}$ decays obtained from steepest descent evaluations. Clearly, at some point in the perturbation procedure these effects "catch up" and require a modification of the theory. Recent papers by Dupree **[24]** and Aamodt **[25]** present new perturbation techniques designed to remedy this situation. At the present time these have not been investigated in sufficient detail to evaluate their success.

V. **Conclusions.** We can conclude this summary of the present state of plasma kinetic theory in the following way:

(1) The equilibrium plasma exhibits three different regions of physical and mathematical behavior. The theory seems to be well understood and offers no blindingly obvious problems in the form of divergences, etc.

(2) The nonequilibrium stable plasma theory has all the problems of the usual kinetic and transport theories, but significant progress has been made. It is quite clear that in addition to the regions that appear in physical space which must be treated by differing asymptotic expansions, different regions both in velocity and time are introduced. A complete picture of phase space for all time has certainly not been found to date.

(3) The nonequilibrium unstable plasma theory is clearly still in its infancy, since in addition to the problems that arise in the nonequilibrium case, nonlinear effects now must be considered. It is in this area that a connection should be made with the vast body of literature on plasma stability theory in spatially inhomogeneous situations and complicated magnetic geometries. At the present time, however, only a glimmering of such a connection has been made. This is a major problem for the future.

REFERENCES

1. E. Frieman, *The present state of plasma kinetic theory*, Proc. Sympos. Appl. Math. Vol. 18, Amer. Math. Soc. Providence, R.I., 1967, p. 257.

2. Y. L. Klimontovich, *Method of second quantization in phase space*, Soviet Physics JETP. **6** (1958), 753.

3. T. H. Dupree, *Kinetic theory of plasma and the electromagnetic field*, Phys. Fluids **6** (1963), 1714.

4. F. H. Tappert, Ph.D. Thesis, Princeton University, 1967, unpublished.

5. E. Frieman, *On a new method in the theory of irreversible processes*, J. Mathematical Phys. **4** (1963), 410.

6. G. Sandri, *The foundations of nonequilibrium statistical mechanics*, I, Ann. Physics **24** (1963), 332.

7. R. Balescu, *Irreversible processes in ionized gases*, Phys. Fluids **3** (1960), 52.

8. A. Lenard, *On Bogoliubov's kinetic equation for a spatially homogeneous plasma*, Ann. Physics **10** (1960), 390.

9. R. Guernsey, Ph.D. Thesis, University of Michigan, 1960, unpublished.

10. E. Frieman and D. L. Book, *A convergent classical kinetic equation for a plasma*, Phys. Fluids **6** (1963), 1700.

11. N. Rostoker and K. Matsuda, *Kinetic theory of particles and waves*, *Plasma physics and controlled nuclear fusion research* (IAEA, Vienna, 1966), Vol. 1, p. 747.

12 A. Kaufmann, *Extension of the plasma kinetic equation to small wave numbers*, Phys. Rev. Letters **17** (1966), 1127.

13. A. Rogister and C. Oberman, *On the kinetic theory of stable and weakly unstable plasma*. Part I, J. Plasma Phys. **2** (1968), 33.

14. R. Cohen, L. Spitzer, Jr., and P. McR. Routly, *The electrical conductivity of an ionized gas*, Phys. Rev. **80** (1950), 230.

15. J. Dorfman and E. Cohen, *On the density expansion of the pair distribution function for a dense gas not in equilibrium*, Phys. Letters **16** (1965), 124.

16. E. Frieman and R. Goldman, *Propagation of correlations in a Boltzmann gas*, J. Mathematical Phys. **7** (1966), 2153.

17. R. Goldman and E. Frieman, *On the logarithmic density behavior of a nonequilibrium Boltzmann gas*, J. Mathematical Phys. **8** (1967), 1414.

18. R. Aamodt, O. Eldridge and N. Rostoker, *Kinetic theory of radiation in a plasma*, Phys. Fluids **7** (1964), 1952.

19. H. Hsuan, Ph.D. Thesis, Princeton University, 1967, unpublished.

20. A. Simon and E. Harris, *Kinetic equations for plasma and radiation*, Phys. Fluids **3** (1960), 245.

21. T. Dupree, *Theory of radiation emission and absorption in plasma*, Phys. Fluids **7** (1964), 923.

22. R. Davidson, *Weak turbulence in a homogeneous plasma*, Princeton Univ. Plasma Physics Lab. MATT-496 (1966).

23. T. O'Neil, *Collisionless damping of nonlinear plasma oscillations*, Phys. Fluids **8** (1965), 2255.

24. T. Dupree, *A perturbation theory for strong plasma turbulence*, Phys. Fluids **9** (1966), 1773.

25. R. Aamodt, Report Center for Plasma Physics, Univ. of Texas, 1966.

PLASMA PHYSICS LABORATORY, PRINCETON UNIVERSITY,
PRINCETON, NEW JERSEY

Author Index

Roman numbers refer to pages on which a reference is made to a work of the author. *Italic numbers* refer to pages on which a complete reference to a work by the author is given. **Boldface numbers** indicate the first page of an article in this volume.

Subject Index